Middle School Geometry for the Common Core

Middle School Geometry for the Common Core

Cataloging-in-Publication Data is on file with the Library of Congress.

Printed in the United States of America

9 8 7 6 5 4 3 2 1

ISBN 978-1-61103-025-9

For more information or to place an order, contact LearningExpress at:
80 Broad Street
4th Floor
New York, NY 10004

Or visit us at:
www.learningexpress.com

Contributor

Kimberly Stafford majored in mathematics and education at Colgate University in upstate New York. She taught math, science, and English in Japan, Virginia, and Oregon before settling in Los Angeles. Kimberly began her work in Southern California as an educator in the classroom but soon decided to launch her own private tutoring business so she could individualize her math instruction. She believes that a solid foundation in math empowers people by enabling them to make the best work, consumer, and personal decisions. Kimberly is unfazed by the ubiquitous student gripe "When am I going to use this in real life?" She stresses that the mastery of math concepts that are less applicable to everyday life helps teach a critical skill—problem solving. The very ability to apply a set of tools to solve new and complex problems is an invaluable skill in both the workforce and personal life. Kimberly believes that mathematics is a beautiful arena for developing organized systems of thinking, clear and supported rationale, and effective problem solving.

Contents

Contents

Contents

Getting to Know the Common Core State Standards

Does even thinking about the Common Core State Standards make your palms sweat a bit and your heart rate elevate slightly? Well, you are not alone. New requirements and standards are always a little nerve-racking—and not just for teachers but for parents, students, and school districts alike. There's been a lot of buzz about the Common Core State Standards, along with a lot of misinformation. People say that the standards were developed without input from teachers, that they're a narrow "script" that teachers cannot deviate from, and that they keep kids from connecting concepts in the real world. But all of these claims are untrue. The fear of change drives much of the misinformation out there, so let's take a closer look at the rationale, development, and goals of the Common Core State Standards.

Much of the backlash against the Common Core State Standards is fueled by misconceptions about who developed the standards, why they were developed, how they are implemented in schools, and what they mean for your school district's teachers and children.

Who?

The idea of having a body of common standards that would cross state lines was introduced by the National Governors Association Center for Best Practices (NGA) along with the Council of Chief State School Officers. Contrary to what some people believe, the standards were not designed by the federal government but instead developed by a diverse group of educators and experts from around the country. Along with teachers, scholars, assessment developers, and parents, the following well-respected professional organizations provided feedback on the standards:

- National Education Association (NEA)
- American Federation of Teachers (AFT)
- National Council of Teachers of Mathematics (NCTM)
- National Council of Teachers of English (NCTE)

The CCSS were developed by well-respected professional organizations along with the help of teachers, scholars, assessment developers, and parents.

Why?

The United States educational system has not kept pace with the skills gained by students in many other industrialized countries. Although the U.S. used to rank first for high school graduation, it has slipped to 22nd among the top 27 industrialized countries.[1] College retention rates are also plummeting. When students successfully make it into the halls of higher education, university educators are noticing that their preparedness levels are declining. The Common Core State Standards aim to raise the bar of the material presented to students in K–12. Not only are the CCSS aligned across state lines, but they're internationally benchmarked to help students acquire the skills they need to compete in the global market. The desired outcome of the CCSS is to give students the applied skills they need for college, work, and life.

[1] BANCHERO, STEPHANIE. "High-School Graduation Rate Inches Up." *The Wall Street Journal*. Accessed February 27, 2014, http://online.wsj.com/news/articles/SB10001424127887323301104578256142504828724

How?

The Common Core State Standards are a collection of learning goals that outline the skills students should attain at each grade level. Covering mathematics and English language arts/literacy from kindergarten through 12th grade, they give students and parents a clear understanding of the knowledge kids need to be college- and career-ready when they graduate from high school.

The CCSS for mathematics signify a shift away from a body of skills that is "a mile wide and an inch deep." Instead of having students move rapidly from one concept to the next, the standards present a more focused set of skills that involve higher-level reasoning and problem solving.

It's important to know that the standards are:

1. based on thorough research and evidence.
2. formulated so that they can be clearly understood by educators, parents, and students alike.
3. focused on rigorous content that requires students to apply higher-level reasoning and problem-solving skills.
4. a dynamic byproduct of the state standards that have been most effective and a revision of state standards that have fallen short.
5. in alignment with the expectations of colleges and employers.
6. in alignment with the educational standards of top-performing countries so that U.S. students can successfully compete when they join the global workforce.

The CCSS for mathematics signify a shift away from a body of skills that is "a mile wide and an inch deep." Instead of having students move rapidly from one concept to the next, the standards present a more focused set of skills that involve higher-level reasoning and problem solving.

What?

The Common Core State Standards should not be thought of as a narrow curriculum that cannot be deviated from. They are a broad collection of skills, not a list of every single skill that a student should learn. How school districts and teachers present the standards is determined by how they choose to design their curricula. There are no strict lesson plans to teach from, and educators are encouraged to tailor their lessons to best meet the specific needs of their student body.

When?

By the early 1990s, state education standards had become popular, and it wasn't too long before every state had its own learning standards. Since the definition of "proficiency" differed from state to state, kids who excelled according to the standards in one area might not meet the benchmarks set in another.

Educators saw a need for a cross-state body of learning standards, and work on the Common Core State Standards began in 2009. The standards were introduced to states for adoption in 2010, and states are independently implementing the CCSS.

Where?

The Common Core State Standards are not a federal mandate forced on schools. The standards have been voluntarily adopted by 43 states. Each state that has adopted the CCSS works to deliver the same general content that other states do at the same grade level. Now, for example, a student who finishes 7th grade in Idaho and enters 8th grade in Florida the following school year will be at the same level as his peers. Previously, students changing school districts across state lines were in jeopardy of having their education compromised because of gaps in the material.

You can learn more about the standards at the official site for the CCSS: www.corestandards.org/about-the-standards.

What's the Big Deal about Geometry?

Now that you know more about the rationale behind the Common Core State Standards, they're a little less scary, right? Maybe you even think that it sounds like a good idea to create a unified set of sequential skills that students need for the college classroom or the workforce. Now let's take a closer look at what the standards mean for middle school geometry—and why middle school geometry is so important in the first place.

You probably already know that one of the most challenging aspects of *high school* geometry is writing proofs. Proofs weave theorems and postulates together in a careful progression of statements to show that a conclusion is valid. Sounds really boring? It's not. Proofs are in fact an exciting and demanding mental workout that forces a student's brain to do the equivalent of squat thrusts and lunges. Kids who are accustomed to "going through the motions" to solve math problems are drawn out of their shell, prompted to think logically and originally on their own. Although it may be frustrating at first, constructing proofs is an invaluable way to learn how to reason and support one's assertions with solid evidence—two critical life skills. After all, who doesn't want to be able to craft a convincing argument for why Mom or Dad should pay for a spring-break trip to Miami Beach?

High school geometry is a great arena in which to learn and improve higher-level reasoning skills, and having a solid foundation of geometry in middle school is essential to getting there.

Constructing proofs is an invaluable way to learn how to reason and support one's assertions with solid evidence—two critical life skills.

The CCSS for Middle School Geometry

Remember, an important aspect of the Common Core State Standards is the increased focus they bring to the topics at hand. In the educational community in recent years, there's been a movement away from teaching through rote memorization and teaching a greater number of concepts with less depth. In keeping with this, the CCSS promote a more vigorous approach to a refined set of concepts.

When it comes to math, the standards challenge students to have a more grounded understanding of the origins and applications of theorems, algorithms, and formulas and to use mathematical ways of thinking to solve real-world problems. Here's a brief summary of the geometry skills that middle school students learn with the CCSS, broken down by grade:

- Sixth grade geometry standards get students comfortable with mathematical and real-world problems involving area, surface area, and volume. Topics include:
  - drawing shapes on the coordinate plane.
  - calculating area, surface area, and volume of 2D and 3D shapes using reasoning skills and formulas.
  - learning how 2D figures, called nets, can be transformed into 3D prisms.

- In seventh grade, students should come with their protractors because they are going to start making constructions and deepening their understanding of angle measurements as they relate to geometric shapes and parallel lines. Topics include:
  - using ratios and proportions to construct geometric figures of varying scales.
  - recognizing special angle relationships and using them to set up equations that can be used to solve for unknown angles in a figure.
  - exploring area and circumference of circles and how these formulas are related.
  - solving real-world problems involving area, volume, and surface area of various 2D and 3D shapes.

- Students should arrive to eighth grade with their colored pencils because they'll be manipulating shapes on and off the coordinate plane. But that's not all they will be doing before heading to high school. Topics in eighth grade include:
  - transforming various geometric shapes to better understand the coordinate plane, similarity, and congruence.
  - learning a proof of the Pythagorean theorem to gain a deeper knowledge of triangles.
  - applying the Pythagorean theorem to solve for unknown sides of right triangles in both mathematical and real-world scenarios.
  - calculating the volume of cones, cylinders, and spheres.

How to Use This Book

We know you're revved up now and excited to get down to geometry as soon as possible, but before jumping into the first chapter, familiarize yourself with the standards and notice how they build upon one another over time. The table at the end of this introduction presents all of the Common Core State Standards for middle school geometry in a manner that's approachable even if you're not on the fast track to becoming a Math Olympian. It also shows what lessons relate to each standard.

Throughout this book, chapter introductions will point out the standards that each chapter pertains to, as well as the critical foundations it covers. You will notice that some of the lessons in this book review concepts that are now part of the CCSS for grades four and five. Remember, before this state-aligned set of standards, skills that were taught in fifth grade in one state may have been presented in sixth or seventh grade in another state. We have intentionally included some foundational concepts to ensure that students have the prerequisite skills they need to be successful with the more rigorous standards in grades six through eight.

Since the book is organized sequentially—each lesson builds on the previous one—it's a good idea to make sure that you fully understand the material in one lesson before you proceed to the next. Throughout the lessons, you'll find examples, illustrations, and boxed theorems and postulates as well as Standard Alert boxes that point out what standard is being covered and why it is valuable. You'll also find two to four sets of practice problems per lesson; the answers are listed at the end of each chapter.

Before diving into Chapter 1, take the Pretest and check your answers to see where your skill level is now. Then, after making your way through all of the lessons in the book, take the Posttest and review any sections that are still tricky for you. If you invest 15 minutes of focused study time on each lesson per day, you should be able to move through this collection of lessons within a month—and, as a result, feel confident in your understanding of geometry and your ability to use it to solve real-world problems.

Standard #	Description	Lessons to Review
6.G.A.1	Find the area of 2D shapes by breaking them down into triangles and rectangles.	Lessons 8, 12, and 15
6.G.A.2	Find the volume of rectangular 3D shapes that have fractional side lengths.	Lesson 18
6.G.A.3	Draw polygons in the coordinate plane using (x,y) points as the corners and find the length of the shapes' vertical and horizontal sides.	Lessons 19, 23, and 24
6.G.A.4	Recognize the connection between 2D shapes and 3D prisms by creating 2D shapes that can be folded to create 3D prisms. Understand the connection between area of 2D shapes and surface area of 3D prisms.	Lesson 17
7.G.A.1	Use the concept of scale to calculate actual side lengths and distances from scale drawings. Reproduce figures of varying sizes by appropriately applying scale.	Lessons 13, 22, 23, and 24
7.G.A.2	Construct geometric shapes with specific angle and side length conditions and understand what sets of conditions would create 0, 1, or more than 1 triangle.	Lessons 2, 6, and 7
7.G.A.3	Define the 2D figures that result from slicing 3D prisms.	Reference B
7.G.B.4	Find the area and circumference of circles and present the relationship between the circumference and area formulas.	Lessons 14 and 15
7.G.B.5	Learn how to identify special pairs of angles and use their relationships to set up and solve for unknown angles in a figure.	Lessons 3, 4, 7, and 11
7.G.B.6	Solve real-world problems involving area, surface area, and volume of 2D and 3D shapes.	Lessons 8, 12, 16, 17, and 18
8.G.A.1	Understand how the appearances of shapes change or remain the same as they are rotated, reflected, and shifted.	Lesson 20
8.G.A.2	Demonstrate that a 2D figure is congruent to another 2D figure if the second figure can be replicated by rotating, reflecting, and shifting the first figure.	Lesson 21
8.G.A.3	Use coordinate pairs (x,y) to create rules for the changes to points and 2D shapes that are dilated, shifted, rotated, and reflected.	Lesson 20
8.G.A.4	Demonstrate that a 2D figure is similar to another 2D figure if the second figure can be replicated by dilating, rotating, reflecting, and shifting the first figure.	Lesson 21
8.G.A.5	Deepen understanding of angle relationships through investigating angles in triangles as well as the angle pairs formed by a line that intersects a pair of parallel lines.	Lessons 5, 6, 7, and 22
8.G.B.6	Demonstrate a thorough understanding of the Pythagorean theorem by explaining its proof.	Reference A

8.G.B.7	Use the Pythagorean theorem to find the unknown sides of right triangles.	Lessons 9 and 10
8.G.B.8	Use the Pythagorean theorem to find the distance between two sides in the coordinate plane.	Lesson 19
8.G.B.9	Apply the formulas for the volumes of cones, cylinders, and spheres to solve problems.	Lesson 18

Middle School Geometry for the Common Core

Pretest

This pretest will test your knowledge of basic geometry. The 24 questions are presented in the same order that the topics they cover are presented in this book. The book builds skills from lesson to lesson, so skills you will learn in earlier lessons may be required to solve problems found in later lessons. This means that you may find the problems toward the beginning of the test easier to solve than the problems toward the end of the test.

For each of the 24 multiple-choice questions, circle the correct answer. The pretest will show you which geometry topics are your strengths and which topics you might need to review—or learn for the first time. After taking the pretest, you might realize that you need only to review a few lessons. Or you might benefit from working through the book from start to finish.

Pretest

1. In the first figure, all of the following are acceptable ways to name the shaded angle EXCEPT

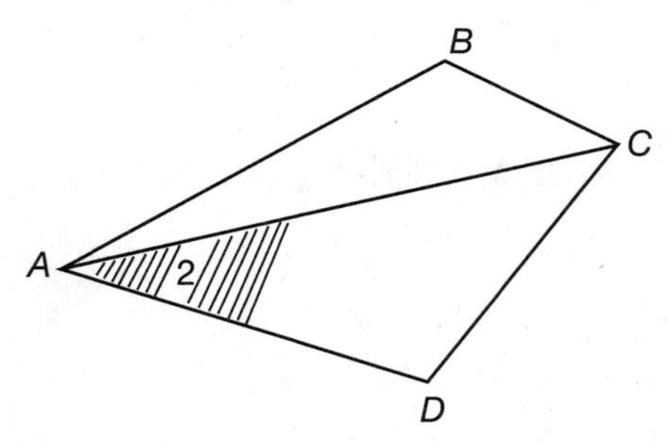

a. $\angle A$
b. $\angle 2$
c. $\angle CAD$
d. $\angle DAC$

2. In the next figure, D bisects $\overline{NA}$ and N bisects $\overline{IA}$. If $\overline{DA} = 2x - 3$ units, what is the length of $\overline{IA}$?

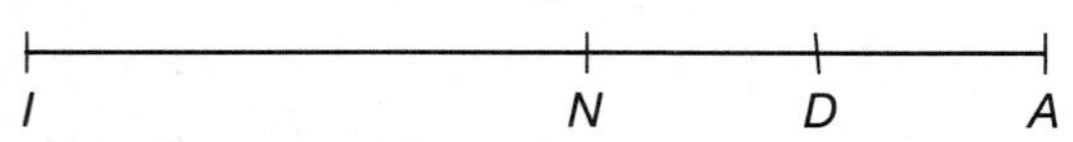

a. $4x - 6$
b. $6x - 9$
c. $8x - 12$
d. $4x + 9$

3. $\angle K$ is complementary to $\angle M$. If $m\angle M = (2x)°$, then what is the measure of $\angle K$ in terms of x?
a. $90° + (2x)°$
b. $90° - (2x)°$
c. $800° + (2x)°$
d. $800° - (2x)°$

4. Solve for x.

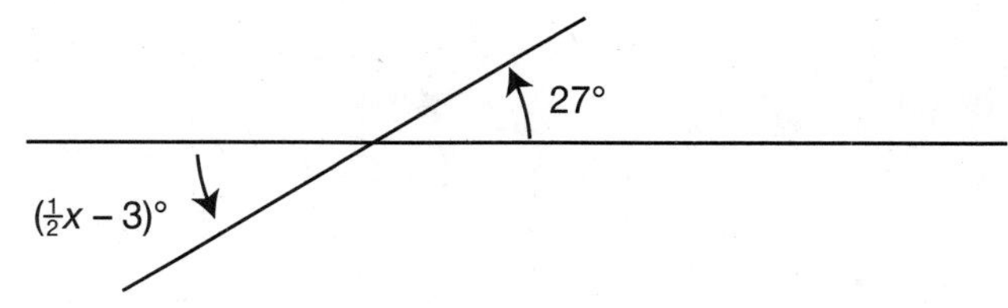

a. 27
b. 30
c. 54
d. 60

5. In the next figure, $l \parallel k$. $\angle A$ and $\angle B$ are what type of angles?

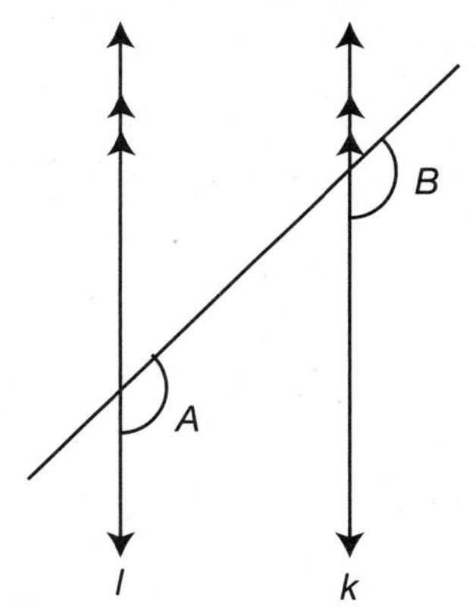

a. Corresponding angles
b. Complementary angles
c. Alternate exterior angles
d. Same-side interior angles

6. In ΔCAT, $\overline{CA} < \overline{AT} < \overline{TC}$. If $\overline{CA} = 12$ and $\overline{TC} = 30$, then what could be a possible length for side $\overline{AT}$?
a. 17
b. 18
c. 19
d. Cannot be determined.

7. In an isosceles triangle ΔLEX, $\overline{LE} \cong \overline{LX}$. If $m\angle LEX = 27°$ find the $m\angle ELX$.

a. 126°
b. 153°
c. 63°
d. 54°

8. Which expression is equal to the area of the given triangle?

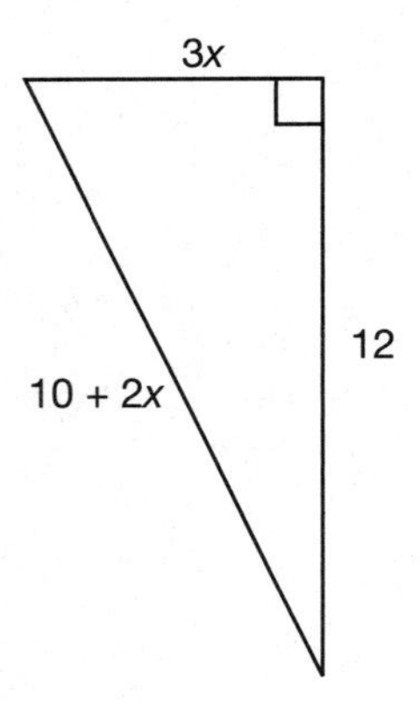

a. $5x + 22$
b. $18x$
c. $120x + 24$
d. $30x + 6x^2$

9. In a right triangle, the lengths of the two legs are 8 cm and 15 cm. Find the length of the hypotenuse.

a. 23 cm
b. 19 cm
c. 7 cm
d. 17 cm

10. ΔPIN is a 30°-60°-90° triangle. If $\overline{PI}$ measures 14 m, as shown in the accompanying figure, what will be the length of $\overline{PN}$?

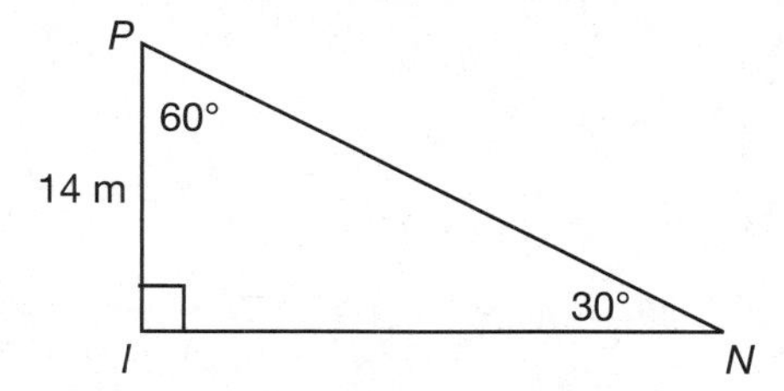

a. 28 m
b. 21 m
c. 14 m
d. 14 m

11. Quadrilateral *TOYS* has the following angle measurements: $m\angle T = (y)°$, $m\angle O = (2y)°$, $m\angle Y = (4y)°$, $m\angle Y = (5y)°$. What is the measure of the largest angle in quadrilateral *TOYS*?
a. 30°
b. 130°
c. 150°
d. 160°

12. What is the area of the right polygon that follows?

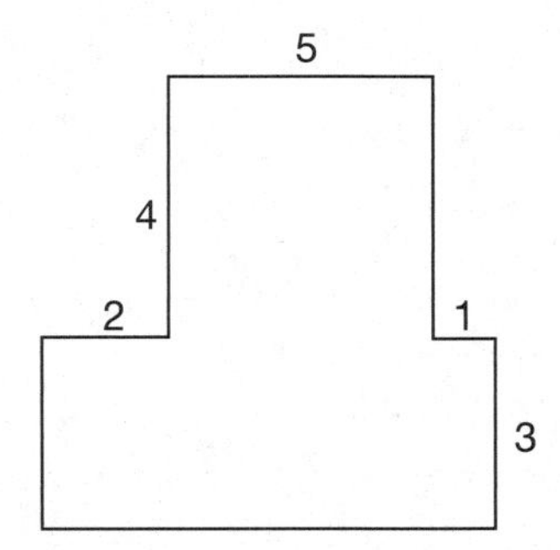

a. 15 units
b. 20 units
c. 30 units
d. 44 units

13. Solve for g: $\frac{15}{24} = \frac{g}{16}$
- **a.** 10
- **b.** 12
- **c.** 11
- **d.** 13

14. What is the circumference, in terms of π, of a circle with a radius of 6 inches?
- **a.** 6π in.
- **b.** 36π in.
- **c.** 3π in.
- **d.** 12π in.

15. What is the area, in terms of π, of circle H if the diameter is 16?

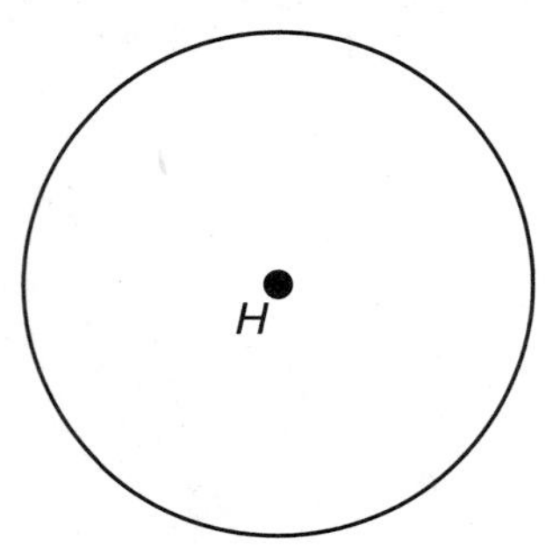

- **a.** 16π cm^2
- **b.** 64π cm^2
- **c.** 32π cm^2
- **d.** 4π cm^2

16. Nina purchases 64 feet of fencing to protect her organic garden project from deer. If she plans on making a rectangular garden with a width of 8 feet, how many feet long can her garden be?
- **a.** 8 ft.
- **b.** 16 ft.
- **c.** 20 ft.
- **d.** 24 ft.

17. What is the surface area of a rectangular prism that is 8 units long, 6 units wide, and 5 units tall?

a. 228 units^2
b. 236 units^2
c. 240 units^2
d. 264 units^2

18. What is the volume of the cylinder in the figure? Round your answer to the nearest tenth.

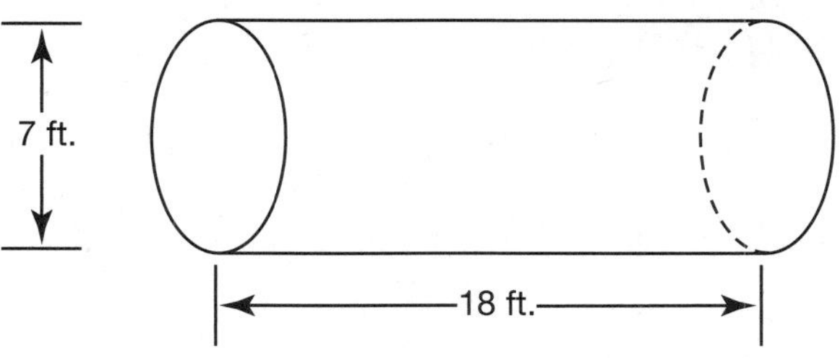

a. 395.6 ft.^{2}
b. 791.3 ft.^{2}
c. 692.4 ft.^{2}
d. 2,769.5 ft.^{2}

19. What is the midpoint between points $R(-6,-5)$ and $H(4,-3)$?

a. (–5,1)
b. (–1,–4)
c. (–1,1)
d. (–2,–4)

20. What transformation could be used to create image point $D'(3,-4)$ from preimage point $D(-3,-4)$?

a. A clockwise rotation of 90°.
b. A reflection over the x-axis.
c. A translation using $T_{-1,0}$
d. A reflection over the y-axis.

21. Which transformations determine similarity in the shaded figures in the following logo?

a. translation then rotation
b. rotation
c. reflection then translation
d. all of the above

22. $\Delta ABC \sim \Delta XYZ$. Find the length of side $\overline{XZ}$.

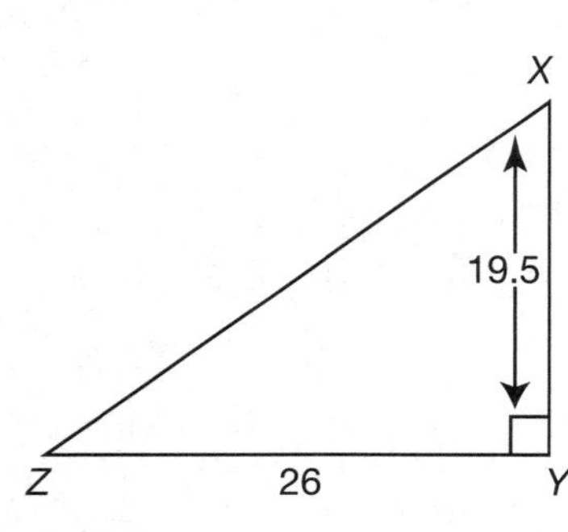

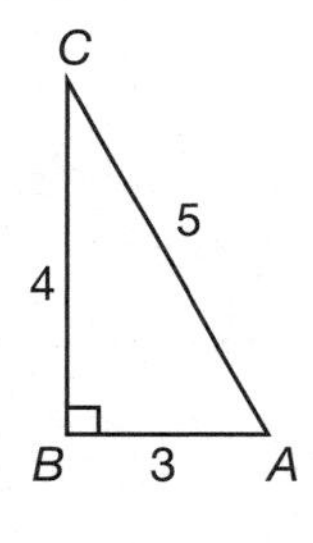

a. 28
b. 29.5
c. 31
d. 32.5

23. A vertical line has a slope that is
a. Undefined
b. Positive
c. Zero
d. Negative

24. Which equation is *not* linear?

a. $5x + 4 = \frac{1}{2}y$

b. $3x = -12 - 4y$

c. $\frac{8x}{y} = 12$

d. $-5 = 20xy$

Answers

For any questions that you may have missed, pay special attention to the lesson listed.

1. a. Lesson 1
2. c. Lesson 2
3. b. Lesson 3
4. d. Lesson 4
5. a. Lesson 5
6. c. Lesson 6
7. a. Lesson 7
8. b. Lesson 8
9. d. Lesson 9
10. a. Lesson 10
11. c. Lesson 11
12. d. Lesson 12
13. a. Lesson 13
14. d. Lesson 14
15. b. Lesson 15
16. d. Lesson 16
17. b. Lesson 17
18. c. Lesson 18
19. b. Lesson 19
20. b. Lesson 20
21. d. Lesson 21
22. d. Lesson 22
23. a. Lesson 23
24. d. Lesson 24

1 Foundational Geometry Concepts

Without geometry life is pointless.
—Anonymous

In this lesson you will become acquainted with the basic definitions of the building blocks of geometry such as points, lines, line segments, rays, planes, and angles.

STANDARD SNEAK PREVIEW

Just as a cook needs to know what a *cup*, *teaspoon*, and *tablespoon* are in order to follow a recipe, we need to know what *points*, *line segments*, and *vertices* are before tackling middle school geometry problems. It's not only necessary that we understand the definitions of these types of terms, we must also be fluent in the conventional notations for them. Rather than jumping straight into the standards for grades 6–8, this lesson contains a review of the building blocks of geometry, which are part of the elementary Common Core State Standards. It's a warm-up of sorts! Make sure to focus on naming angles in the second part of this lesson, since that is a common area where beginning geometry students make mistakes.

Points, Lines, Rays, and Planes

A **point** is the most basic geometric unit. It looks like a dot that you would make with your pen and is named by a single capital letter. Although points have no direction, length, or size, they are essential in helping you name lines, rays, planes, and angles. Point A is demonstrated as such:

$.A$

A **line** is a straight collection of points that extend infinitely in both directions, like the horizon over the ocean. Even though it takes only two points to define a line, lines contain an infinite number of points, which means more points than you can even count. Points that are on the same line are called **collinear**. A line is named by writing any of the two points on it and then placing a double-ended arrow above the letters. So, if you saw the symbol $\overleftrightarrow{AB}$, you would read it "line AB." If you saw this symbol, $\overleftrightarrow{BA}$, you would read it "line BA." The order in which you write the letters does not matter:

$\overleftrightarrow{AB}$ and $\overleftrightarrow{BA}$ both define the line.

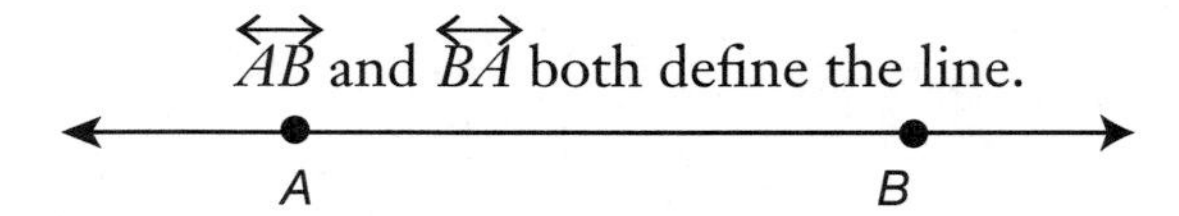

A **line** can also be named by writing a single lowercase letter that has been written beside it:

l can be used to define this line.

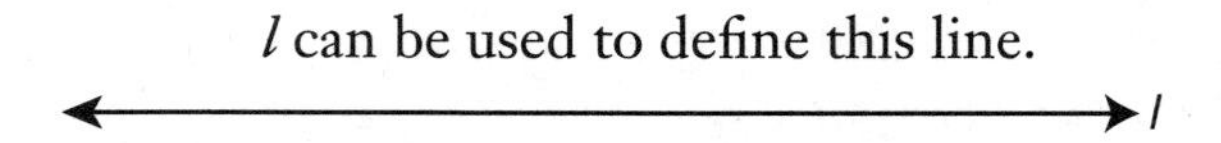

A **line segment** is a piece of a line that does not continue endlessly in both directions. Instead, it is a fixed length that has a beginning point and an endpoint. A ruler is a good example of many line segments. (Line segments still contain a countless number of points.) A line segment is named using the letter endpoints with an arrowless line segment above them.

$\overline{CD}$ defines this line segment.

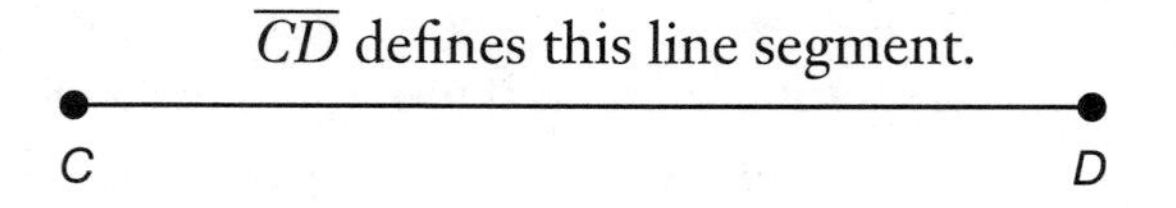

A **ray** is a mix between a line and a line segment. Rays contain one fixed endpoint with the other end of the ray continuing forever. Think of a ray as a shining flashlight. A ray is always named by the endpoint and any other point on the ray, with a one-sided arrow written above.

$\overrightarrow{EF}$ and $\overrightarrow{EG}$ both define this ray.
$\overrightarrow{FG}$ does *not* define this ray.

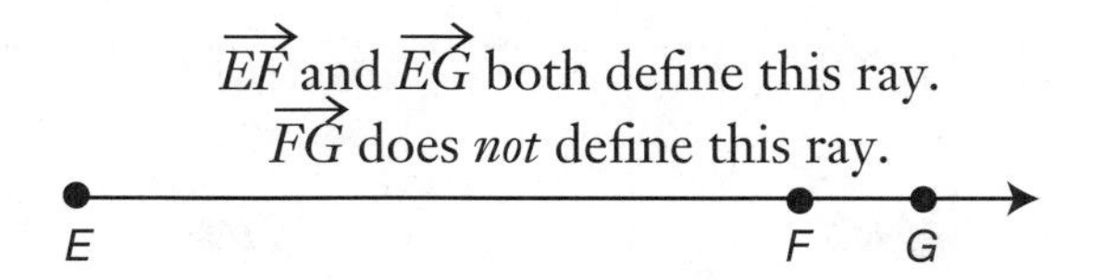

A **plane** is a collection of points that makes a flat surface, which extends continuously in all directions, like an endless wall. Do you remember that a line is created with just two points, and that points that are on the same line are considered to be **collinear**? When points do NOT lie on the same line, they are **non-collinear**. A plane needs a minimum of three non-collinear points to be defined. In geometry, a plane is usually drawn as a four-sided shape with a capital letter in one of its corners, as in the following plane *Q*. Less commonly, planes can be referred to as any three points contained on them, as with the second following plane *XYZ*.

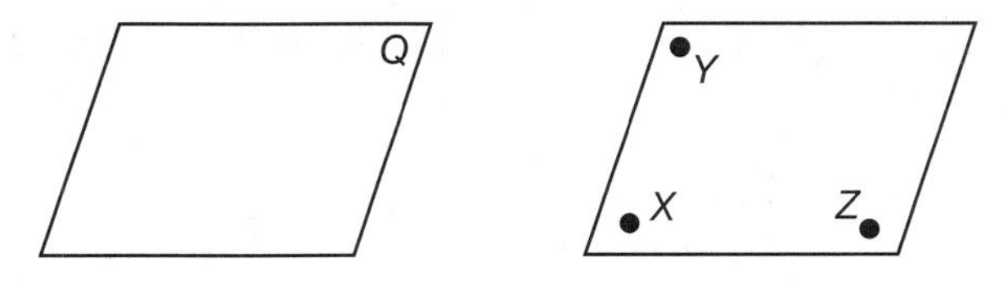

Review the concepts in the table and then complete the practice questions on lines, rays, and planes.

The Basic Building Blocks of Geometry					
Figure	**Name**	**Symbol**	**Read As**	**Properties**	**Examples**
.*A*	Point	.*A*	point *A*	• has no size • has no dimension • indicates a definite location • named with an italicized uppercase letter	• pencil point • corner of a room
A *B* (line with arrows at both ends) *l* ↔	Line	$\overleftrightarrow{AB}$ or $\overleftrightarrow{BA}$	line *AB* or *BA* or line *l*	• is straight • has no thickness • an infinite set of points that extends in opposite directions • one dimension	• highway without boundaries • hallway without bounds
A *B* (ray with arrow at one end)	Ray	$\overrightarrow{AB}$	ray *AB* (endpoint always first)	• is part of a line • has only one endpoint • an infinite set of points that extends in one direction • one dimension	• flashlight • laser beam

A B	Line segment	$\overline{AB}$ or $\overline{BA}$	segment *AB* or *BA*	• is part of a line • has two endpoints • an infinite set of points • one dimension	• edge of a ruler • base board
.B .A C X	Plane	None	plane *ABC* or plane *X*	• is a flat surface • has no thickness • an infinite set of points that extends in all directions • two dimensions	• floor without boundaries • surface of a football field without boundaries

Practice 1

1. What are the different ways to name the line segments in the accompanying figure? (Hint: You should be able to come up with six answers.)

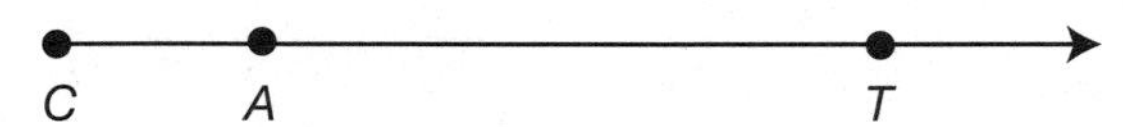

2. In the figure from question 1, what are the different ways to name the ray?

3. What are three ways to name the following line?

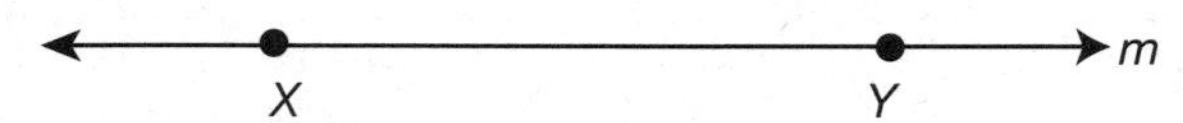

4. What is the fewest number of points needed to create a plane?

5. How many points does a line have?

6. True or false: When you're naming a ray, it does not matter which letter comes first.

7. A hose spraying water is most similar to a line, a line segment, or a ray?

Angles

An **angle** is created when two rays come together at one common endpoint. This endpoint is called the **vertex** of the angle. The angle is the space between the two rays, or sides, and is measured in degrees. The number of degrees an angle has identifies what type of angle it is. You will learn more about that in the next lesson. Either the ∠ symbol or the ∡ symbol are used when discussing angles. An angle can be named in three different ways including:

1. By its vertex (if there are *only* two rays that connect at the vertex).
2. By using three letters, starting with a point on one of the sides, followed by the vertex point, and ending with a point on the remaining side. (The order does not matter as long as the vertex is in the center.)
3. By a number drawn inside the angle to identify it.

The following figure shows the three ways to name an angle.

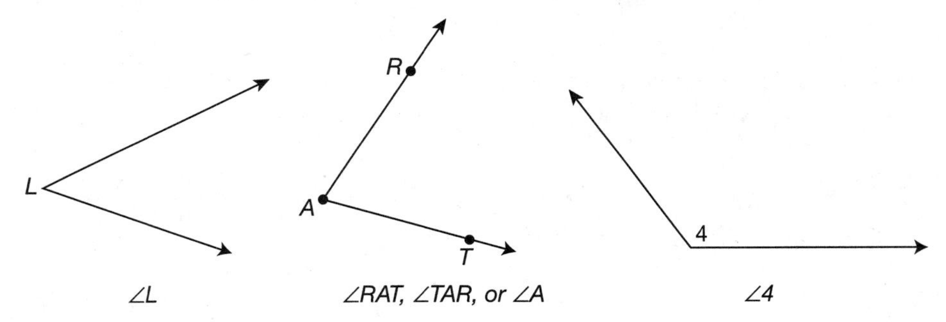

The following figure shows a situation where vertex O is actually the vertex for three different angles: $\angle DOT$, $\angle TOG$, and $\angle DOG$

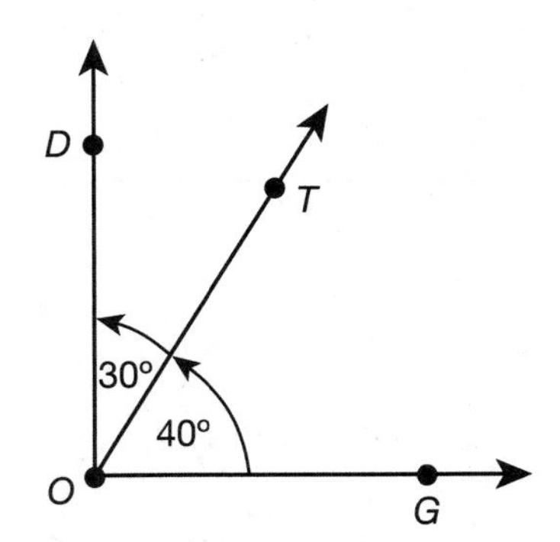

Since more than one angle has its vertex at O, it is incorrect to refer to any of the three angles listed above as $\angle O$:

- The 30° angle must be named $\angle DOT$ or $\angle TOG$
- The 40° angle must be named $\angle TOG$ or $\angle GOT$
- The 70° angle must be named $\angle DOG$ or $\angle GOD$

Practice 2

1. What are three different ways that the following angle can be named?

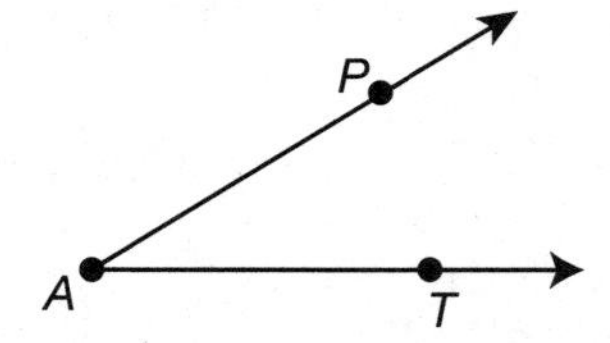

2. An angle is formed when two ________________ come together at a common endpoint.

3. How many different angles in the following figure have ray $\overrightarrow{XB}$ as one of the sides? Name them.

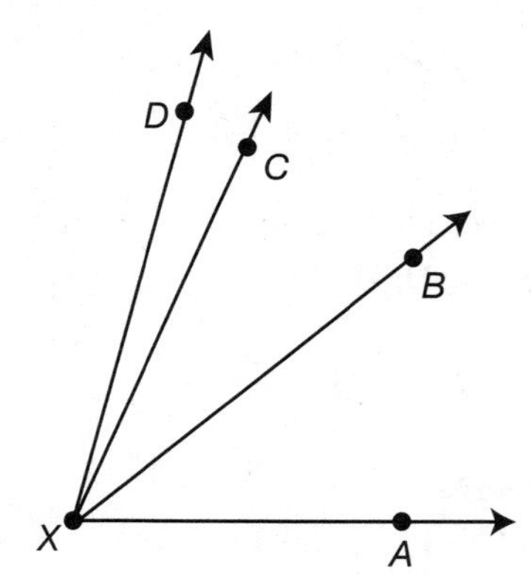

4. Point A in the following figure is called the ________________ of the angle.

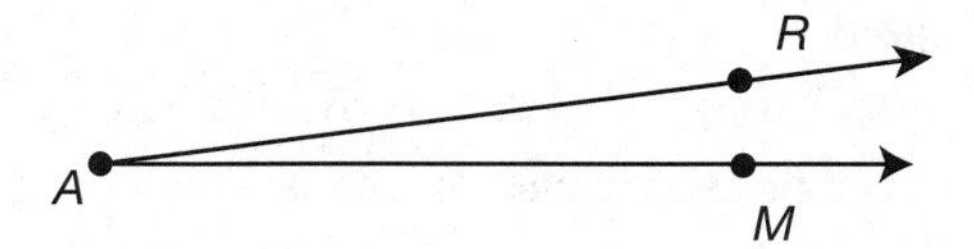

5. Why is it that W in the following figure cannot be used to name the angle?

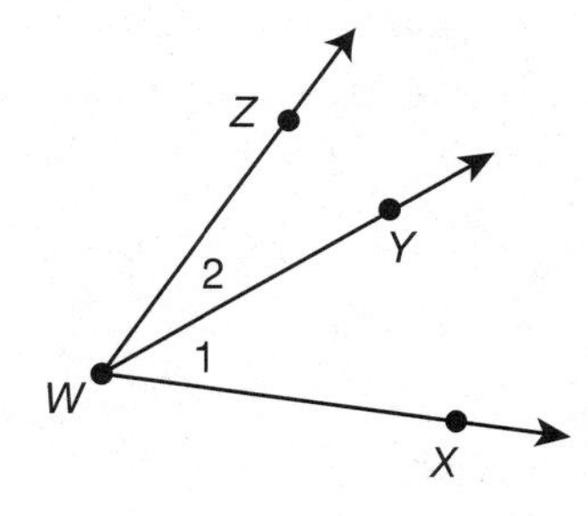

6. Using the figure above, list five ways to name the given angles.

7. True or false: An angle can have two vertices.

Answers

Practice 1

1. The line segments can be named $\overline{CA}$, $\overline{AC}$, $\overline{CT}$, $\overline{TC}$, $\overline{TA}$, or $\overline{AT}$ since the order of the letters does not matter.
2. The ray can only be named $\overrightarrow{CA}$ or $\overrightarrow{CT}$ since the endpoint, C, must come first.
3. $\overleftrightarrow{XY}$, $\overleftrightarrow{YX}$, or m
4. It takes three non-collinear points to make a plane.
5. A line has an infinite, or countless, number of points.
6. False: When you are naming rays, the endpoint must always come first.
7. A hose spraying water is most similar to a ray since there is a definite endpoint.

Practice 2

1. $\angle PAT$, $\angle TAP$, $\angle A$
2. rays
3. Three angles have $\overrightarrow{XB}$ as a side: $\angle AXB$, $\angle BXC$, and $\angle BXD$.
4. vertex
5. W cannot be used to name either of the angles because more than two rays come together at this vertex.
6. $\angle XWY$ or $\angle 1$, $\angle YWZ$ or $\angle 2$, and $\angle XWZ$ (or $\angle ZWX$)
7. False. Angles have only one vertex, where the two sides, or rays, come together.

2

Introduction to Angles and Segments

Where there is matter, there is geometry.
—Johannes Kepler

In this lesson you will learn how to use a protractor to measure and create angles. You will also learn the basics of measuring line segments.

STANDARD SNEAK PREVIEW

Lots of interesting careers require the accurate drawing of angles and geometric shapes to create miniature models of real-world products. Engineers, mechanical drafters, and cartographers (mapmakers) all need to be familiar with how to construct angles, and the Common Core State Standards want to make sure you can, too! In this lesson we are going to lay the foundation for *Standard 7.G.A.2, which requires students to construct geometric shapes with specific angle and side length conditions.* We are going to focus on how to do this by hand, but as you can imagine, many careers now use technology to draw shapes and render plans. If you have access to a computer, you should be able to find some online tools that can be used to explore the construction of angles and shapes. There are some highly sophisticated computer programs on the market that offer affordable student rates, and some are even free.

Using a Protractor to Measure Angles

A **protractor** is a special ruler that is used to measure and draw angles. Unlike distances, which can be measured in various units like inches, feet, or miles, angles are measured only in **degrees**. Degrees are always notated with the symbol °. The expression "$m\angle B = 40°$" is read, *The measure of an angle B is 40 degrees.*

Most protractors have two scales along their arc. The lower scale, starting with zero on the right, is used to **measure angles** that have a "bottom" ray pointing out to the right and open in a counterclockwise arc. To measure a counterclockwise angle, place the vertex in the circular opening at the base of the protractor and extend the bottom ray through the 0° mark out to the right. Then read the angle's measure using the bottom scale. In the following figure, the counterclockwise angle drawn is 15°.

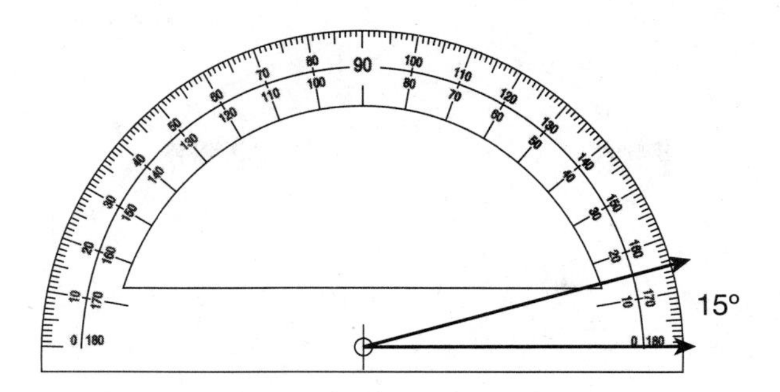

The second scale along the top, which begins with zero on the left, measures angles that have a "bottom" ray pointing out to the left and open in a clockwise arc. To measure a clockwise angle, place the vertex in the circular opening at the base of the protractor and extend the bottom ray through the 0° mark out to the left. Then read the angle's measure using the top scale. In the following figure, the clockwise angle drawn is 120°.

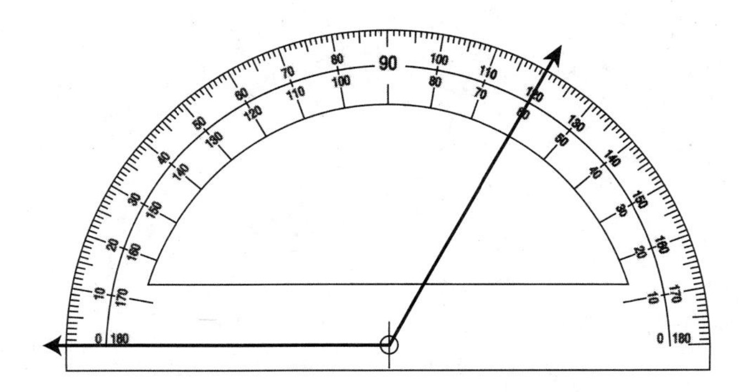

Notice that whether you're reading the top or bottom scale, that 90° is directly in the center. It's good to check that you're using the correct scales by making sure that angles less than 90° are tighter and smaller than this midway point, and that angles greater than 90° are wider and more open. Sometimes, you will need to measure an angle that does not pass through 0°, but instead passes through two different measurements. When this happens, subtract the smaller measurement from the larger measurement in order to identify the number of degrees in the angle. For example, say that in the following figure, you need to find the $m\angle BXC$. $\overrightarrow{XB}$ goes through 30° and $\overrightarrow{XC}$ goes through 110°, so in order to find $m\angle BXC$, subtract 30° from 110° to get 80°. Therefore, $m\angle BXC = 80°$.

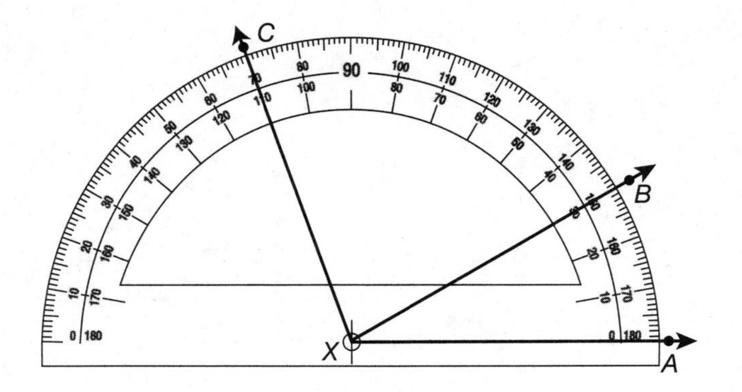

Practice 1

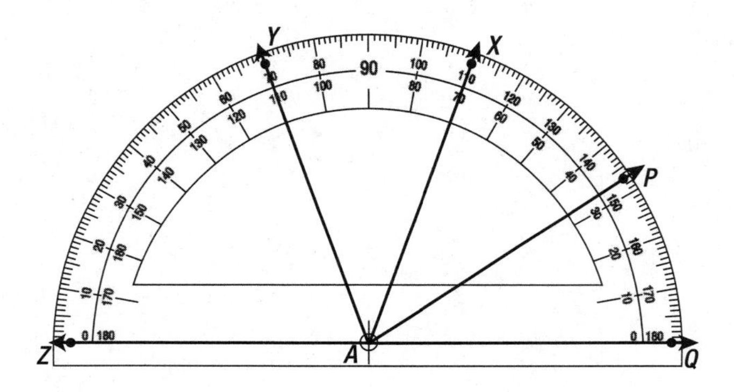

Use the protractor to find the measure of the given angles.

1. $m\angle QAP =$

2. $m\angle PAX =$

3. $m\angle QAY =$

4. $m\angle ZAX =$

5. $m\angle ZAP =$

6. $m\angle PAY =$

7. $m\angle QAZ =$

Using a Protractor to Draw Angles

STANDARD ALERT!

Standard 7.G.A.2 requires students to construct geometric shapes with specific angle and side length conditions. Get ready to learn your way around a protractor!

Using a protractor to draw angles is easy once you understand how to measure angles. First, draw a ray pointing out toward the right. Place the protractor's circular opening over the point that will be the angle's vertex. Then align that ray with the 0° mark on the right. Remember to use the lower scale when you find the degree of the angle you are drawing. Make a dot there that marks the endpoint of the second ray. Last, remove the protractor and connect the vertex with your drawn endpoint. In the following figure, we are ready to make a 75° angle.

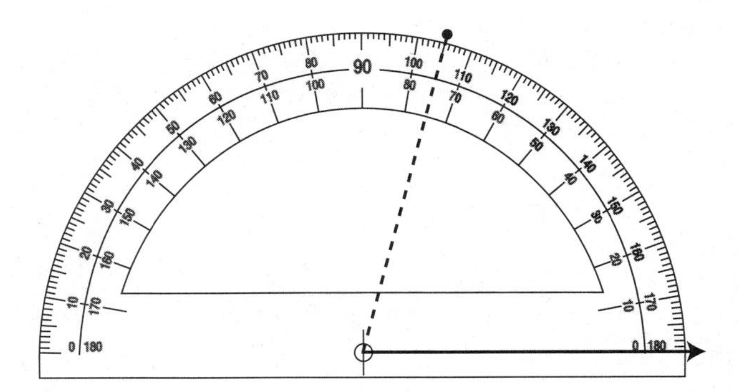

In the previous protractor illustration, the starting ray for the 75° angle was a right-facing ray. Therefore, the lower scale was used when making the dot and accompanying line segment to complete the angle. Remember that when beginning with a ray that is pointing out to the 0° mark on the *left*, the scale along the *top* of the protractor must be used. In the following protractor illustration, a dot has been drawn to create a 105° angle. Notice that if this angle were combined with the 75° angle in the previous illustration, together they would make a straight line. What do these two angles sum to? We will revisit this special relationship in the next lesson.

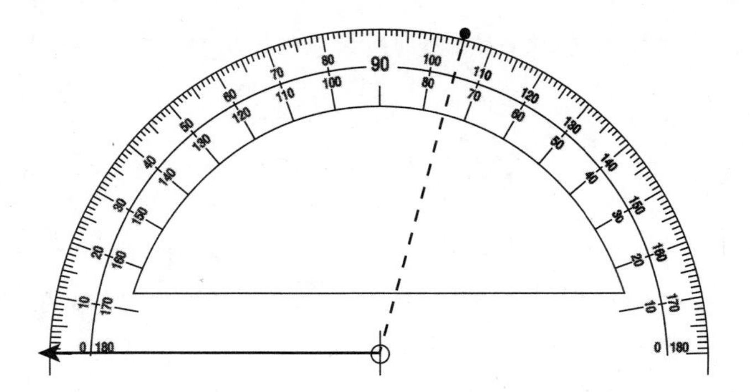

Practice 2

Beginning with a right-facing ray, use a protractor to construct angles with the given measures.

1. 45°

2. 75°

3. 100°

4. 125°

5. 32°

Beginning with a left-facing ray, use a protractor to construct angles with the given measures.

6. 170°

7. 80°

8. 5°

9. 90°

10. What do you notice about the 90° angle formed with a right-facing ray and the 90° angle formed with a left-facing ray?

11. What do you notice about the 100° angle formed with a right-facing ray in question 3 and the 80° angle formed with a left-facing ray in question 8?

Measuring Line Segments

Line segments are easy to measure when you have a ruler, but often in geometry you will be asked to determine the length of a missing segment with only limited information. In order to indicate the length of a line segment, two letters defining the segment are written with a small line over them. For example, $\overline{BC}$ can be read, *The length of line segment* $\overline{BC}$.

As long as points are all on the same line, the sum of all of the non-overlapping segments will equal the length of the entire line segment. For example, notice that in the following figure, $\overline{AB} = 5$ and $\overline{CD} = 6$, but that segment $\overline{BC}$ has no given measurement. If you were given that $\overline{AD} = 18$, you could determine that $\overline{BC}$ measures 7 units, by subtracting 5 and 6 from 18.

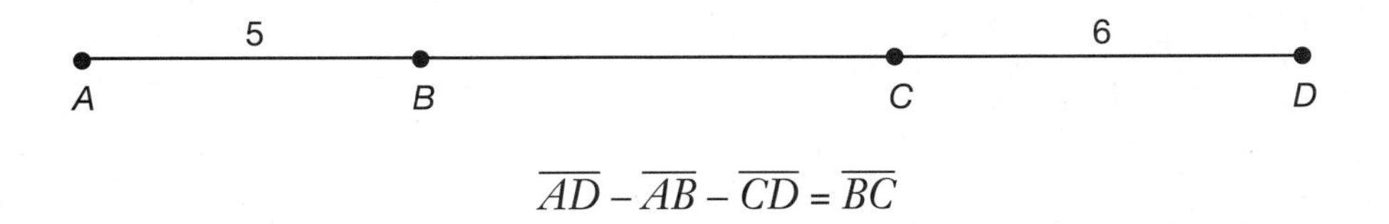

$$\overline{AD} - \overline{AB} - \overline{CD} = \overline{BC}$$

Therefore since 18 – 5 – 6 = 7, BC = 7 units.

Determining segment length gets more challenging when some of the given segments are overlapping. For example, consider this figure.

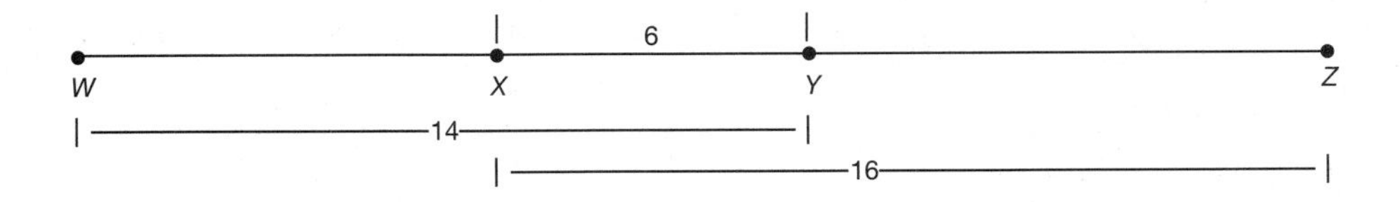

If you were given that $\overline{WY} = 14$, $\overline{XY} = 6$, and $\overline{XZ} = 16$, how would you determine the length of $\overline{WZ}$? You cannot add all three segments together, since $\overline{XY}$ would be included three times, since it is part of both $\overline{WY}$ and $\overline{XZ}$. Therefore, the correct way to do this is to add $\overline{WY}$ and $\overline{XZ}$, and then subtract $\overline{XY}$. This would look like: (14 + 16) – (6) = 24 units.

Midpoints and Bisectors

Line segments can have both **midpoints** and **bisectors**. A **midpoint** is a point that divides a line segment into two segments of equal length. A **bisector** is a line, ray, or segment that goes through a line segment's midpoint, dividing the segment into two equal parts. In the next figure, point I is the midpoint of $\overline{JM}$. In the same illustration, $\overline{BN}$ is the bisector of $\overline{JM}$ since it passes through midpoint I.

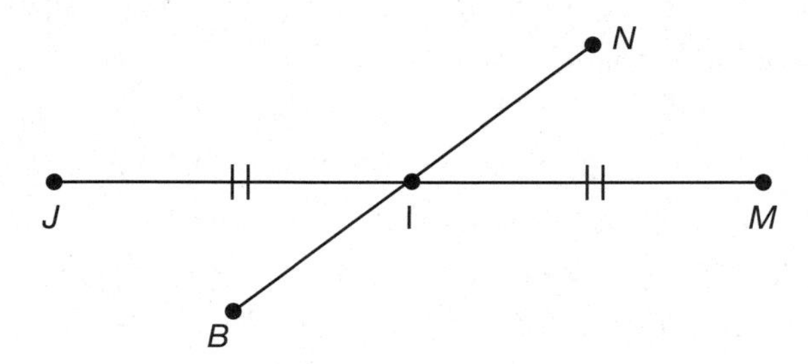

Similar to a line segment bisector, an angle bisector is a ray that passes through the vertex of an angle and divides it into two equal angles. In the next figure, $\overrightarrow{YW}$ bisects $\angle XYZ$ and therefore $m\angle XYW = m\angle WYZ$. When two angles have equal measures, it is illustrated by drawing arcs with identical hash marks through them. You will learn more about this in the following lessons.

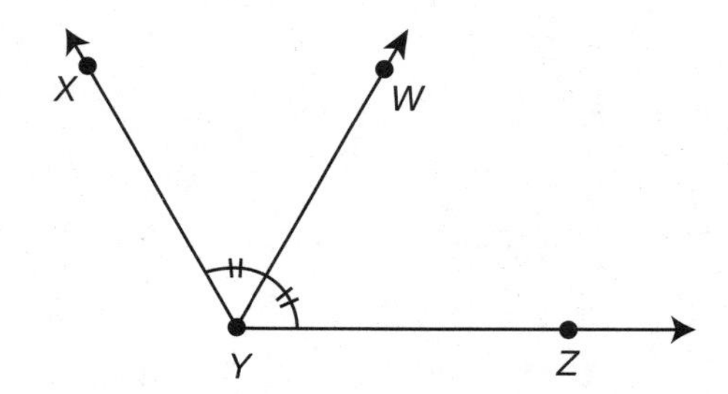

Practice 3

Use the following figure to answer questions 1 through 5. Note: The figure is NOT drawn to scale.

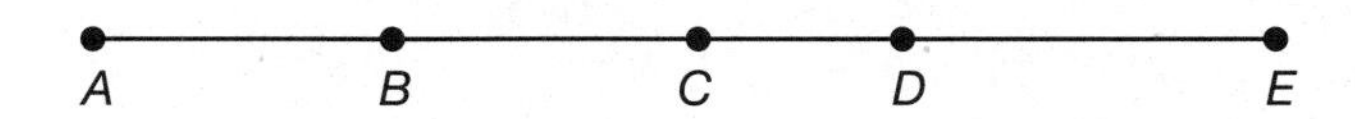

1. If $\overline{AD} = 8$, $\overline{CE} = 6$, and $\overline{CD} = 2$, what is the length of $\overline{AE}$?

2. If $\overline{AB} = 13$, B is the midpoint of $\overline{AC}$, and C is the midpoint of $\overline{AE}$, what is the length of $\overline{AE}$?

3. If $\overline{AB} = 3.4$, $\overline{BD} = 7.6$, and $\overline{AE} = 14.8$, what is the length of $\overline{DE}$?

4. If B is the midpoint of $\overline{AC}$, and C is the midpoint of $\overline{AE}$, what is the length of $\overline{BC}$ if $\overline{AE} = 56$?

5. If $\overline{AD}$ is twice as long as $\overline{DE}$, and $\overline{AE} = 36$, what is the length of $\overline{AD}$?

Answers

Practice 1

1. 33°
2. 37° (70° – 33° = 37°)
3. 110°
4. 110°
5. 147°
6. 77° (110° – 33° = 77°)
7. 180°

Practice 2

1.

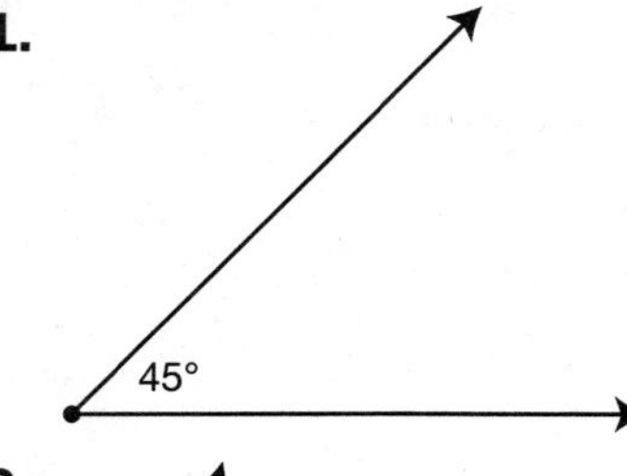

2.

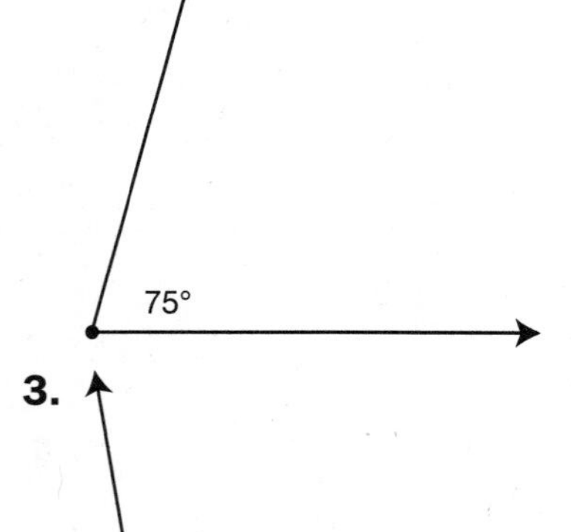

3.

100°

4. 

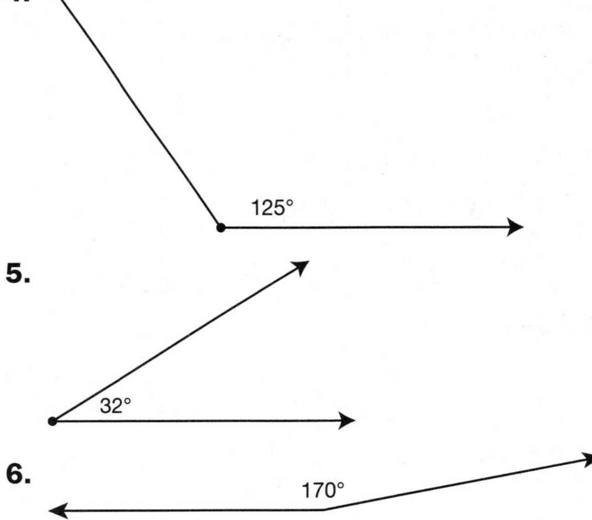

5.

6.

7.

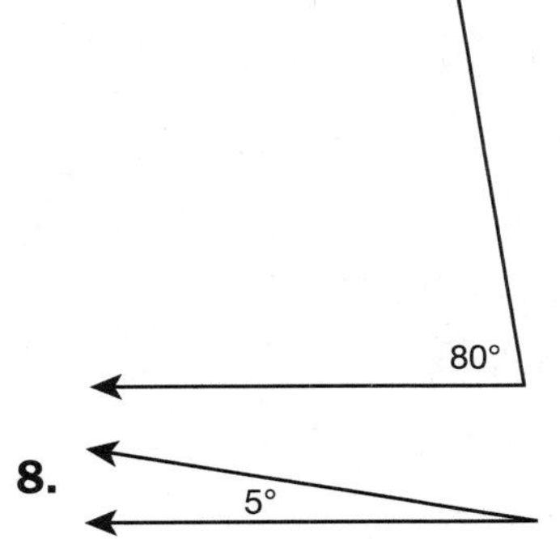

8.

9.

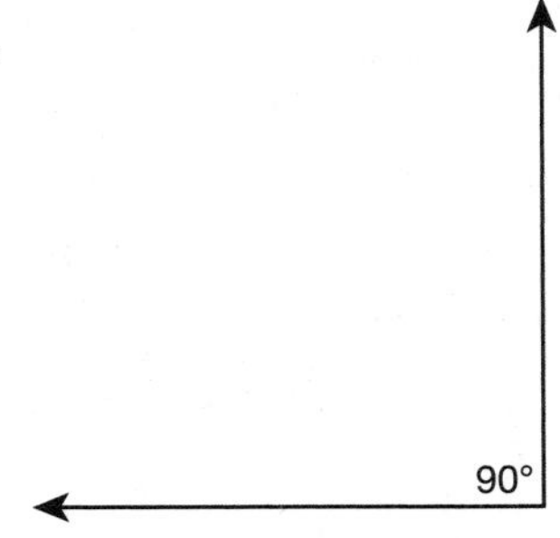

10. The 90° angle formed with a right-facing ray and the 90° angle formed with a left-facing ray look the same, but are just facing opposite directions. If you pressed them together, they would make a straight line.

11. The 80° angle formed with a left-facing ray in question 8 ends at the same ray that ends the 100° angle formed with a right-facing ray in question 3. If you pressed them together, they would make a straight line.

Practice 3

1. $\overline{AE}$ **= 12.** Add $\overline{AD}$ = 8 to $\overline{CE}$ = 6 and subtract $\overline{CD}$ = 2, since $\overline{CD}$ is counted twice where the two segments overlap.
2. $\overline{AE}$ **= 52.** Since $\overline{AB}$ = 13 and B is the midpoint of $\overline{AC}$, then $\overline{AC}$ = 26. Since C is the midpoint of $\overline{AE}$, then $\overline{AE}$ = 52.
3. $\overline{DE}$ **= 3.8.** Subtract the smaller segments from the whole: 14.8 – 7.6 – 3.4 = 3.8.
4. $\overline{BC}$ **= 14.** Since $\overline{AE}$ = 56 and C is its midpoint, $\overline{AC}$ = 28. Since B is the midpoint of $\overline{AC}$, then $\overline{BC}$ =14.
5. $\overline{AD}$ **= 24.** Since $\overline{AD}$ is twice as large as $\overline{DE}$, let $\overline{AD}$ = $2x$ and $\overline{DE}$ = x and then write the formula, $2x + x = 36$. Solving for x you see that $x = 12$, so $\overline{AD}$ is twice that.

3

Angles in the Plane

Geometry is just plane fun.
—Anonymous

In this lesson you will learn how to classify angles in one of four categories: acute, right, obtuse, and straight. You will also learn about complementary and supplementary angle relationships and perpendicular lines.

STANDARD SNEAK PREVIEW

A talented sports coach carefully looks for skillful relationships between her players so that she can use them in pairs to work well together. In geometry, it is similarly important to seek out special pairs of angles that work together to fit a certain criterion. Knowing how to name, measure, and construct angles accurately are important skills we focused on in the past two lessons and now we are going to build upon this foundation. ***Standard 7.G.B.5 requires students to use facts about supplementary and complementary angles in order to write and solve equations for unknown angles in figures.*** Although we'll save our equation writing and solving skills for Lesson 4, get ready to become an expert on supplementary and complementary angles in the second half of this lesson!

Defining Angles

Angles are defined by their relationships to the degree measures of 90° and 180°. It's key to remember that 90° angles look like the corner of a page from this book, and 180° angles look like a flat line. Although geometric illustrations are not always perfectly drawn to scale, you should be able to tell which of the following four categories an angle falls into: acute, right, obtuse, or straight.

Acute Angles

An angle that is larger than 0° but smaller than 90° is called an **acute angle**. (A clever way to remember this is to think, "Oh, isn't that small angle so *cute*!")

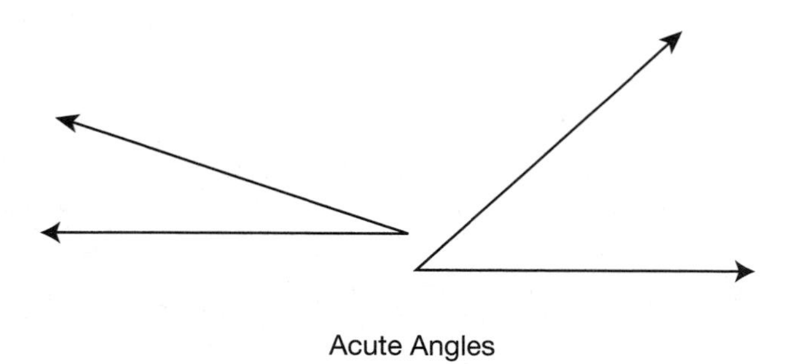

Acute Angles

Right Angles

An angle that measures 90° is called a **right angle**. Since right angles are extra special angles, they have their own symbol to show they are 90°. A small square is drawn at the vertex of right angles to note their unique nature, as shown in the following figure. *It is critical to remember that unless an angle has this special mark, or it is otherwise a given that it is a 90° angle, you cannot assume it is a right angle.* Kind of like how in our justice system, people are "innocent until proven guilty," it should be said that "angles are not 90° until proven to be right angles!" We'll learn about the special properties of right angles in the following lesson. (It might help you to remember that a right angle has a side that rises directly up from its base by thinking, "Oh, isn't that angle so up*right*!")

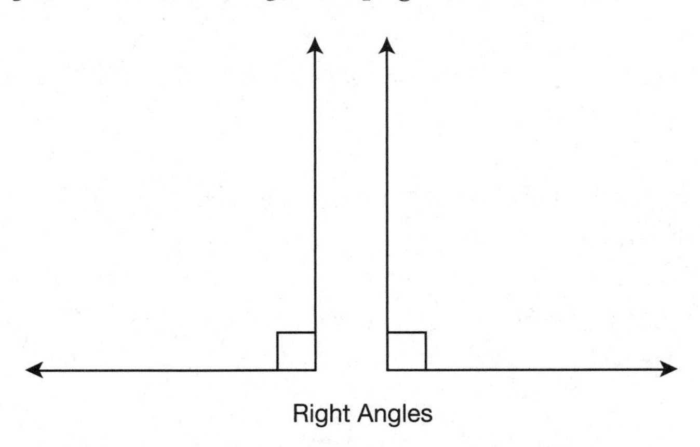

Right Angles

Obtuse Angles

An angle that is larger than 90° but smaller than 180° is called an **obtuse angle**. (Do you like rhymes? "That *obtuse* is as big as a *moose*!")

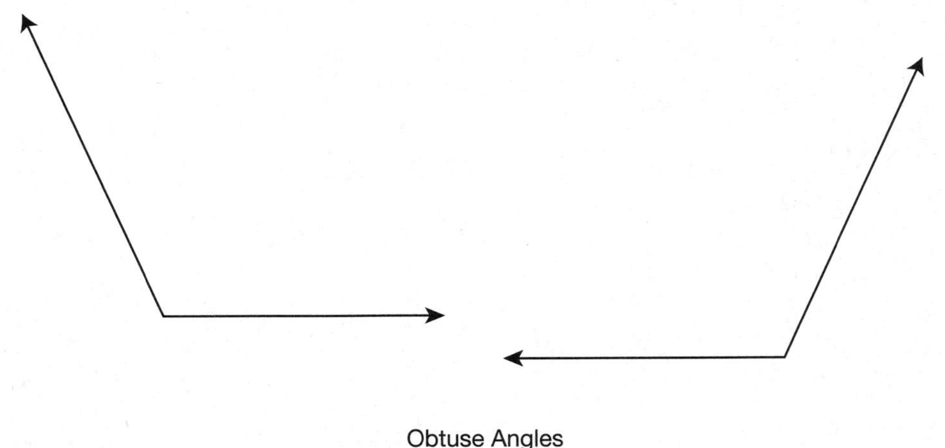

Obtuse Angles

Straight Angles

An angle that is 180° looks like a straight line and is often called a **straight angle**. Take note that there are no names for angles larger than 180°.

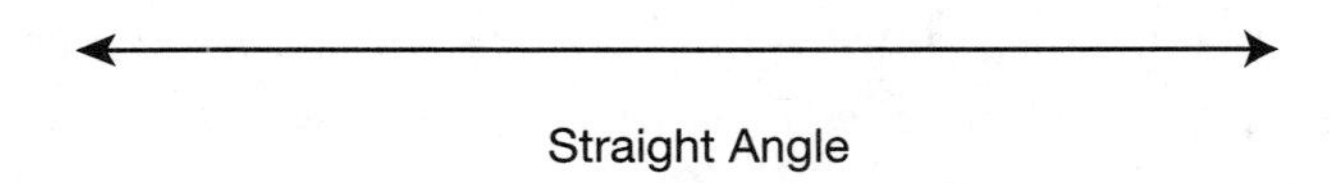

Straight Angle

Practice 1

Answer the following questions about angle classifications.

1. If a straight angle is bisected, the two angles that are created are ______________.

2. If a straight angle is broken into three angles of equal measure, then each of those angles is ____________________.

3. True or false: An acute angle plus an acute angle will always give you an obtuse angle.

4. A straight angle minus an obtuse angle will give you an acute angle. Is this statement sometimes, always, or never true?

5. A right angle plus an acute angle will give you an obtuse angle. Is this statement sometimes, always, or never true?

6. An obtuse angle minus an acute angle will give you an acute angle. Is this statement sometimes, always, or never true?

7. Which of the following angles is not acute, obtuse, right, or straight?
   a. 180°
   b. 167°
   c. 195°
   d. 3.7°

Use the following images to complete questions 8 through 11.

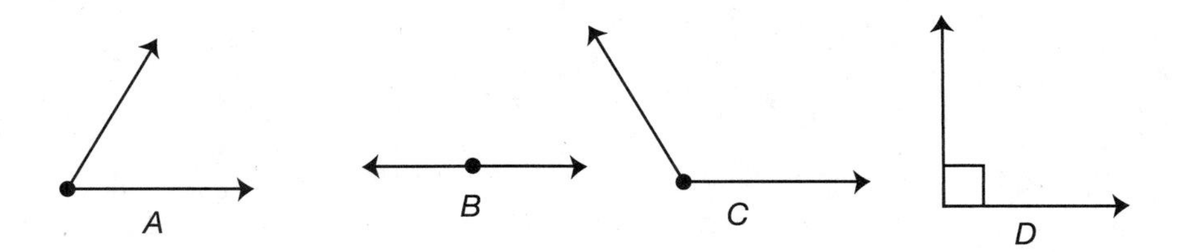

8. Which angle is acute?

9. Which angle is a right angle?

10. Which angle is a straight angle?

11. Which angle is obtuse?

Supplementary and Complementary Angles

STANDARD ALERT!

Standard 7.G.B.5 requires students to use facts about supplementary and complementary angles in order to write and solve equations to solve for an unknown angle in a figure. Make sure you are completely clear on these definitions!

Now you know how to classify angles based on their specific degree measurement. Next, you will learn about the special relationships that can exist between two or more angles.

Complementary Angles

When two angles have measurements that sum to 90°, we say that the angles are **complementary**. The next figure shows two pairs of complementary angles. Notice that complementary angles can be two completely separate angles, as with $\angle 1$ and $\angle 2$. Also, complementary angles can share a common side and form a right angle together, as with ∠3 and ∠4. Whenever acute angles form a right angle, they are complementary.

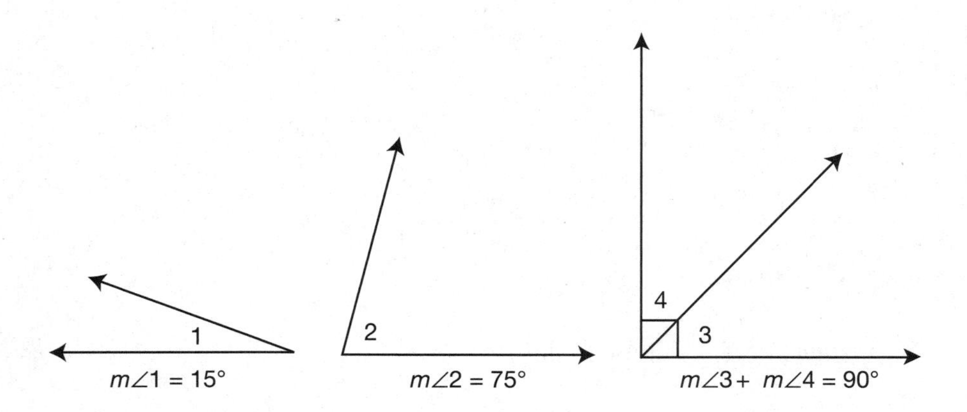

Supplementary Angles

When two angles have measurements that sum to 180°, we say that the angles are **supplementary**. The next figure shows two pairs of supplementary angles. Supplementary angles can be two completely separate angles, as with ∠5 and ∠6. Supplementary angles can also share a common side and form a straight angle, as is the case with ∠7 and ∠8. When supplementary angles are adjacent and form a straight angle, they are called a **linear pair**.

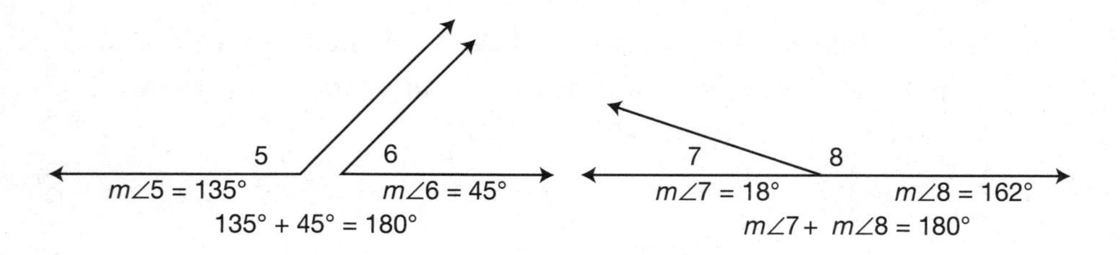

Practice 2

1.

1

2

Are angles 1 and 2 supplementary, complementary, or neither?

For questions 2 through 5, find the measure of the complement and the supplement for each of the given angles.

2. $\angle 2 = 15°$: complement = _____ ; supplement = _____

3. $\angle 3 = 78.5°$: complement = _____ ; supplement = _____

4. $\angle 4 = 137°$: complement = _____ ; supplement = _____

5. $\angle 5 = 180°$: complement = _____ ; supplement = _____

For questions 6 through 9, use the following illustration:

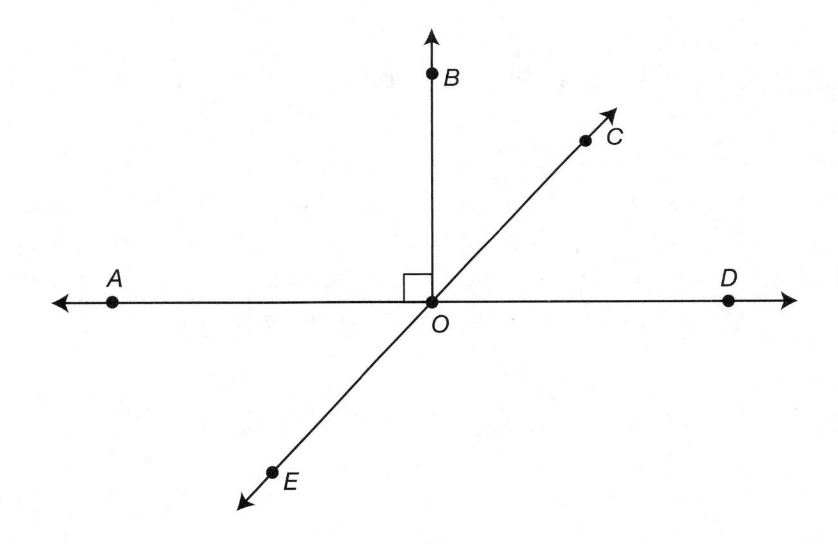

6. Name two supplementary angles of $\angle DOE$.

7. Name a pair of complementary angles.

8. Name two right angles.

9. Name two straight angles.

For questions 10 through 14, state whether the following statements are true or false.

10. Complementary angles must be acute.

11. Supplementary angles must be obtuse.

12. Two acute angles can be supplementary.

13. Any two right angles are supplementary.

14. Two acute angles are always complementary.

Right Angles and Perpendicular Lines

Earlier in this lesson you learned that a right angle contains 90°. When two lines intersect to form a 90° angle, those lines are called **perpendicular**. The symbol $\perp$ is used to indicate that two lines are perpendicular. Another way to show that two lines are perpendicular is to draw a small square in the angle where the lines meet, as in the following figure.

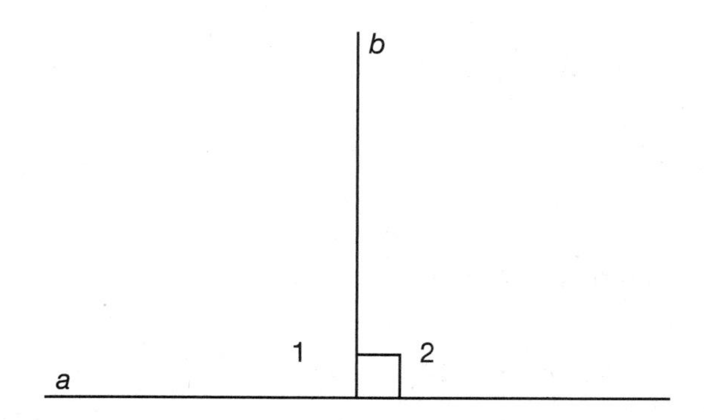

It would be correct here to write, "$a \perp b$." Notice that $\angle 1$ and $\angle 2$ make a straight angle that measures 180°. Although only $\angle 2$ is marked as being 90°, since $\angle 1$ and $\angle 2$ are supplementary, it must be true that $m\angle 1$ is also equal to 90°.

TIP: When two adjacent angles form a straight angle, if one of the angles is a right angle, then the second angle is also 90°.

Remember, in geometry illustrations it is not guaranteed that figures will be drawn to scale. Therefore, you can never assume that two lines are perpendicular unless you see a square in the angle, the symbol ⊥, or you are told that the angle is 90°!

Practice 3

Use the figure below to answer the following questions.

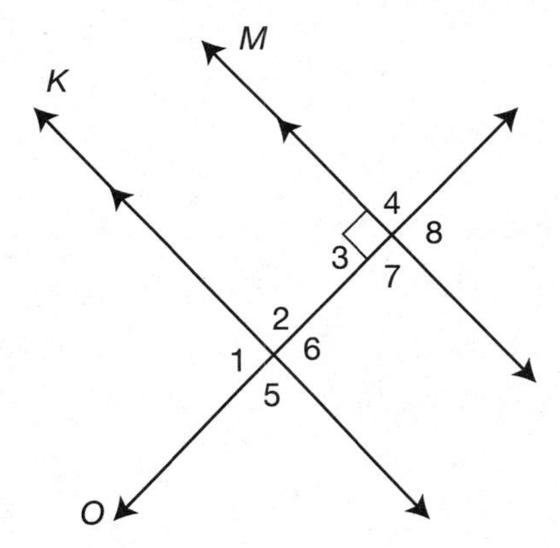

1. If $m\angle 7 = (3x - 9)°$, find the value of x.

2. True or false: Angles 3 and 4 are complementary.

3. True or false: You cannot assume that $\angle 4$ is a right angle since it does not have the box drawn in.

4. If $m\angle 1 = (5x + 10)°$ and $m\angle 4 = (8x - 2y)°$, find the values of x and y.

Answers

Practice 1

1. right angles $\frac{180°}{2} = 90°$
2. an acute angle $\frac{180°}{3} = 60°$
3. False. Two small acute angles do not have to have a sum greater than 90°.
4. Always. Such as: 180 – 102 = 78
5. Always
6. Sometimes. 170° – 10° = 160°, but 100° – 60° = 50°
7. **c.** Remember, an obtuse angle must be *between* 90° and 180°.
8. A
9. D
10. B
11. C

Practice 2

1. supplementary
2. complement = 75°; supplement = 165°
3. complement = 11.5°; supplement = 101.5°
4. complement = does not exist; supplement = 43°
5. 180° is a straight angle and it does not have a complement or a supplement.
6. $\angle AOE$ and $\angle COD$
7. $\angle BOC$ and $\angle COD$
8. $\angle AOB$ and $\angle DOB$
9. $\angle AOD$ and $\angle EOC$
10. True
11. False—There are two cases: Either one angle must be obtuse and the other angle must be acute *or* both angles must be right angles.
12. False
13. True
14. False—Acute angles *can* be complementary, but they do not *have* to be.

Practice 3

1. $x = 33$
2. False; they are supplementary.
3. False; since $\angle 3$ and $\angle 4$ form a straight line, and $m\angle 3 = 90°$, then $m\angle 4 = 90°$.
4. $x = 16$ and $y = 19$

4

Special Pairs of Angles

Geometry is the foundation of all painting.
—Albert Dürer

In this lesson you will learn how to identify adjacent and vertical angles and then you will gain an understanding of the relationships between these types of angles.

STANDARD SNEAK PREVIEW

Do you recall the sports coach in the introduction to Lesson 3, who was looking for pairs of players to work together smoothly? She will probably be looking for several pairs of athletes to work together to fulfill various roles—offense, defense, team leadership, etc. Now that you're proficient with supplementary and complementary angles, we want to show you the rest of *Standard 7.G.B.5: students will use facts about supplementary, complementary, vertical, and adjacent angles in order to write and solve equations for unknown angles in a figure*. So now, not only are you going to get your feet wet with writing and solving equations for unknown angles, you're also going to learn about a few other special pairs of angles. Ready to brush off those algebra skills?

Adjacent Angles

Adjacent angles are angles that share a common vertex, have one common side, and have no interior points in common. In the next figure, $\angle ZWY$ and $\angle YWX$ are adjacent since they share side WY. Since they are not overlapping, they also have no common interior points.

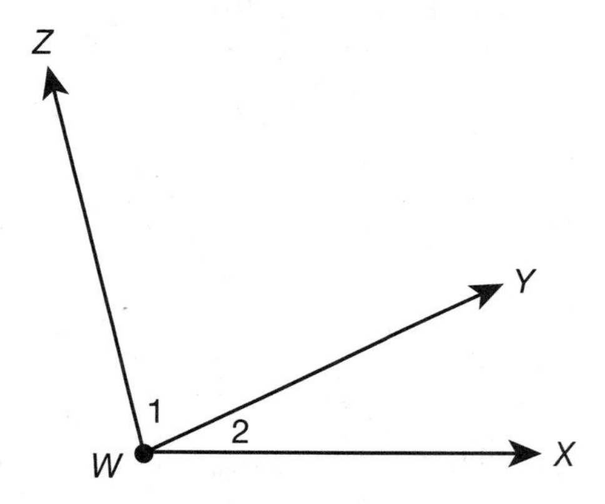

It is important to understand that in this figure, $\angle ZWY$ and $\angle ZWX$ are NOT adjacent since the interior of $\angle ZWY$ is included in $\angle ZWX$. Remember, adjacent angles do not have any interior points in common. The word *adjacent* means *next to*, so adjacent angles are next to each other, but cannot overlap with one another.

Just as with numbers and variables, addition and subtraction can be used to combine or separate measures of angles. Addition and subtraction are especially helpful when you are working with adjacent angles. For example, in the given figure, $m\angle 1 + m\angle 2 = m\angle ZWX$.

Another way to look at this could be to use subtraction in the following way: $m\angle ZWX - m\angle 1 = m\angle 2$.

TIP: The sum of two adjacent angles equals the measure of the larger angle formed by their non-common sides.

You should remember from the last lesson that two angles are supplementary when they sum to 180°. When two adjacent angles are supplementary, they are referred to as a **linear pair** and they form a straight angle, as in the following figure:

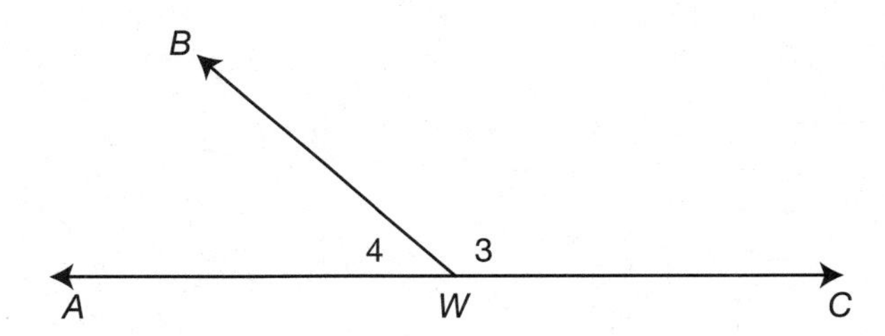

In this figure, subtraction can be used to form the following relationships:

$$180° - m\angle 3 = m\angle 4$$
$$180° - m\angle 4 = m\angle 3$$

STANDARD ALERT!

The two equations above will help you solve for unknown adjacent angles, which is part of Standard 7.G.B.5. Be sure to apply what you previously learned about supplementary and complementary angles to solve the problems in Practice 1. Several of these questions will require you to apply your algebra skills to write and solve equations with variables.

Practice 1

Use this figure and the given angle measurements to answer questions 1 through 5.

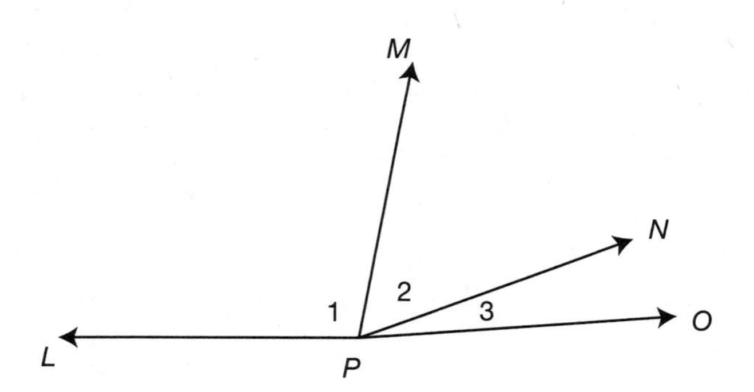

1. $m\angle 1 = 93°$
$m\angle 2 = 48°$
Find $m\angle LPN$.

2. $m\angle MPO = 78°$
$m\angle 2 = 42°$
Find $m\angle 3$.

3. $m\angle 1 = 3x$
$m\angle 2 = 2x$
$m\angle 3 = x$
Find $m\angle LPO$ in terms of x.

4. $m\angle LPO = 175°$
$m\angle 3 = 33°$
$m\angle 1 = 95°$
Find $m\angle 2$.

5. $m\angle 1 = (4y + 15)°$
$m\angle 2 = (3y - 6)°$
$m\angle LPN = 149°$
Find $m\angle 1$ and $m\angle 2$.

Use the following figure and the given angle measurements to answer questions 6 through 11. Note that for some of the angle measurements given below, the figure is not drawn to scale.

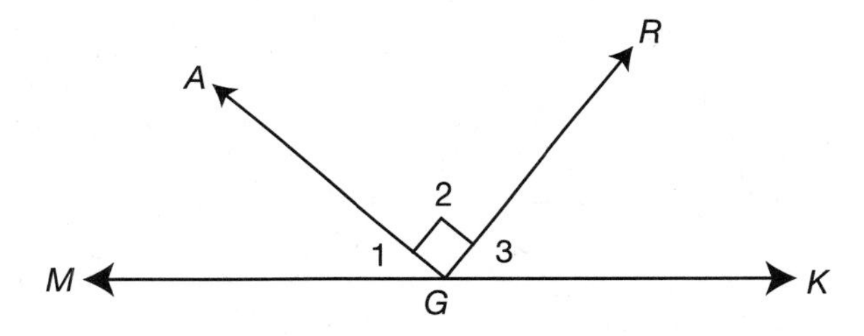

6. $m\angle 1 = 37.5°$
Find $m\angle MGR$.

7. $m\angle 3 = (2x + 10)°$
$m\angle AGK = 144°$
Find the value of x.

8. If $m\angle 1$ is one-half the $m\angle 3$, find the measure of angles 1 and 3. (*Remember:* $\angle AGR$ is a right angle and $\angle MGK$ is a straight angle.)

9. True or false: Angles 1 and 3 are adjacent angles.

10. Definition: Angles 1, 2, and 3 are three adjacent angles that form a ________________ angle.

11. $m\angle 1 = (3x + 3)°$
$m\angle 3 = (6x - 3)°$
Find $m\angle 1$, $m\angle 3$, and $m\angle AGK$.

Vertical Angles

Draw a thin and tall letter X on a piece of paper. Do you notice that the two wider angles formed on the left and right sides of the X appear to be the same size? And do you notice that the smaller angles formed above and below the X also appear to be the same size? The types of angles we are investigating here are called **vertical angles**. **Vertical angles** are the non-adjacent angles formed when any two line segments or lines intersect. (Non-adjacent angles do not share a common side, so they are the angles that are across from each other, rather than next to each other.) **Vertical angles** are **congruent**, which means they are equal in size and have the same angle measurement. When two line segments cross, two pairs of vertical angles are always formed. Vertical angles are sometimes referred to as **opposite angles**, since they are opposite each other, but this name is not as commonly used.

TIP: When two rays, line segments, or lines intersect, two pairs of vertical angles are formed.

In the following figure, angles 1 and 3 are one pair of congruent vertical angles. Angles 2 and 4 are another pair of congruent vertical angles. Notice how vertical angles are not adjacent, since they do not share a common side, but instead, they are opposite each other.

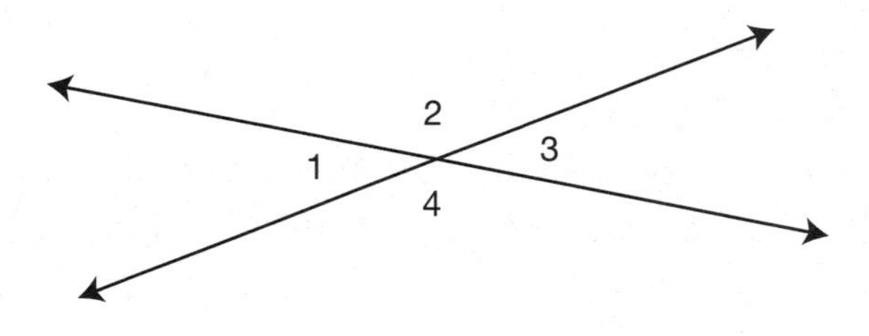

TIP: Vertical angles are congruent.

Most of the time, intersecting lines will form one pair of obtuse vertical angles and one pair of acute vertical angles, as in the preceding figure. Angles 1 and 3 are a pair of congruent acute angles, while angles 2 and 4 are a pair of congruent obtuse angles. In this case, the obtuse vertical angles will be congruent and the acute vertical angles will also be congruent.

Vertical Right Angles

Another case is when two lines are perpendicular and intersect to form a 90° angle. In this case, the vertical angles are also right angles, so whenever one angle between intersecting lines is 90°, then all four of the resulting angles are also 90°. Can you think of how you could prove that this statement is true by using what you know about vertical, adjacent, and supplementary angles? Vertical right angles are illustrated in the following figure, where $g \perp h$ and $\angle 1 = \angle 2 = \angle 3 = \angle 4 = 90°$.

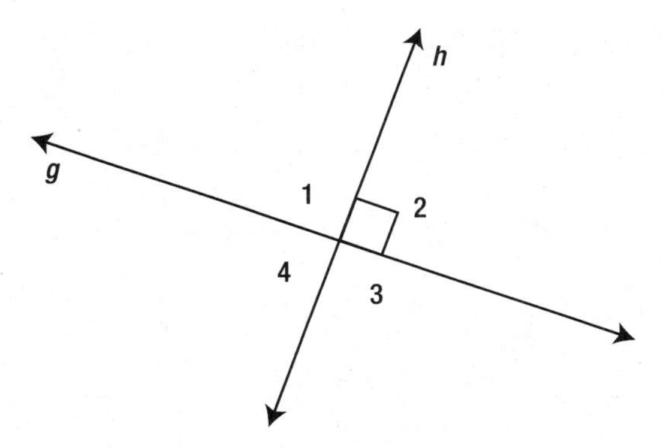

TIP: When two intersecting lines are perpendicular, four right angles are formed.

STANDARD ALERT!

Now that you understand the special relationship between vertical angles, combine that with your knowledge of supplementary, complementary, and adjacent angles to solve for unknown information in the complex illustration in Practice 2. Standard 7.G.B.5 is a very important one, so you'll have more opportunities to practice it in following lessons, but make sure you're confident with these concepts before moving ahead!

Practice 2

Use this figure to answer questions 1 through 5.

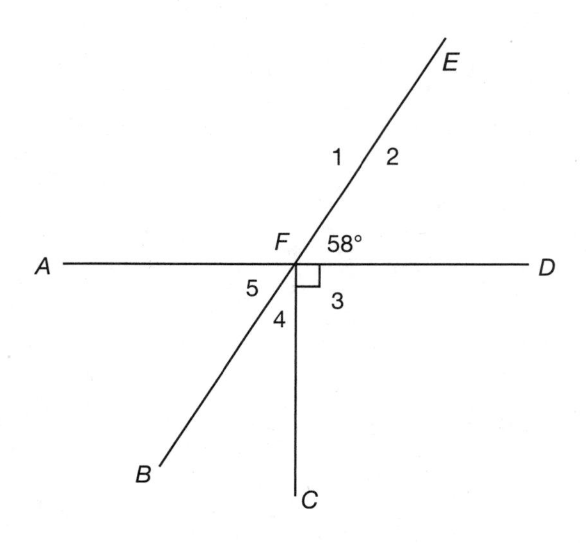

1. Name one pair of vertical angles.

2. Find $m\angle CFE$.

3. Find $m\angle 5$.

4. Find $m\angle 4$.

5. True or false: Angles 1 and 4 are vertical angles.

State whether the following statements are true or false.

6. A pair of vertical angles can be complementary.

7. A pair of vertical angles can be supplementary.

8. Vertical angles are always acute.

Use this figure to answer questions 9 through 12.

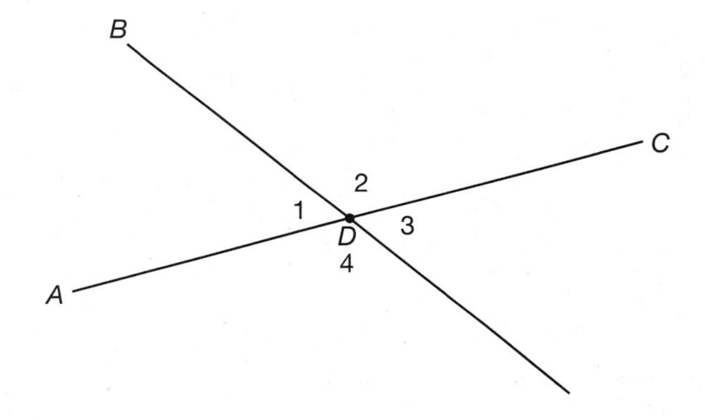

In the figure, $m\angle 1 = 72°$, $m\angle 2 = (2x + 6)°$.

9. Find $m\angle 3$.

10. Find $m\angle 2$.

11. Find $m\angle ADC$

12. Find the value of x.

Answers

Practice 1

1. $m\angle LPN = 141°$
2. $m\angle 3 = 36°$
3. $m\angle LPO = 6x$
4. $m\angle 2 = 47°$
5. $m\angle 1 = 95°$ and $m\angle 2 = 54°$
6. $m\angle MGR = 127.5°$
7. $x = 22$
8. $m\angle 1 = 30°$ and $m\angle 3 = 60°$
9. False
10. Straight
11. $m\angle 1 = 33°$, $m\angle 3 = 57°$, and $m\angle AGK = 147°$

Practice 2

1. Angles 5 and 2 are vertical angles.
2. $m\angle CFE = 148°$
3. $m\angle 5 = 58°$
4. $m\angle 4 = 32°$
5. False
6. True. Two vertical angles can each equal 45°.
7. True. Two vertical angles can each equal 90°.
8. False
9. $m\angle 3 = 72°$
10. $m\angle 2 = 108°$
11. $m\angle ADC = 180°$
12. $x = 51$

5

Angles Formed by Parallel Lines

If parallel lines meet at infinity—infinity must be a very noisy place with all those lines crashing together.
—Anonymous

In this lesson you will learn about transversals, parallel lines, and the relationships that exist between the angles formed by these lines.

STANDARD SNEAK PREVIEW

What do sidewalks, airplane landing strips, railroad tracks, and basketball courts all have in common? They are all made out of parallel lines that never touch! City planners, landscape designers, and construction workers use parallel lines in their plans and daily hands-on work to design things we use every day. Even though you don't have a career yet, it's impossible to look around your room without seeing endless examples of parallel lines. *Standard 8.G.A.5 asks students to learn about the special angles created when parallel lines are cut by a transversal*, and that is precisely what we're going to do!

Transversals

A transversal is a line that passes through two or more lines at two separate points on each line. In the following figure, line M is a transversal through lines J and F. Line N is not a transversal since it crosses lines J and F at the same point.

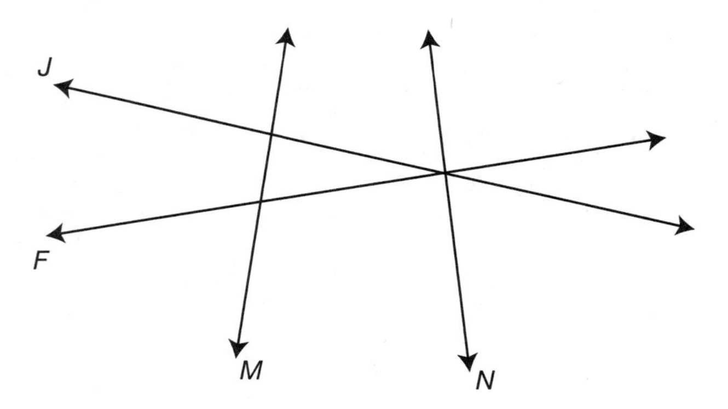

Transversals can sometimes be difficult to recognize, but they are very important since they define specific relationships between the angles they form. Use the following practice questions to become familiar with transversals.

Practice 1

Use the following figure to answer questions 1 through 4.

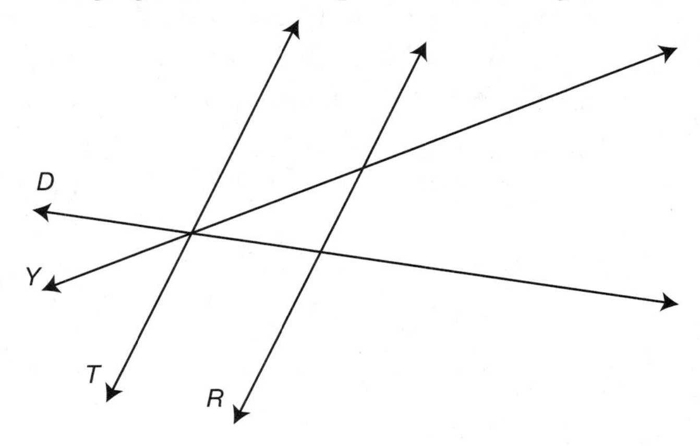

1. Is line D a transversal? Why or why not?

2. Is line Y a transversal? Why or why not?

3. Is line T a transversal? Why or why not?

4. Is line R a transversal? Why or why not?

Parallel Lines

If you were to draw a pair of coplanar lines (lines in the same plane), there are three ways to do this. One way is with the two lines intersecting at one and only one point. The point where they cross each other is called the *point of intersection*. These are seen in the following figure.

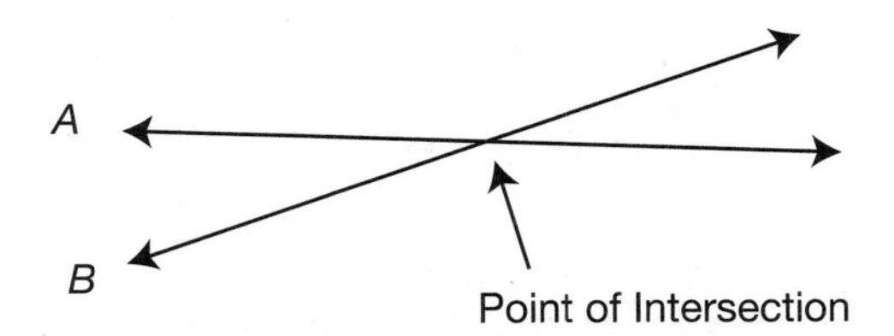

A second way to draw two lines is to construct them on top of one another. In this case, they share all of their points in common, and are essentially the same line.

The third way to draw two lines is so that they will never intersect. In this case, they are called **parallel** lines. In order to indicate that two lines are parallel, they are drawn with single-arrow or double-arrow markings on them, which you can see in the following figure. The symbol ‖ is used to indicate parallel lines. In order to write that line *K* is parallel to line *S*, the notation *K* ‖ *S* is used.

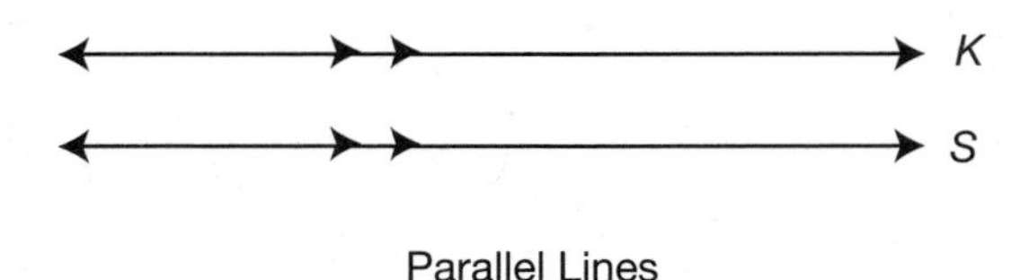

Parallel Lines

Transversals through Parallel Lines

When a transversal intersects two parallel lines, special angle relationships are formed. In the next figure, lines *P* and *Q* are parallel and line *X* is a transversal that forms eight numbered angles.

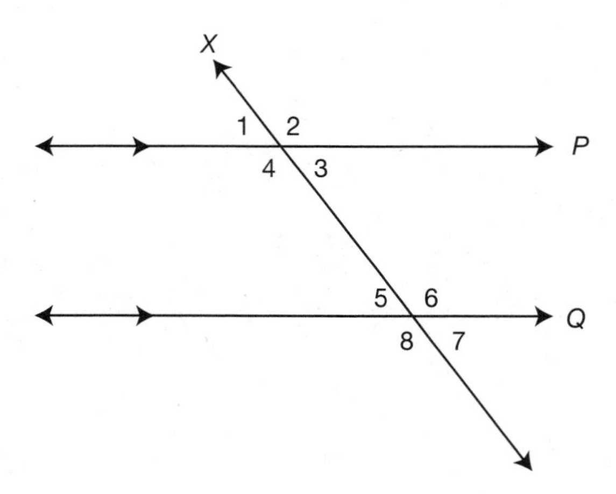

You can probably tell that all of the odd-numbered angles are acute angles: $\angle 1$, $\angle 3$, $\angle 5$, and $\angle 7$. All of the even-numbered angles are obtuse angles: $\angle 2$, $\angle 4$, $\angle 6$, and $\angle 8$. When a transversal intersects parallel lines, all of the acute angles are congruent to one another. Similarly, all of the obtuse angles are congruent to one another. Is "congruent" a new word for you? **Congruent** is an important term in geometry used when two angles or shapes have the same measurements.

TIP: When a transversal intersects parallel lines, all of the acute angles formed are congruent and all of the obtuse angles are congruent.

Special Angle Relationships

Additionally, there are special names and relationships for the pairs of angles formed by a transversal passing through parallel lines.

STANDARD ALERT!

Now that you know that along a transversal, intersecting parallel lines create congruent acute and congruent obtuse angles, you're ready to dig deeper into Standard 8.G.A.5. There's a lot of new vocabulary regarding the special angles created by a transversal, but these are very important building blocks for geometric reasoning and proofs, so you should take time to thoroughly understand them.

Corresponding Angles

Corresponding angles have the same relative position in respect to the parallel line and the transversal. For example, in the figure on the following page, $\angle 2$ and $\angle 6$ are both the top right angle formed by the transversal and the parallel line. Therefore, we would say that $\angle 2$ and $\angle 6$ are corresponding angles. (See the shaded angles in this figure.) Corresponding angles are always congruent. Can you find another pair of corresponding angles? Try to find three more pairs of corresponding angles and then check your answers under the "Tip" box that follows.

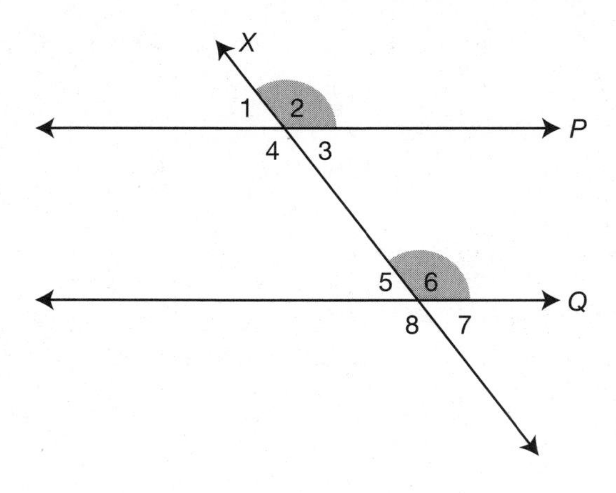

TIP: Corresponding angles are congruent.

Hopefully you were able to recognize that the following pairs of angles are corresponding: ∡1 and ∡5, ∡4 and ∡8, ∡3 and ∡7.

Alternate Exterior Angles

Exterior means *outside*, and *alternate* means *other* or *opposite*. **Alternate exterior angles** are the angles *outside* the parallel lines and on the *opposite* sides of the transversal. For example, ∠2 is on the *right* side of transversal *X*, and is located on the outside of parallel line *P*. ∠8 is on the *left* side of transversal *X*, and is located on the outside of parallel line *Q*. Therefore, ∠2 and ∠8 are *alternate exterior* angles. (See the shaded angles in the following figure.) Alternate exterior angles are always congruent. Can you find another pair of alternate exterior angles? Check your answer under the "Tip" box that follows.

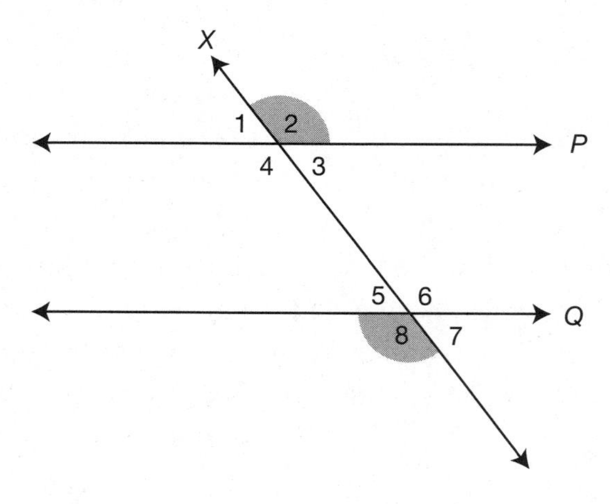

TIP: Alternate exterior angles are congruent.

Another pair of alternate exterior angles is ∠1 and ∠7.

STANDARD ALERT!

Standard 8.G.A.5 doesn't just want you to know that alternate exterior angles are congruent; it also wants you to be able to explain why they are congruent. A sample argument could look like this:

- *∠2 is congruent to ∠6 because they are corresponding angles*
- *∠8 is congruent to ∠6 because they are vertical angles*
- *Therefore, ∠2 must be congruent to ∠8 because they are both congruent to ∠6*

Alternate Interior Angles

Interior means *inside*, and we already know what *alternate* means. **Alternate interior angles** are the angles *inside* the parallel lines and on *opposite* sides of the transversal. For example, ∠3 is on the *right* side of transversal *X*, and is located inside the parallel lines *P* and *Q*. ∠5 is on the *left* side of transversal *X*, and is also located on the inside of parallel lines *P* and *Q*. Therefore, ∠3 and ∠5 are *alternate interior angles*. (See the shaded angles in the next figure.) Alternate interior angles are congruent. Can you find another pair of alternate interior angles? Hopefully by now you know where you can check your answer.

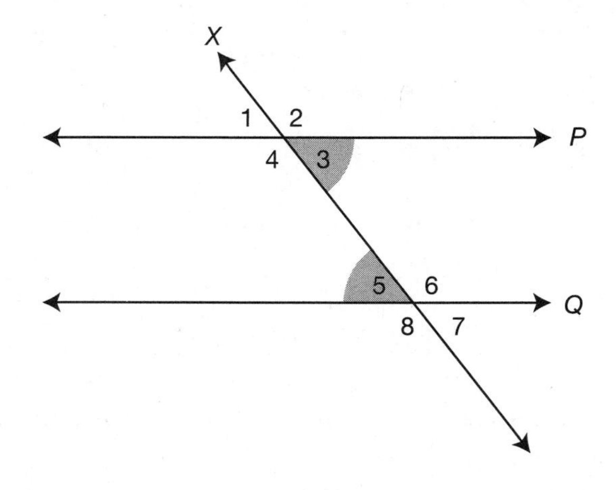

TIP: Alternate interior angles are congruent.

Angle 4 and angle 6 represent another pair of alternate interior angles.

STANDARD ALERT!

In the last Standard Alert box we demonstrated ***why*** *alternate exterior angles are congruent. Standard 8.G.A.5 wants you to also come up with an argument for why alternate interior angles are congruent. Notice how the facts regarding corresponding angles and vertical angles are used together to demonstrate why alternate exterior angles are congruent:*

- $\angle 3$ is congruent to $\angle 7$ because they are corresponding angles
- $\angle 5$ is congruent to $\angle 7$ because they are vertical angles
- Therefore, $\angle 3$ must be congruent to $\angle 5$ because they are both congruent to $\angle 7$

Same-Side Interior Angles

Same-side interior angles are the angles *inside* the parallel lines, on the *same* side of the transversal. For example, $\angle 3$ and $\angle 6$ are both on the *right* side of transversal X. They are also both located on the inside of parallel lines P and Q. Therefore, $\angle 3$ and $\angle 6$ are *same-side interior angles*. (See the shaded angles in the following figure.) Same-side interior angles are supplementary and add up to 180°. Can you find another pair of same-side interior angles?

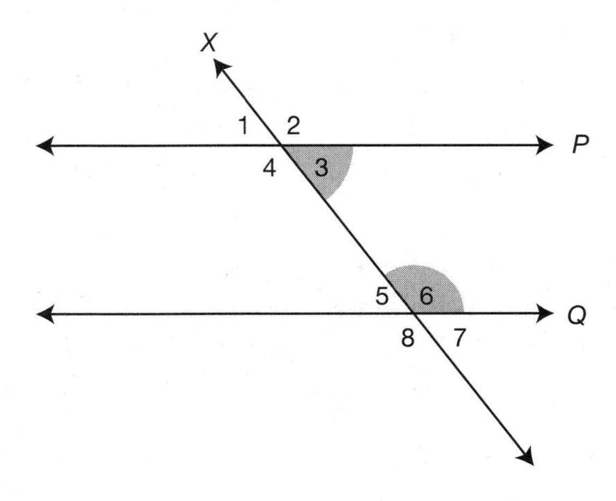

TIP: Same-side interior angles are supplementary.

STANDARD ALERT!

In the past two Standard Alert boxes we've used previous geometric facts to demonstrate why the alternate exterior and alternate interior angle special relationships exist. Now see if you can fill in the appropriate angles to create an argument for why same-side interior angles are supplementary. Check your answers below.

- *$\angle 5$ is congruent to $\angle 3$ because they are ______________*
- *$\angle 5$ is ______________ to $\angle 6$ because they are adjacent angles that make a straight angle*
- *Therefore, $\angle 3$ must be supplementary to $\angle 6$ because they are both ______________ to $\angle 5$*

Answers: alternate interior, supplementary, supplementary

The property regarding same-side interior angles can be extended to include all pairs of acute and obtuse angles formed with parallel lines and transversals. When two parallel lines are intersected by a transversal, any pair of one acute and one obtuse angle is supplementary.

TIP: When a transversal intersects parallel lines, any pair of one obtuse and one acute angle will be supplementary.

Note: Even when a transversal intersects two non-parallel lines, the same angle relationship names can be used. The properties of congruence will not hold true if the intersected lines are not parallel, but it is still correct to refer to angles as "corresponding," "alternate exterior," and "same-side interior."

Practice 2

Use the non-parallel lines in the next figure to practice identifying relationships between angles in questions 1 through 6.

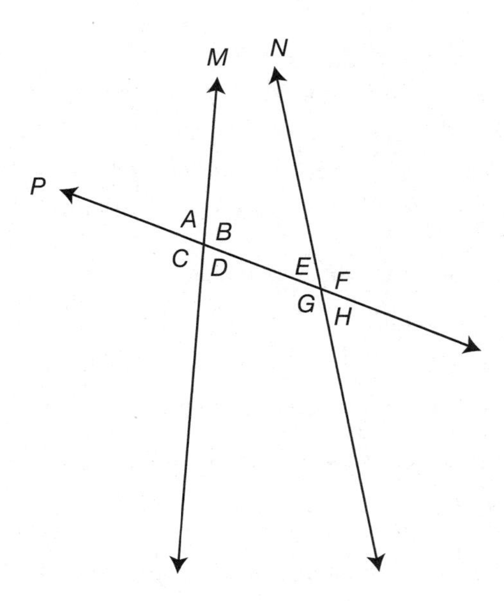

1. Name two pairs of *alternate exterior angles.*

2. Name two pairs of *same-side interior angles.*

3. Name four pairs of *corresponding angles.*

4. Name two pairs of *alternate interior angles.*

5. Complete the following statement and explain why the lines would be parallel: If $\angle D$ were congruent to $\angle$_____ or to $\angle$_____ , then line M and line N would be parallel.

6. Complete the following statement and explain why the lines would be parallel: If $\angle F$ were congruent to $\angle$_____ or to $\angle$_____ , then line M and line N would be parallel.

Review of Angle Relationships

Review the facts about special angle relationships in the box and then finish the last set of practice problems.

TIP: When a transversal intersects parallel lines:

- Corresponding angles are congruent.
- Alternate interior angles are congruent.
- Alternate exterior angles are congruent.
- Same-side interior angles are supplementary.
- Any pair of acute and obtuse angles are supplementary.

STANDARD ALERT!

Now that you are familiar with all the special angle relationships formed by transversals and parallel lines, you'll get to practice Standard 8.G.A.5 along with Standard 7.G.B.5 to solve for unknown angles in Practice 3. Get ready!

Practice 3

Use the parallel lines in the next figure to practice identifying relationships between angles in questions 1 through 8.

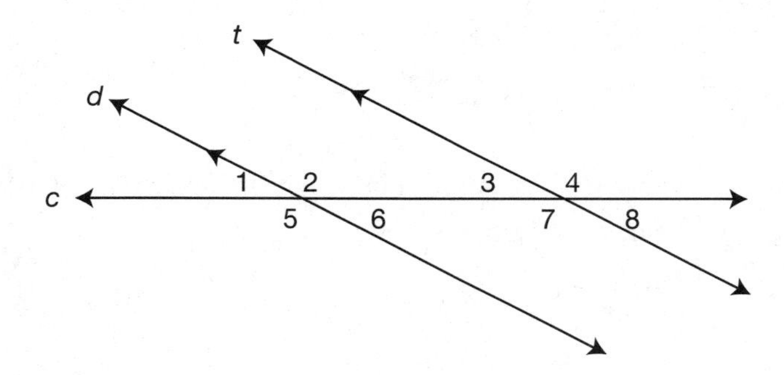

1. Which angles have the same measure as $\angle 2$?

2. $\angle 1$ is an alternate exterior angle to which angle?

3. What is the relationship between $\angle 6$ and $\angle 7$?

4. If $m\angle 2 = (4x + 10)°$ and $m\angle 3 = (2x - 10)°$, find the measure of $\angle 2$ and $\angle 3$.

5. If the measure of $\angle 5$ is 20 less than four times the measure of $\angle 8$, find the $m\angle 5$ and $m\angle 8$.

6. If $m\angle 7 = (5x + 26)°$ and $m\angle 4 = (7x - 18)°$, find $m\angle 7$ and $m\angle 4$.

7. $\angle 3$ and $\angle 8$ are called ________________ angles.

8. $\angle 4$ is supplementary to which four angles?

Perpendicular Transversals through Lines

An important case is when a transversal is perpendicular to a pair of parallel lines as in the next figure:

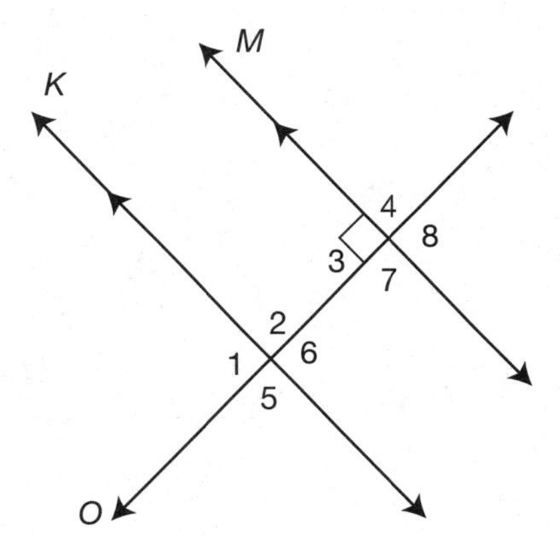

In this case, only $\angle 3$ is marked as 90° with the box at the intersection of the lines. You learned that when two lines are perpendicular, all four of the angles formed are 90°. Therefore, you know that $\angle 3 = \angle 4 = \angle 7 = \angle 8 = 90°$. Remembering that $\angle 3$ is congruent to its alternate interior angle, $\angle 6$, you can conclude that $\angle 6$ is also a right angle. Once certain that $\angle 6 = 90°$, you can conclude that angles 1, 2, and 5 are also right angles. Therefore, when a transversal intersects a pair of parallel lines to form one perpendicular angle, all eight angles formed are right angles.

TIP: When a transversal is perpendicular to two parallel lines, it forms eight right angles.

Practice 4

Use the following figure to answer questions 1 through 10.

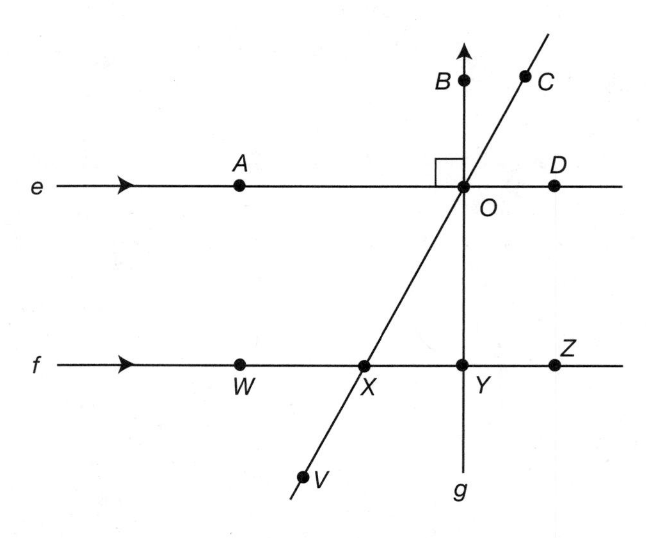

$e \parallel f$
$g \perp f$ and $g \perp e$
$m\angle COD = 62°$

1. Find $m\angle BOC$.

2. Find $m\angle AOC$.

3. Find $m\angle WXO$.

4. Find $m\angle WXV$.

5. Find $m\angle OYZ$.

6. True or false: $\angle WXV$ is an alternate exterior angle to $\angle COD$.

7. True or false: $\angle BOC$ and $\angle AOX$ are vertical angles.

8. True or false: $\angle DOC$ and $\angle XOA$ are vertical angles.

9. If $m\angle AOX = (4x - 26)°$ and $m\angle XOY = (x + 6)°$, find the value of x.

10. $\angle XYO$ is a/an ____________ angle.

Answers

Practice 1

1. Line D is a transversal through lines T and R only since it intersects them at two different points.
2. Line Y is a transversal through lines T and R only since it intersects them at two different points.
3. Line T is not a transversal since it intersects lines D and Y at the same point.
4. Line R is a transversal through lines D and Y only since it intersects them at two different points.

Practice 2

1. ($\angle A$ and $\angle H$) and ($\angle C$ and $\angle F$) are two pairs of *alternate exterior angles.*
2. ($\angle B$ and $\angle E$) and ($\angle D$ and $\angle G$) are two pairs of *same-side interior angles.*
3. ($\angle A$ and $\angle E$), ($\angle B$ and $\angle F$), ($\angle C$ and $\angle G$), and ($\angle D$ and $\angle H$) are four pairs of *corresponding angles.*
4. ($\angle D$ and $\angle E$) and ($\angle B$ and $\angle G$) are two pairs of *alternate interior angles.*
5. $\angle E$ (alternate interior angles) or $\angle H$ (corresponding angles)
6. $\angle C$ (alternate exterior angles) or $\angle B$ (corresponding angles)

Practice 3

1. $\angle 5$, $\angle 4$, $\angle 7$
2. $\angle 8$
3. $\angle 6$ and $\angle 7$ are same-side interior angles, and they are supplementary.
4. Same-side interior angles are supplementary, so $m\angle 2 + m\angle 3 = 180$:

   $(4x + 10) + (2x - 10) = 180°$

   $6x = 180°$, so $x = 30$

   $m\angle 2 = (4(30) + 10) = 130°$

   $m\angle 3 = (2(30) - 10) = 50°$

   $m\angle 2 = 130°$ and $m\angle 3 = 50°$
5. $m\angle 5$ is supplementary to $m\angle 6$, and $m\angle 6 = m\angle 8$, so $m\angle 5 + m\angle 8 = 180°$:

   $m\angle 8 = x$

   $(4x - 20) + x = 180°$

   $5x - 20 = 180°$

   $5x = 200°$

   $x = 40°$

   $m\angle 8 = 40°$ and $m\angle 5 = 140°$

6. Angles 7 and 4 are vertical, which means $m\angle 7 = m\angle 4$:
   $(5x + 26)° = (7x - 18)°$
   $26 + 18 = 7x - 5x$
   $44 = 2x$, so $x = 22$
   $(5(22) + 26) = 136$ and $(7(22) - 18) = 136$
   $m\angle 7 = m\angle 4 = 136°$
7. vertical (or opposite)
8. $\angle 1$, $\angle 3$, $\angle 6$, and $\angle 8$ are all supplementary to $\angle 4$

Practice 4

1. $m\angle BOC = 28°$
2. $m\angle AOC = 118°$
3. $m\angle WXO = 118°$
4. $m\angle WXV = 62°$
5. $m\angle OYZ = 90°$
6. True
7. False. $\angle BOC$ and $\angle XOY$ are vertical angles.
8. True
9. $x = 22$
10. right / 90°

6

Side and Angle Theorems for Triangles

The only angle from which to approach a problem is the TRY-Angle.
—Anonymous

In this lesson you'll learn about the special relationships of interior and exterior angles of triangles. You'll also perform an experiment to determine the rule regarding the side lengths needed to form a triangle.

STANDARD SNEAK PREVIEW

If engineers had to choose just one shape to design with, they would certainly select the triangle. One reason is because triangles are the strongest geometric shape. Look around and you'll notice that they are in roofs of houses, bridges, and the infrastructures of skyscrapers. Not only are they strong, but triangles are clever, too! They can be used to measure distances, locate yourself on a map in the wilderness, and pinpoint your location in a massive city on a smartphone map! (Cell phones rely on satellites, which use triangulation to determine location.) Since triangles are important in so many different ways, it's no wonder that the CCSS has them in lots of different standards.

We are going to begin with ***Standard 7.G.A.2, which wants students to know about certain side-length requirements that are needed in order to create a triangle.*** Then we'll switch from investigating a triangle's sides to its angles as we cover ***Standard 8.G.A.5 with informal arguments that will reveal facts about the sum of a triangle's interior angles, as well as a special relationship involving a triangle's exterior angle***. The concepts presented in this lesson are important foundations for your future work with triangles, so take your time as you make your way through the following content.

Naming and Labeling Triangles

You may remember that triangles are named using their three vertices and a small triangle is drawn in front of the vertex names. The following triangle can be named ΔLMN, ΔMNL, ΔNML, or any other order or direction of the three vertices.

It is standard practice to name the angles in triangles with capital letters. The sides opposite each angle are named with the accompanying lowercase letter. For example, in the following figure, notice how side l is opposite $\angle L$, side m is opposite $\angle M$, and side n is opposite $\angle N$.

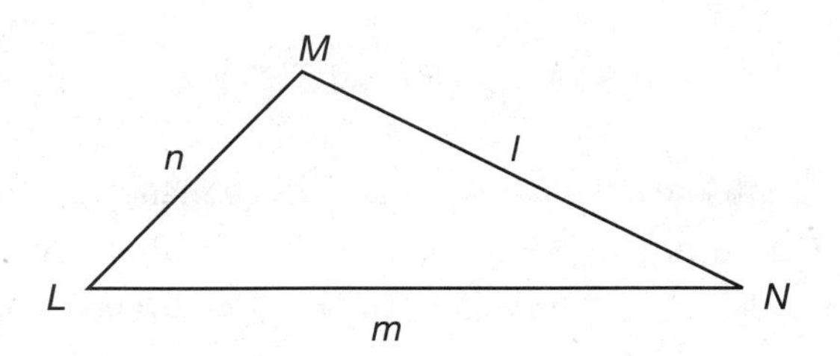

TIP: In triangles, the angles are named with capital letters and each opposite side is named with the corresponding lowercase letter.

Relationships between Sides and Angles in Triangles

There is a special relationship between the relative length of a side and the relative measurement of its opposite angle. (The word "relative" used in this way means how it relates to the other parts of the triangle.) The longest side is always opposite the largest angle, the shortest side is always opposite the smallest angle, and the middle-length side is always opposite the middle-sized angle. For example, if the previous figure were drawn to scale (with side m the longest, side n the shortest, and side l in between m and n), since side m is the longest side, we could conclude that $\angle M$ is the largest angle. Similarly, since side n is the shortest side, we could conclude that $\angle N$ is the smallest angle.

TIP: The largest angle in a triangle is always opposite its longest side. The smallest angle is always opposite the shortest side. The middle-sized angle is always opposite the middle-sized side.

STANDARD ALERT!

Standard 7.G.A.2 asks students to construct triangles with certain given conditions and to know when those conditions will create a unique triangle, more than one triangle, or no triangle. Do the following activity to investigate this concept and see if you can figure out a condition regarding side lengths that must be met in order for the sides to create a triangle. You will need a ruler, a pair of scissors, and a scrap piece of paper for this activity.

How Long Can You Go?

1. Using your ruler, scrap paper, and scissors, cut out three different side lengths that measure 2", 3", and 5". Label each edge with its side length.
2. Use the 2" side, 3" side, and the edge of your ruler as the base, and see if you can:
   - Make a triangle with a 6" base. (Circle **y** or **n**)
   - What about a 5.5" base? (Circle **y** or **n**)
   - Lastly, use your ruler to attempt a 5" base (Circle **y** or **n**).
   - What did you notice about all of these suggested base lengths?
3. Using the 2" side, the 3" side, and the edge of your ruler as the base, what is the shortest possible length that can be used to create a triangle? _______
4. Make two different triangles in the same manner, and after measuring their two different bases write them down here: _______ and _______.
5. What is the sum of the two fixed side lengths you were just using? _______
6. What do you notice about the length of the longest possible third side and how it compares to the sum of the other two sides? ____

The **triangle inequality theorem** is a fancy name for the rule that defines the relationship between the side lengths of a triangle. Hopefully, in the previous activity you noticed that *any two sides of a triangle must have a sum* *that is greater than the length of the remaining side*. When you used the 3" and 5" side, it was impossible to have a 5" base because the two sides just sat flatly on the base and didn't make a triangle. Notice that the two base lengths you were able to create were both less than 8".

TIP: The sum of any two sides of a triangle must be greater than the length of the remaining third side.

Using the triangle inequality theorem, if two sides of a triangle are 7 and 10, you can determine that the third side must be less than 17 units. One way to remember this theorem is to think that any two sides of a triangle must be "stronger" (longer) than the third side. In real-world terms, normally two people together can carry more weight than a third solo person, which is an idea similar to the triangle inequality theorem. In the following set of problems you will get to test your understanding of standard 7.G.A.2.

Practice 1

ΔNPR in the following figure is drawn to scale and should be used to answer questions 1 through 6.

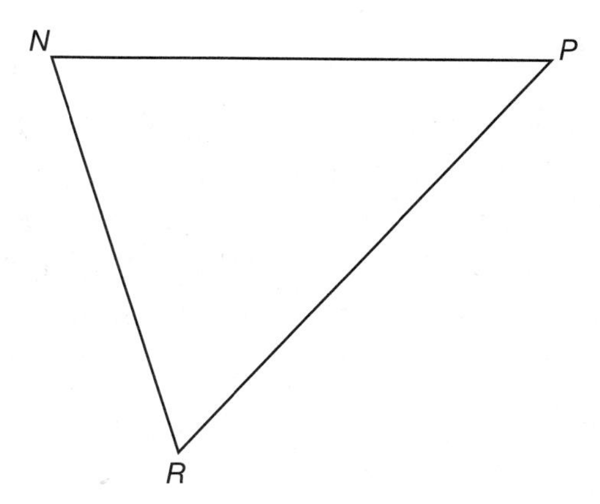

1. What would be another way to name side $\overline{NP}$?

2. What would be another way to name side $\overline{PR}$?

3. Given that ΔNPR is drawn to scale, which angle is the smallest and why?

4. If side $r = 17$ mm and side $p = 12$ mm, what is the largest even-numbered side length that could be possible for side n?

5. Given that $k > 0$, side $r = 6.5k + 0.5$, $p = 6.2k + 0.3$, and side $n = 6.7k + 0.8$, which angle of ΔNPR must have the largest measure?

6. True or false: The side lengths for ΔNPR could be $n = 0.9$ cm, $r = 1.6$ cm, and $p = 0.7$ cm.

Angle Sum Theorem

STANDARD ALERT!

Standard 8.G.A.5 asks students to use informal arguments to establish facts about the sum of a triangle's interior angles. Rather than just memorizing a fact presented in text or by a teacher, students should be able to independently support why this fact is true, so follow along in the next activity and see if you can present this reasoning to a sibling or parent in your home.

One of the most common characteristics used when problem solving with triangles relates to the sum of a triangle's interior angles. We are going to explore a triangle's interior angles by manipulating ΔABC, but before we begin, write down an *estimate* you have for each of the angles, just by looking at the triangle—don't use your protractor skills from lesson 2!

$\angle A =$ ________, $\angle B =$ ________, $\angle C =$ ________

Now add your estimates together and write your estimated sum here: _________

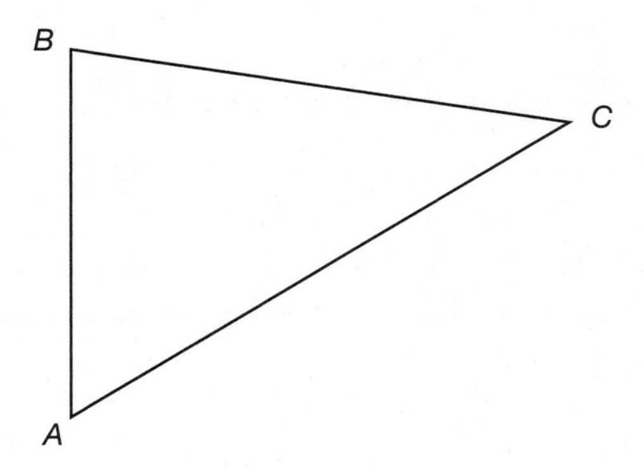

Next, notice how the triangle was manipulated in the following illustration. If you were to tear off the three corners of ΔABC and place their untorn edges together, you would end up with a straight line, as is shown.

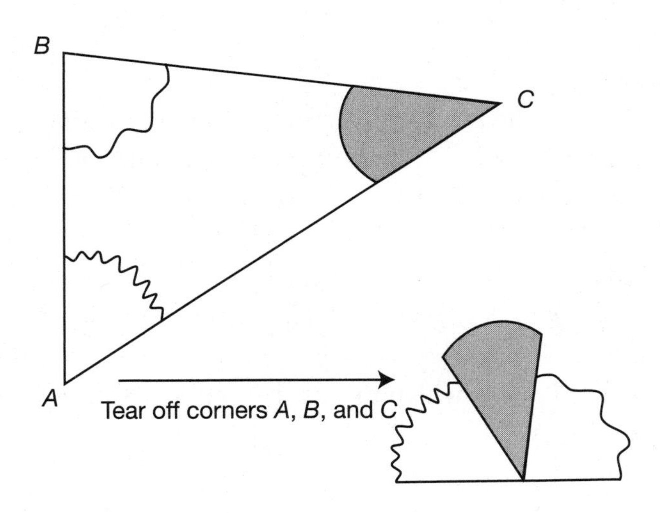

Do you think this was a one-time fluke? Now try this exercise yourself: Using a pair of scissors, cut out a triangle that looks very different from ΔABC. Next, carefully tear off the three corners and see what happens when you line up their straight edges. Recalling the measure of a straight angle, what does this experiment tell you about the sum of the three interior angles of a triangle? Hopefully your answer matches the theorem that is listed in the following "Tip" box. How close was your estimate above to the answer you got here? Now you can grab your protractor, measure each angle in ΔABC, and verify that the sum of the three angles is 180°.

TIP: The sum of the interior angles of a triangle is 180°.

The 180° sum of interior angles can be used to determine the measure of a missing angle from a triangle. Use this figure to understand the following two examples.

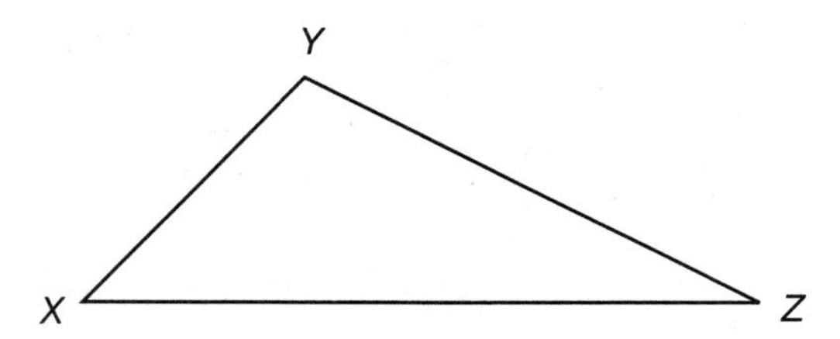

Example 1
If $m\angle X = 45°$ and $m\angle Y = 100°$, then the $m\angle Z$ can be determined by setting up $\angle X + \angle Y + \angle Z = 180°$. From this equation we see that $45° + 100° + \angle Z = 180°$. Therefore $\angle Z$ must equal 35°.

Example 2
If $m\angle X = j$, $m\angle Y = 3j$, and $m\angle Z = 2j$, then the following equation can be set up: $\angle X + \angle Y + \angle Z = j + 3j + 2j = 6j = 180°$. From this equation we see that $j = 30°$, which means that $m\angle X = 30°$, $m\angle Y = 90°$, and $m\angle Z = 60°$.

STANDARD ALERT!

Now that you have established a useful fact about the sum of a triangle's interior angles as part of Standard 8.G.A.5, combine this knowledge with 7.G.B.5 to solve for missing angles in the following practice!

Practice 2

Use the following figure and the given angle information to answer questions 1 through 5.

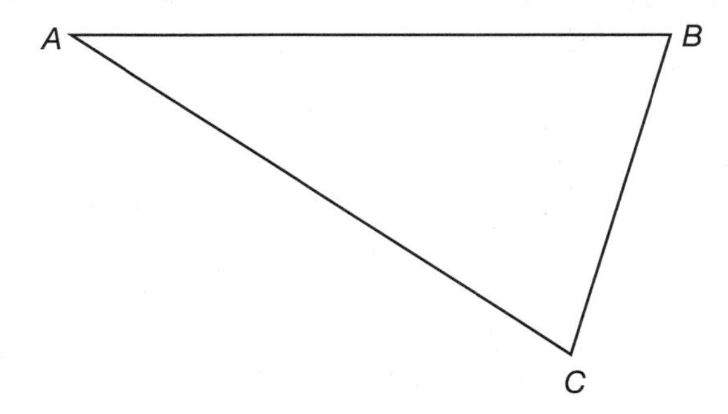

1. Find $m\angle A$ if $m\angle B = 72.3°$ and $m\angle C = 74.5°$.

2. Find the value of y if $m\angle A = (y + 9)°$, $m\angle B = (3y + 17)°$, and $m\angle C = (4y - 6)°$.

3. Using the information provided in question **2**, what is the measure of angles A, B, and C?

4. Find the measure of all the angles, given the following: The measure of angle A is five less than the measure of angle B. The measure of angle C is 15 less than twice the measure of angle B. (*Note:* Figure not drawn to scale.)

5. Assume that angle C is a right angle and angle A is one-half the measure of angle B. Find the measure of all the angles.

Exterior Angles in Triangles

Now that you know the relationship between the interior angles of a triangle, it's time to investigate the *exterior* angles of a triangle. First, let's figure out what an exterior angle is! Triangles have angles that occur outside of the triangle when one of the sides is extended. In the following figure, $\angle 1$ is defined as the **exterior angle** of ΔABC. Since $\angle 2$ is next to $\angle 1$, we call this the **adjacent interior angle**. $\angle 3$ and $\angle 4$ are not adjacent to $\angle 1$, so they are considered the **remote interior angles** to $\angle 1$.

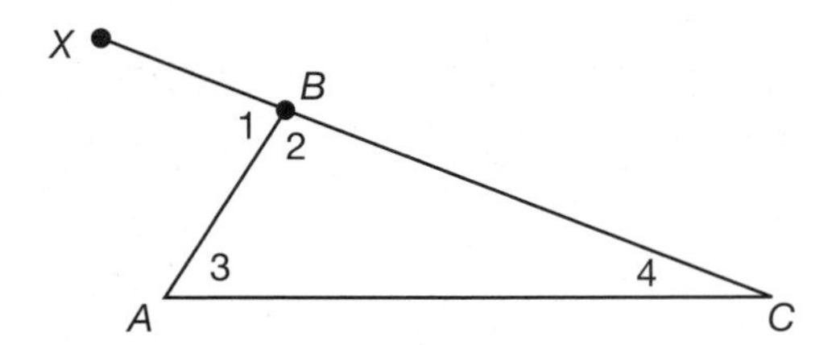

STANDARD ALERT!

Standard 8.G.A.5 asks students to use informal arguments to establish facts about a triangle's exterior angles. Pay close attention to the following argument, mark this page in your book, and see if you can re-create this informal proof the next time you pick up this book!

The exterior angle of a triangle has a special relationship to its remote interior angles. Let's use what we know about supplementary angles, straight angles, and the interior angles of a triangle to figure out what that relationship is:

- $m\angle 2 + m\angle 1 = 180°$ since they form a straight angle
- $m\angle 2 + m\angle 3 + m\angle 4 = 180°$ since they are the interior angles of a triangle
- Since both of these equations equal 180°, set them equal to each other to get $m\angle 2 + m\angle 1 = m\angle 2 + m\angle 3 + m\angle 4$
- Subtract the $m\angle 2$ from both sides to reveal a special relationship: $m\angle 1 = m\angle 3 + m\angle 4$

Can you make a hypothesis now of what the special relationship is between a triangle's exterior angle and its remote interior angles? Give it a try and check your answer against the rule stated in the following "Tip" box!

TIP: The measure of a triangle's exterior angle is always equal to the sum of the measures of its two remote interior angles.

STANDARD ALERT!

Now that you know how a triangle's exterior angle relates to its remote interior angles, combine this Standard 8.G.A.5 knowledge with 7.G.B.5 to solve for missing angles in the following practice!

Practice 3

Use the preceding figure to answer questions 1 through 3. Assume that the figure is not drawn to scale.

1. If $m\angle 1 = 88°$ and $m\angle 3 = 49°$, what is the $m\angle 4$?

2. If $m\angle 1 = g°$ and $m\angle 4 = h°$, what algebraic expression would represent $m\angle 4$? (*Hint:* How would you identify the $m\angle 4$? Leave your answer in terms of g and h.)

3. If $m\angle 1 = (3x - 12)°$, $m\angle 4 = (2x - 4)°$, and $m\angle 3 = (0.5x + 8.5)°$, find the measure of all the angles in ΔABC.

4. True or false: A triangle's exterior angle will always be obtuse.

5. An isosceles triangle has a base angle R that measures 37°. An exterior angle is drawn next to $\angle R$. What is the measure of this exterior angle?

Answers

Practice 1

1. side r (the lowercase of the opposite angle)
2. side n (the lowercase of the opposite angle)
3. $\angle P$ would be the smallest because it is opposite the shortest side.
4. Side n must be smaller than the sum of r and p, so the largest even-numbered side length it could be would be 28 mm.
5. Angle N must have the largest measure since it is opposite the largest side.
6. False. The sum of n and p is 1.6, which means that side r must be less than 1.6 cm.

Practice 2

1. $m\angle A = 33.2°$, since $180° - 72.3° - 74.5° = 33.2°$.
2. $y = 20°$, since $(y + 9) + (3y + 17) + (4y - 6) = 180°$.
3. $m\angle A = 29°$, $m\angle B = 77°$, and $m\angle C = 74°$
4. $m\angle A = 45°$, $m\angle B = 50°$, and $m\angle C = 85°$, since $(B - 5) + (B) + (2B - 15) = 180°$.
5. $m\angle A = 30°$, $m\angle B = 60°$, and $m\angle C = 90°$, since $90° + \frac{1}{2}B + B = 180°$.

Practice 3

1. $m\angle 4 = 39°$
2. $m\angle 4 = (g - h)$. ($m\angle 1 = m\angle 3 + m\angle 4$, which means $g = h + m\angle 4$)
3. $x = 33°$, so $m\angle 1 = 87°$, $m\angle 3 = 62°$, $m\angle 4 = 25°$, and $m\angle 2 = 93°$
4. False. A remote angle will be acute when it is adjacent to an obtuse angle, and it will be right when it is adjacent to a 90-degree angle.
5. The exterior angle will measure 143°.

7 Classifying and Problem Solving Special Triangles

Geometry existed before creation.
—Plato

Now that you have learned about the requirements for a triangle's angles and sides, you will learn how triangles are classified according to their angle measures and side lengths. You will also receive some new tools for writing and solving equations for unknown information in these special triangles.

STANDARD SNEAK PREVIEW

You learned some really key triangle theorems in the previous lesson—nice going! In this lesson we'll learn some of the technical language used to classify triangles. You may have noticed that people love to classify things around them. We even classify people we know—"he's a bookworm, she's a jock, he's a runner, she's a musician . . ." It's human nature to want to be able to associate something with a larger group that shares similar traits. When we do this we know more about how something relates to our life and how we should interact with it. If I'm a bookworm and I go to a party by myself, I might look for someone else who seems to be a bookworm so that we can share thoughts on great new novels. In geometry, we classify shapes by common attributes. Since certain types of triangles enjoy special privileges and theorems, it's helpful to be able to quickly recognize these "teacher's pet" triangles! After completing this lesson you'll have a more refined way to discuss triangles' features at the next birthday party you go to! Learning how to classify and work with special types of triangles will help you master *Standard 7.G.A.2, which looks at side-length requirements; Standard 8.G.A.5, which investigates interior and exterior angles relationships in triangles;* and *Standard 7.G.B.5, which focuses on solving for unknown angles in triangles.*

Classifying Triangles by Angle Measurements

One way that triangles can be classified is by the measure of their angles. Do you remember that angles are classified in their relation (larger or smaller) to a 90° angle? The same is true with triangles:

1. **Acute triangles:** Acute triangles have three acute angles (acute = less than 90°).

2. **Right triangles:** Right triangles have one 90° angle.
   - Can a right triangle have two right angles? Use what you know about the sum of a triangle's interior angles to answer this question.

- If one of the angles in a right triangle is 90°, why must the other two angles be complementary?

3. **Obtuse triangles:** Obtuse triangles have one obtuse angle (obtuse = greater than 90°).
   - Use what you know abut the sum of the interior angles of a triangle to see if more than one angle in an obtuse triangle can be obtuse.

Compare the following figures to the previous definitions:

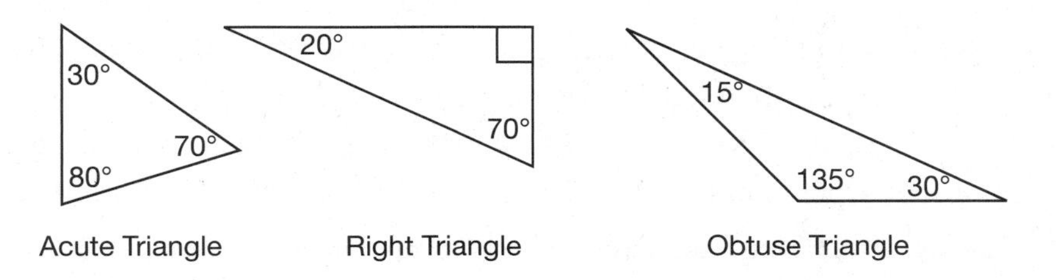

Acute Triangle Right Triangle Obtuse Triangle

Classifying Triangles by Side Lengths

Just like how someone can be a bookworm *and* a jock, triangles can be classified by their angle measurements *and* their side lengths. Triangles can have 0, 2, or 3 congruent side lengths. Hopefully you remember that there is a special relationship between a triangle's side lengths and the angles opposite each side (hint: the longest side is opposite the largest angle).

You will notice in the following definitions that the relationship between side lengths and angles can be extended to congruencies within triangles: Triangles that have 0 congruent sides have 0 congruent angles, triangles that have 2 congruent sides will have 2 congruent angles, and triangles that have 3 congruent sides will have 3 congruent angles. Here are the names given to these classifications of triangles:

Scalene Triangles

Scalene triangles have three side lengths that are all different. This means that their angles are also unique. Scalene triangles can either be **obtuse** or **acute**. See the next figure. Notice that each side length has a different number of hash marks through it. This symbolizes that no two sides are the same length.

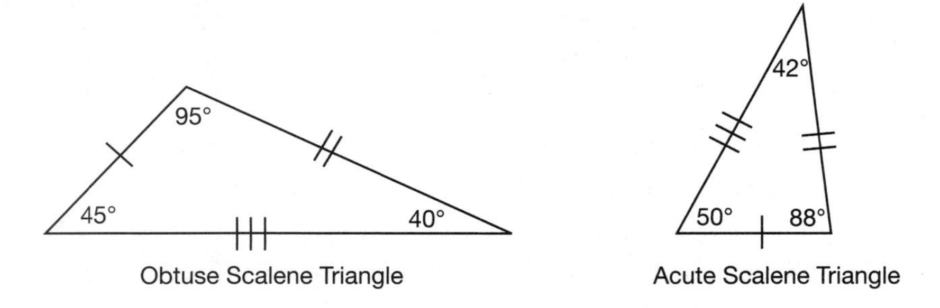

Isosceles Triangles

When a triangle has two angles that are equal in measure, it is called an **isosceles triangle**. The two congruent angles are referred to as the **base angles** and the side included between them is called the **base**. The congruent base angles are opposite a pair of congruent sides, which are called **legs**. The non-congruent angle is called the **vertex angle**. Isosceles triangles can either be obtuse or acute. See the next figure.

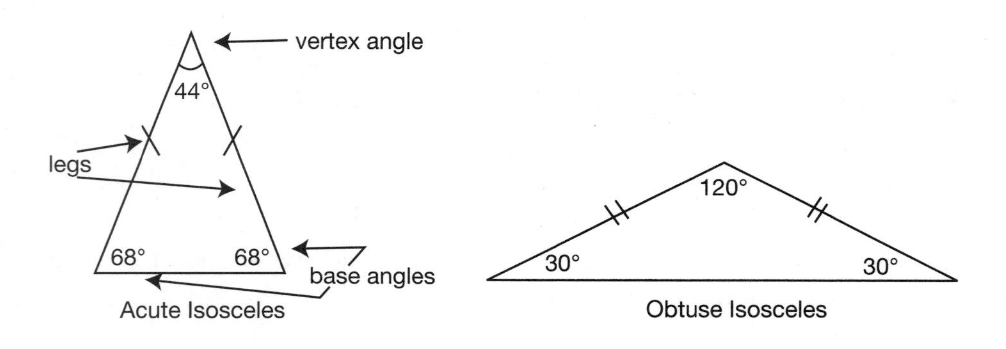

TIP: Isosceles triangles have one pair of congruent sides (called "legs") and one pair of congruent angles (called "base angles").

Equilateral Triangles

Equilateral triangles have three sides that are equal in length. Equilateral triangles also have three congruent angles so they can also be referred to as "**equiangular.**" Since each of the angles in an equilateral triangle has a measure of 60°, these triangles are always acute. Study this important triangle in the following figure.

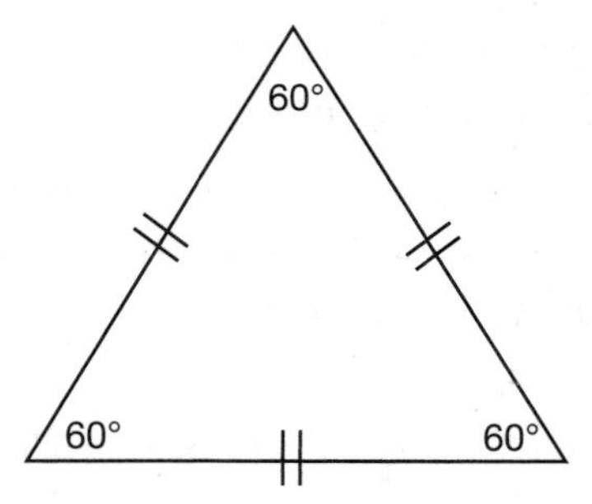

Equilateral or Equiangular Triangle

TIP: The terms *equilateral* and *equiangular* both refer to triangles that have three congruent angles of 60° and three congruent sides.

Notice that with many triangles there can be two classifications—one that describes the side lengths and one that describes the angles. Examples of triangles that fall under more than one classification are *obtuse isosceles*, *acute scalene*, and *right isosceles*. Also, it is critical that you understand that a triangle's angles have the same congruencies as its sides: a triangle with no congruent sides has no congruent angles (scalene), a triangle with two congruent sides has two congruent angles (isosceles), and a triangle with three congruent sides has three congruent angles (equilateral).

Practice 1

1. What is the sum of the two acute angles in a right triangle?

2. In choices **a–e** two angles of a triangle are given. Find the missing angle and then classify each triangle as specifically as possible. (Each triangle should have one classification for its angle measurements and one for its side measurements.)
   a. 70° and 40°
   b. 17° and 73°
   c. 34° and 52°
   d. 45° and 90°
   e. 60° and 60°

3. In ΔTLC, $m\angle T = 48°$. If the measure of $\angle L$ is three times the measure of $\angle C$, find the measures of $\angle L$ and $\angle C$ and classify ΔTLC.

For questions 4 through 8, identify whether the given conditions *sometimes*, *always*, or *never* result in an *acute* triangle.

4. An equiangular triangle with side length of 100 cm.

5. An isosceles triangle with base angles measuring 15°.

6. A scalene triangle with one angle measuring 62°.

7. Isosceles ΔCAT where $\angle A$ is the vertex, and the vertex is three times larger than each base angle.

8. In ΔREM, $\overline{RE} = 10$ cm, $\overline{EM} = 10$ cm, and $\overline{MR} = 10$ cm.

Calculation Tips for Special Triangles

STANDARD ALERT!

Now that you know the special angle relationships in right, isosceles, and equilateral triangles, you are ready for a new set of tools that can be applied to solving for missing angles in figures, which is a key component of Standard 7.G.B.5!

When you know that a triangle is isosceles, equilateral, or right, there are several ways you can identify missing information about its sides or angles.

Problem Solving with Isosceles Triangles

With an isosceles triangle, if you are given only one angle's measurement, it is possible to figure out the remaining two angles. You need only to know if the given angle is a base angle or a vertex angle.

Example 1 (vertex angle given)

Given that the vertex angle of an isosceles triangle is 120°, the remaining two base angles must equally share the remaining 60° of the triangle since 180° – 120° = 60°. In this case, the two base angles would each measure 30°.

Example 2 (base angle given)

Given that the base angle of an isosceles triangle is 75°, you know that the other base angle is also 75°. This would then leave 30° for the vertex of the triangle, since 180° – 2 × (75°) = 30°.

Algebraic Problem Solving with Isosceles Triangles

The same techniques can be used with isosceles triangles when you are given an algebraic expression with variables to represent either the base or the vertex angle of an isosceles triangle.

Example 3 (algebraic expression given)

If given that the vertex angle of an isosceles triangle is 10 more than twice the measure of its base angle, the following could be identified: Each of the base angles could be represented with b and the vertex angle could be represented as $10 + 2b$. Then since the sum of the vertex angle plus two of the base angles is 180°, the following equation can be written: $(10 + 2b) + (b) + (b) = 180°$. Therefore, $10 + 4b = 180°$ and $b = 42.5°$. In this case, each base angle would measure 42.5° and the vertex angle would measure 95°.

Practice 2

1. In an isosceles triangle, one of the base angles measures 37.2°. Find the measure of the vertex angle.

2. What is the measure of each base angle in an isosceles triangle whose vertex measures 117°?

3. In an isosceles triangle, the vertex angle measures $(5x – 10)°$ and each base angle measures $(y)°$. Write an algebraic expression to represent y in terms of x.

4. If the vertex angle of an isosceles triangle is four times as large as each of the base angles, find the measure of each angle of the triangle.

5. The base angle of an isosceles triangle is 30 less than twice the triangle's vertex angle. Find the measure of each of the angles and classify it as an acute or obtuse isosceles.

Problem Solving with Right Triangles

A helpful tidbit to remember about right triangles is that the two remaining acute angles will always be complementary. With this in mind, instead of summing up three angles to 180°, you can just sum the two acute angles to 90°.

Example 4 (right triangle)

In a right triangle, one of the acute angles is 14 less than three times the other acute angle. In order to find the acute angle measurements, let x represent one of the acute angles and let $3x - 14$ represent the larger acute angle. Then after solving the equation $(3x - 14) + (x) = 90°$, you will see that $x = 26°$. Therefore, the smaller acute angle will measure 26° and the larger one will be 64°.

Problem Solving with Equilateral Triangles

With equilateral or equiangular triangles, you can set the sides equal to one another to solve for variables when you are given algebraic expressions to represent the side lengths. The same is true for their angle measures, which can be set equal to one another.

Example 5 (equilateral triangle)

In an equilateral triangle, one of the sides measures $(4x - 20)$ centimeters and another side measures $(130 - 2x)$ centimeters. In order to find the side lengths, let the two sides equal each other in the equation: $4x - 20 = 130 - 2x$. Solving for x, you will get $x = 25$, which results in sides of 80 centimeters.

Practice 3

1. A right triangle has one acute angle that measure 42.75°. Find the measurement of the remaining angle.

2. An equiangular triangle has one angle that measures $(5x - 12y)°$ and another angle that measures $(7x + 18y)°$. Write an expression that represents x in terms of y. (Hint: Write an equation that relates the two angles to each other, and then get x by itself.)

3. Use the following figure to solve for h, u, and f. (Hint: Notice what kind of triangle this is and begin by solving for h.)

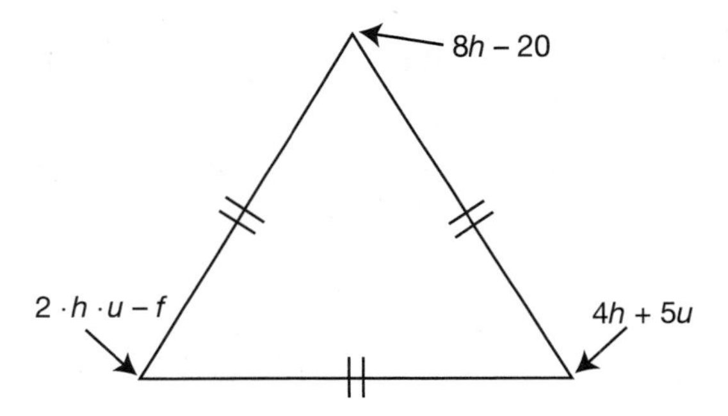

4. In an equiangular triangle, each of the angles equals $3p$. The side lengths are each $(2p - 7)$ inches. Find the measure of the angles and lengths of the sides.

Answers

Practice 1

1. The sum of the two acute angles in a right triangle is always 90°.
2. a. 70°; acute isosceles
   b. 90°; right scalene
   c. 94°; obtuse scalene
   d. 45°; right isosceles
   e. 60°; acute equilateral
3. $\angle L = 99°$; $\angle C = 33°$; and ΔTLC is an obtuse scalene triangle.
4. always (Regardless of its side length, an equiangular triangle will always be acute since its three angles will each measure 60 degrees.)
5. never
6. sometimes
7. never (The base angles would each measure 36° and the vertex angle would measure 118°.)
8. always (This is an equilateral triangle, which is always acute.)

Practice 2

1. 105.6°
2. 31.5°
3. $y = \frac{180 - (5x - 10)}{2} = \frac{190 - 5x}{2}$
4. base angles = 30° and the vertex angle = 120°
5. base angles = 66° and the vertex angle = 48°, so this is an acute isosceles triangle.

Practice 3

1. 47.25° (90° – 42.75° = 47.25°)
2. Solve for x in the equation: $5x - 12y = 7x + 18y$

$$\begin{aligned} 5x - 12y &= 7x + 18y \\ 5x - 7x &= 18y + 12y \\ -2x &= 30y \\ x &= \frac{30y}{-2} \\ x &= -15y \end{aligned}$$

3. $h = 10$ (since $8h - 20 = 60°$), then $u = 4$ and $f = 20$
4. Each angle measures 60° so $p = 20$ and the side lengths are each 33 inches.

8

Finding the Altitudes and Areas of Triangles

Arithmetic! Algebra! Geometry! Grandiose trinity!
Luminous triangle! Whoever has not known you is without sense!
—Comte de Lautreamont

After completing this lesson you'll understand how to find the three different altitudes and bases of a triangle and how to use these two measurements to calculate the area of triangles.

STANDARD SNEAK PREVIEW

Area is the concept that helps you decide which slice of pizza to pick out of a pie—if you're hungry, you'll pick the largest slice, and that's the slice with the biggest area! Now that you have gathered some triangle momentum, you are ready to learn how to work with altitudes and area. Since finding the area of two-dimensional objects is a practical skill that comes up often in personal and work life, it is one of the key concepts of the middle school Common Core State Standards. Area is in several standards, but the first one we'll address in this lesson is ***Standard 6.G.A.1, which asks students to find the area of triangles***. This standard also wants us to be able to find the area of more complex shapes by breaking them down into triangles and other four-sided figures, but we'll tackle that in a later lesson. ***Standard 7.G.B.6 states that students will apply triangle area to solving real-world problems***, so we'll take a look at some examples of practical applications of triangle area. Before we start having fun with area, we need to learn what a triangle's altitude is since the altitude of a triangle is necessary to calculate the area. Read on!

Finding a Triangle's Altitude

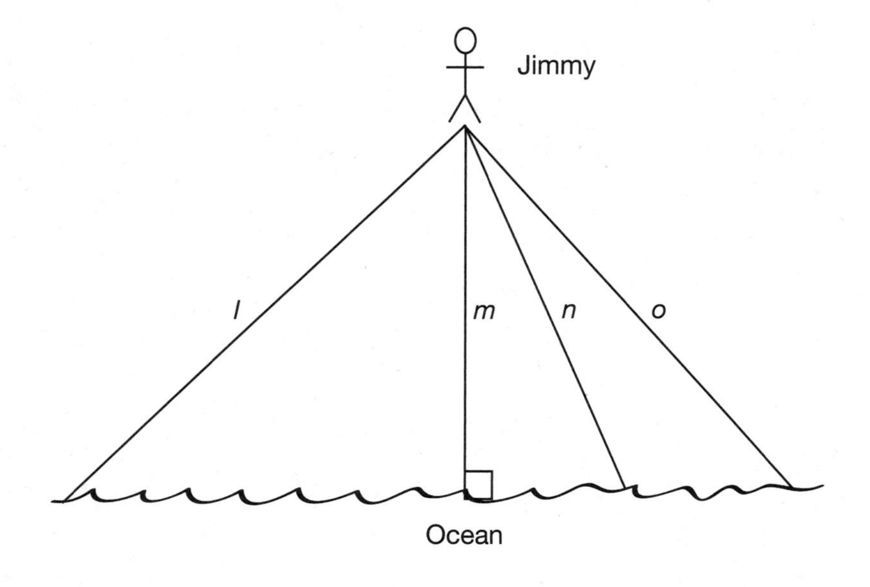

Look at the illustration of Jimmy on the beach. If asked, "How far is Jimmy from the ocean's edge?" which line segment would you use? You'd probably reply with line segment *m* since that is the most direct path to the ocean. We began with this illustration because line *m* is a clear example of the **altitude** of a triangle. The **altitude** of a triangle is the segment that is drawn from a vertex so that it perpendicularly intersects the opposite side. In this illustration, the base is the ocean's edge and the top vertex is Jimmy. Although line segments *l*, *n*, and *o* also connect Jimmy to the ocean, none of them can be considered the altitude because these segments are not perpendicular to the ocean's edge.

TIP: The **altitude** of a triangle is the perpendicular line segment connecting the vertex with the opposite base.

Let's now look at the following figure of ΔABC to illustrate another important term. You can see that $\overline{BY}$ is the perpendicular line segment connecting $\angle B$ to $\overline{AC}$, so it is clear that $\overline{BY}$ is the altitude. The side that is perpendicularly intersected by the altitude is defined as the **base**. In this case, $\overline{AC}$ is the base. Base is a very important concept in triangles!

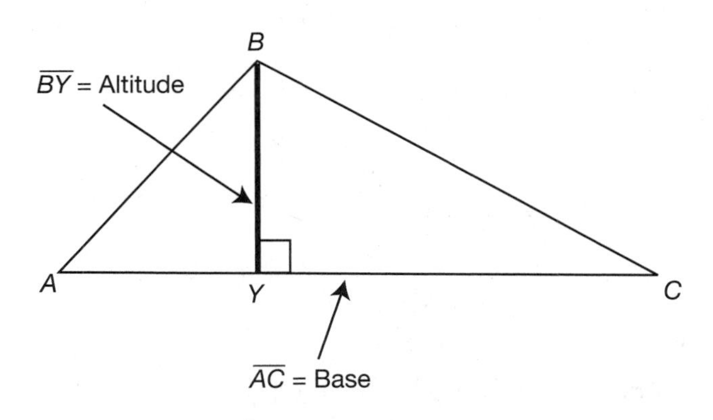

TIP: The **base** of a triangle is the side that is being perpendicularly intersected by the altitude.

Nonstandard Altitudes and Bases

The altitude does not have to be drawn from the top vertex to the bottom side. It can be drawn in different directions as you can see in this figure. Remember though, as the altitude changes, the base changes since they are relative to one another.

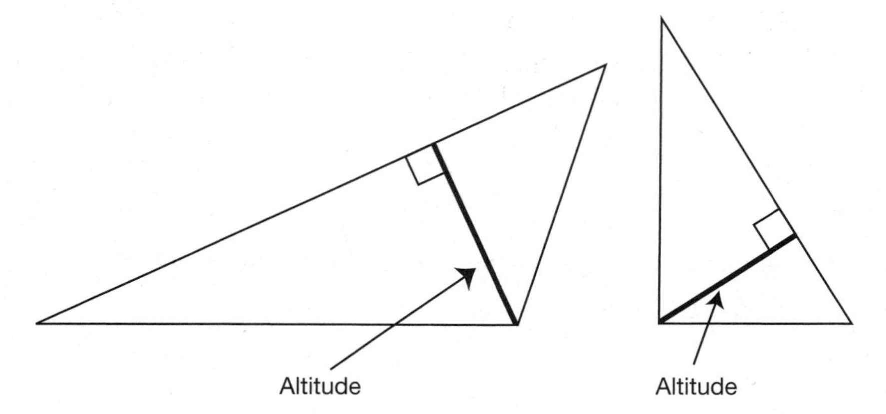

When you are finding the altitude of an obtuse triangle, sometimes the base must be extended so that the altitude can intersect the base in a perpendicular manner. You can see this in the next figure. Notice that in ΔJEN, side $\overline{JN}$ is extended out to point P so that the altitude $\overline{EP}$ can be drawn. In this figure $\overline{JN}$ is the base, not the extended side $\overline{JP}$.

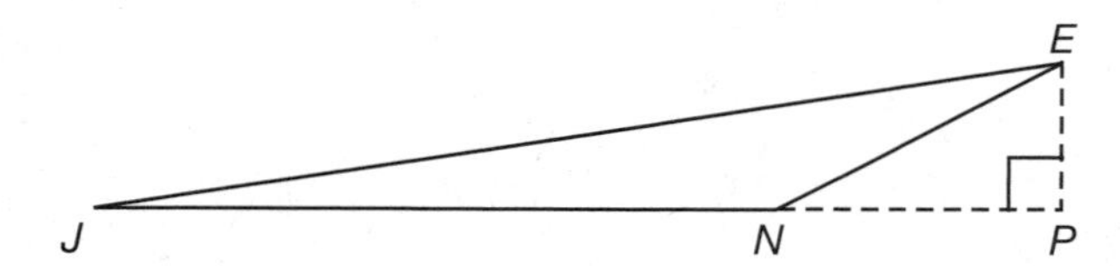

Change Your Altitude, Buddy!

The last important concept to understand about altitudes is that since an altitude can be drawn from each vertex, all triangles always have three altitudes. Therefore, all triangles always have three bases as well: The base will change with the altitude to be the side that the altitude perpendicularly intersects. To visualize this, refer to the following illustration of ΔABC drawn three times. In each copy of ΔABC a different altitude is drawn, which then defines a different base. Look over the following altitude and base relationships and make sure you would be comfortable completing this type of exercise on your own.

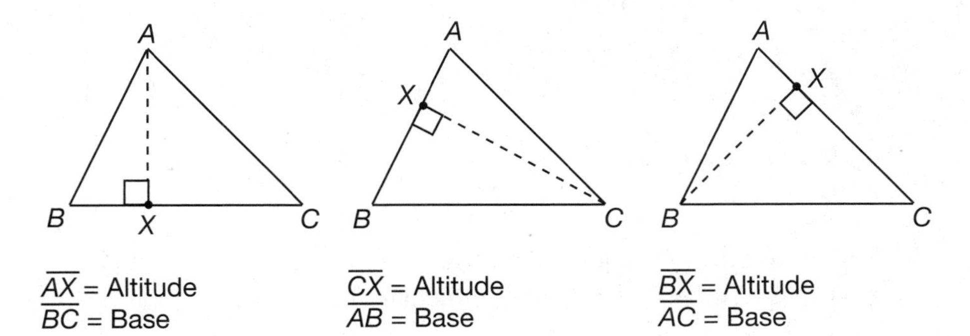

Practice 1

Identify each statement below as true or false.

1. More than one altitude can be drawn from the vertex of a triangle.
2. Some triangles have only one altitude.
3. Sometimes the altitude extends outside of a triangle.
4. The base is the side of the triangle that the altitude intersects at a right angle.
5. In ΔJEN on the previous page, the altitude drawn from vertex J would be inside the triangle.

For questions 6 through 8, draw in the three different altitudes for the acute, right, and obtuse triangle and label them *Alt 1*, *Alt 2*, and *Alt 3* so that they each correspond to *Base 1*, *Base 2*, and *Base 3*.

6.

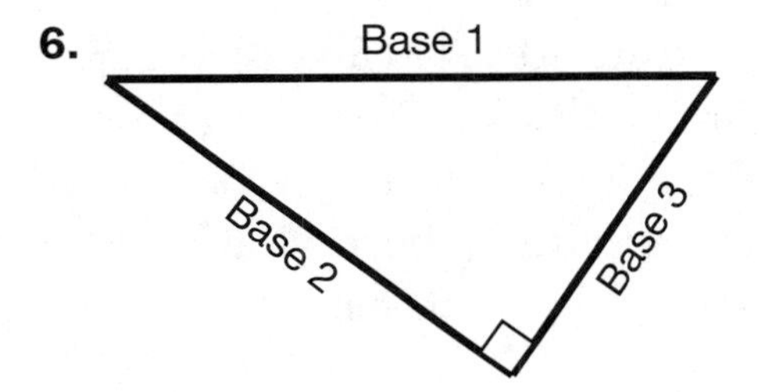

7.

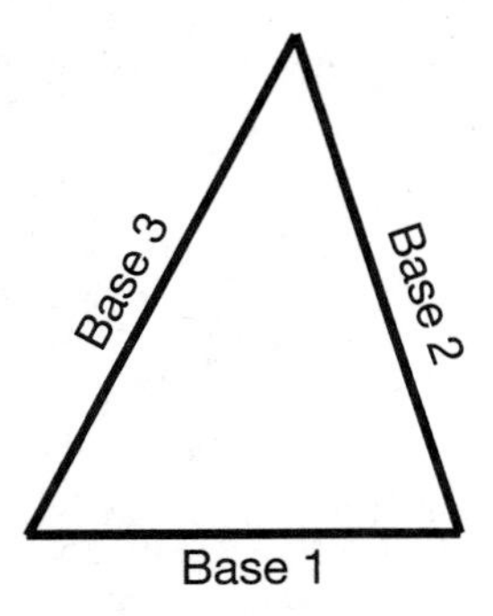

8. Remember that with obtuse triangles you will need to extend two of the sides in order to create perpendicular altitudes.

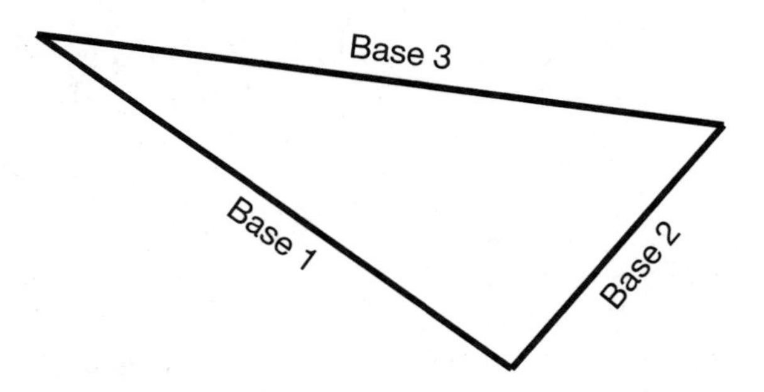

Area

STANDARD ALERT!

Now that you know how to identify the three different altitude and base pairs of a triangle, you are ready for Standard 6.G.A.1: finding the area of triangles!

What Is Area?

A useful and common measure of two-dimensional shapes is area. The **area** of a two-dimensional shape is the amount of space *inside* it. Area is actually a measure of the number of squares that are needed to fully cover a space. To get a visual of what this means, imagine that Lucy is painting a wall. At first, she makes a square on the wall that is 1 foot long by 1 foot tall, and fills it in with paint. The number of these 1-foot squares Lucy needs to paint in order to fully cover the wall is the *area* of the wall.

How Is Area Measured?

Although area is a measure of the number of squares needed to cover a shape, how do we know the size of each of the squares? Similar to how length can be measured in inches, feet, or miles, depending on what we're measuring, area can be measured in different sized squares. If no unit of measurement is given for a shape, the squares are written as basic *units*. They can also be square inches, square feet, or square miles depending on the general size of a shape. For example, a huge plot of land might be measured in square miles, but it makes more sense to measure a front lawn in square yards, and a garden box in square feet.

How Is Area Written?

The correct way area is written is with a squared exponent following the unit of measure, such as 25 in.^{2}. The squared exponent is referring to the inches and the 2 is NOT applied to the 25; 25 in.^{2} is said 25 *square inches*. An area of 25 in.^{2} means that 25 one-by-one-inch squares are needed to fill a certain space.

TIP: ***Area*** **is the measurement of the amount of space contained inside a figure. It is measured in square units and written with an exponent of 2 following the units.**

Finding the Area of a Triangle

In the first half of this lesson, you learned how to identify the altitude and base of a triangle. A triangle's altitude is the same thing as its height.

Every triangle has three different altitudes and as the altitude changes, the perpendicular base changes along with it. The formula for calculating the area of a triangle is $A = \frac{1}{2}bh$, where A is the area, b is the length of the base, and h is the height or altitude. In a later lesson we will investigate where this formula comes from.

TIP: The area of a triangle is given by $A = \frac{1}{2}$(base)(height)

Note that in the next figure, by using different sides as the *base*, there are three different ways that the area can be correctly calculated:

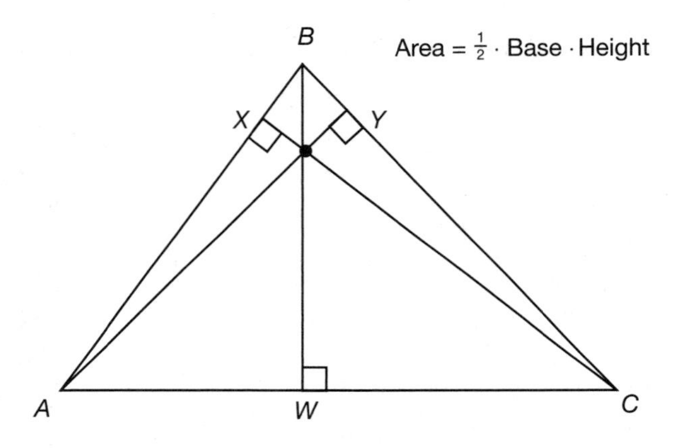

$A = \frac{1}{2}\,\overline{AC} \cdot \overline{BW}$ (using $\overline{AC}$ as base)
or
$A = \frac{1}{2}\,\overline{AB} \cdot \overline{CX}$ (using $\overline{AB}$ as base)
or
$A = \frac{1}{2}\,\overline{BC} \cdot \overline{AY}$ (using $\overline{BC}$ as base)

Finding the Area of an Obtuse Triangle

When you are working with obtuse triangles, the same formula for area is used, but take special care to choose the correct pairing of altitude and base. You already know that in obtuse triangles, sometimes the base must be extended so that a perpendicular altitude can be drawn from the opposite vertex. It is important to remember to use the length of the external altitude as your height, and the *non*-extended base as your base. For example, in the following figure, $\overline{JK}$ is the *non*-extended base that you will use to calculate the area (NOT $\overline{JE}$).

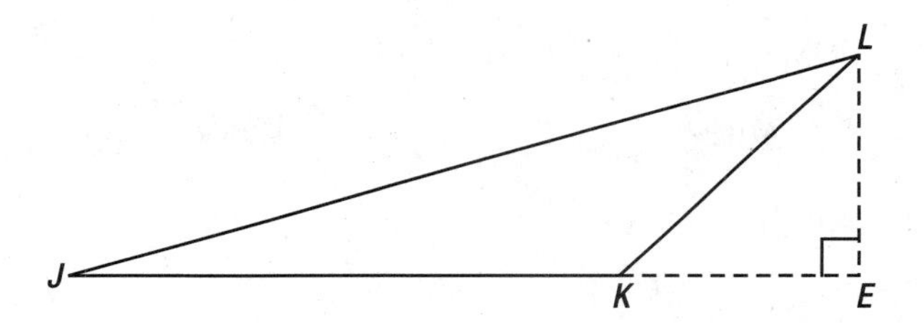

$A = \frac{1}{2}\overline{JK} \cdot \overline{LE}$

STANDARD ALERT!

You've had practice identifying the altitude and base of a triangle and now you have the area formula for triangles, so the following set offers you some practice applying area of triangles to real-world problems, which is part of Standard 7.G.B.6.

Practice 2

Use ΔABC to answer questions 1 through 3.

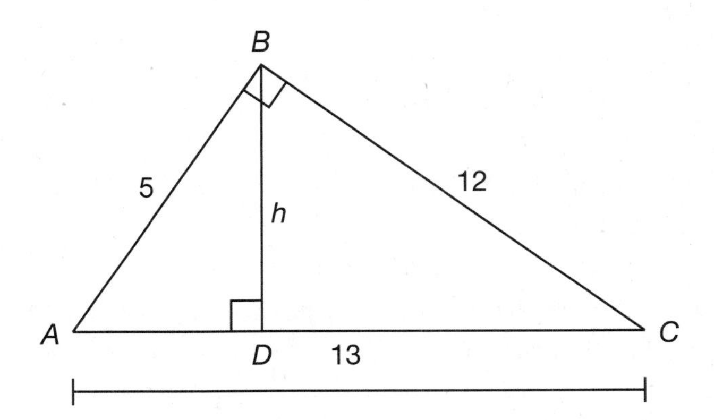

1. Determine the corresponding altitude for each base:
a. If $\overline{AC}$ is the base, then _____ is the altitude
b. If $\overline{BC}$ is the base, then _____ is the altitude
c. If $\overline{AB}$ is the base, then _____ is the altitude

2. What is the area of ΔABC?

3. Use your answer from question 2 to determine the length of altitude h.

Use ΔRSB to answer questions 4 and 5.

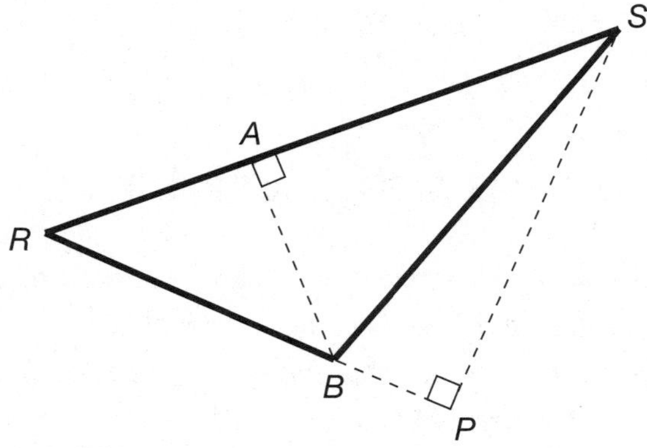

In Δ *RSB*:

$\overline{RP} = 5$ $\overline{SP} = 4$
$\overline{RB} = 3$ $\overline{BS} = 4.5$
$\overline{AB} = 2$

4. Determine which of the given information should be used to determine the area and calculate it.

5. Determine the length of base $\overline{RS}$.

Use the following information and figure to answer questions 6 through 8.

Ryan is the garden coordinator at his school and he has decided to use a 30-yard by 10-yard garden plot to plant succulents and vegetables. He wants to diagonally divide the space as shown.

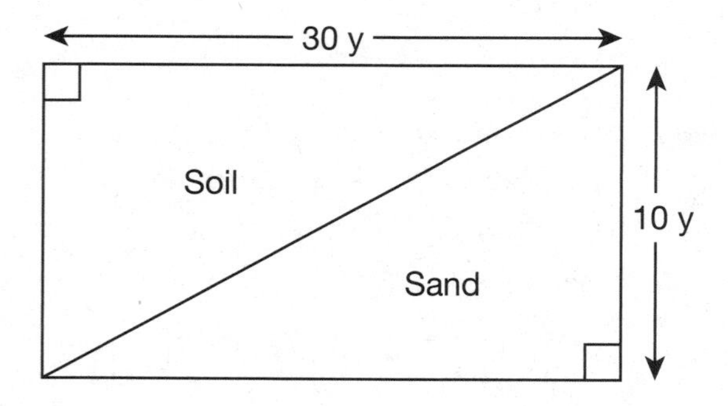

6. If Ryan fills the bottom triangle with sand for the succulents and it costs $1.20 per square foot, how much will he spend on sand?

7. Ryan wants to use an organic manure for the vegetable section of the garden, but it costs $2.00 per square foot. If he doesn't want to spend more than half of his $500 budget on the soil, is this a purchase he should make?

8. Use the two triangles to find the combined area of the rectangular space. At a nearby university campus Ryan is offered a plot that is twice as long and twice as wide, so the dimensions are 60 yards by 20 yards. Use the area you found for a 30-yard by 10-yard plot to make a hypothesis for what the area of this larger plot will be. Redo your calculations to find the area of the larger plot and compare your answer to your hypothesis.

State whether the following statements in questions 9 through 11 are true or false.

9. Area can be measured and expressed in unit squares, unit circles, or unit triangles.

10. One triangular chicken coop has 12 square feet of space for the chickens and another has 16 square feet of space. The coop with the smaller square footage will need a shorter fence to contain the chickens.

11. The base of a triangle is always longer than its corresponding altitude.

Answers

Practice 1

1. False. Each vertex has just one altitude, but a triangle has three altitudes.
2. False
3. True
4. True
5. False
6. 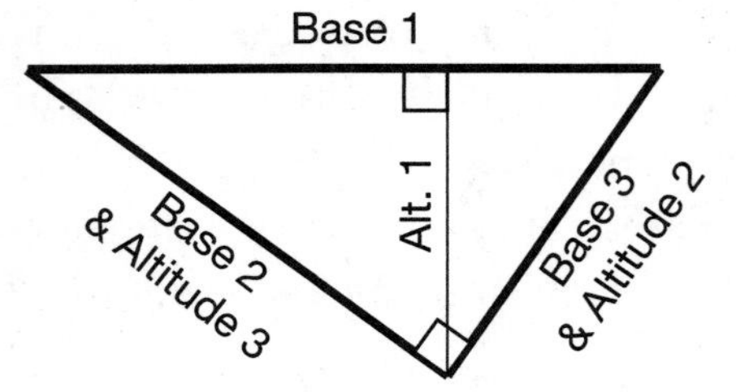

7. 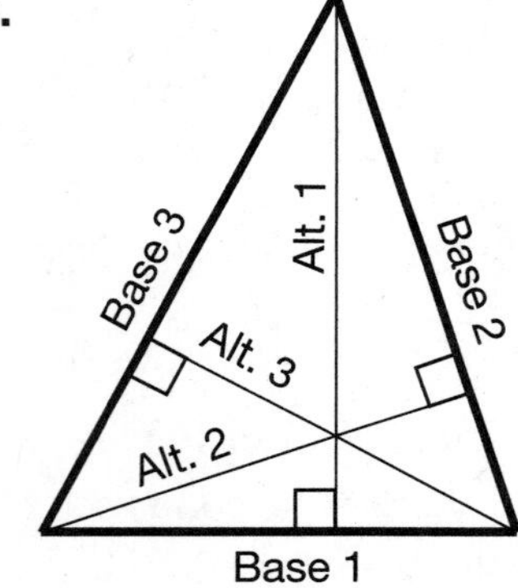

8. 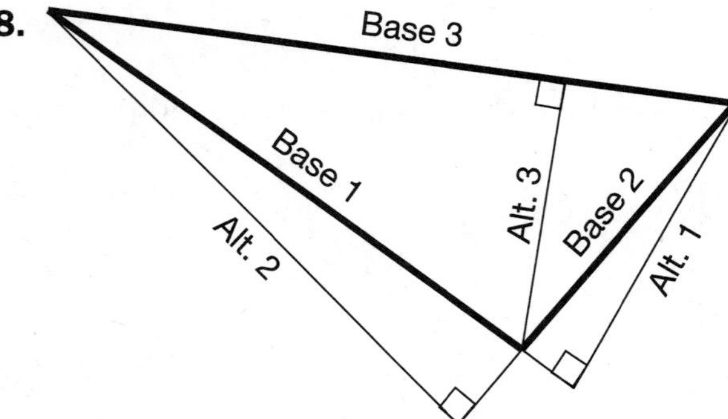

Practice 2

1. Determine the corresponding altitude for each base:
   a. If $\overline{AC}$ is the base, then $\overline{BD}$ is the altitude
   b. If $\overline{BC}$ is the base, then $\overline{AB}$ is the altitude
   c. If $\overline{AB}$ is the base, then $\overline{BC}$ is the altitude
2. Area of $\Delta ABC = 30$ units^2
3. $A = \frac{1}{2}bh$ so sub in the area and the base to get $30 = \frac{1}{2}(13)h$. $h = \frac{60}{13} \approx 4.6$
4. Base = $\overline{RB} = 3$. Altitude = $\overline{SP} = 4$. $A = \frac{1}{2}bh$, so $A = \frac{1}{2}(3)(4) = 6$ units^2
5. $A = \frac{1}{2}bh$, so sub in the $A = 6$ and the $h = \overline{AB} = 2$ to get $6 = \frac{1}{2}(b)(2)$. $b = 6$
6. $A = \frac{1}{2}bh = \frac{1}{2}(30)(10) = 150y^2$. $150 \times \$1.20 = \$180$
7. $A = \frac{1}{2}bh = \frac{1}{2}(30)(10) = 150y^2$. $150 \times \$2.00 = \$300$. This will cost more than half of his budget.
8. Area of two 30×10 triangles: $\frac{1}{2}bh + \frac{1}{2}bh = \frac{1}{2}(30)(10) + \frac{1}{2}(30)(10) = 150 + 150 = 300$

   Area of two 60×20 triangles: $\frac{1}{2}bh + \frac{1}{2}bh = \frac{1}{2}(60)(20) + \frac{1}{2}(60)(20) = 600 + 600 = 1{,}200$. So, when the dimensions are twice as long, instead of the combined area being doubled, it is 4 times bigger.
9. False: Area is only measured and expressed in unit squares.
10. False: The chicken coop with 12 square feet of space might be really long and flat, meaning that it could need more fencing than a more evenly shaped coop with 16 square feet of space.
11. False: The altitude can be longer than the base.

9 The Pythagorean Theorem

Do not worry too much about your difficulties in mathematics;
I can assure you that mine are still greater.
—Albert Einstein

In this lesson you will learn about the widely used Pythagorean theorem and its application to right triangles.

STANDARD SNEAK PREVIEW

We've spent a few lessons acquainting ourselves with theorems that enable us to solve for unknown *angles* in triangles. Now we're going to learn how to apply an important theorem that will empower you to solve for unknown *sides* in right triangles. This theorem was proven more than 2,000 years ago by a Greek mathematician named Pythagoras; it's such a major theorem that thousands of years later it appears in three separate Common Core State Standards! (Wouldn't that be cool if you discovered something that people were still writing about thousands of year from now?) Let's not get ahead of ourselves just yet! For now you're going to *learn how to apply the Pythagorean theorem to solve for the missing side lengths of right triangles, which is Standard 8.G.B.7*. When you're done with this lesson, check out Reference A at the back of this book. There, we explain the proof of the Pythagorean theorem, which covers Standard 8.G.B.6.

A Closer Look at Right Triangles

So far you have learned that a right triangle has one right angle and two acute angles that sum to 90°. Before discussing the Pythagorean theorem, you must become familiar with the standard way to label the sides and angles of right triangles. In a right triangle, the two sides that form the right angle are referred to as the **legs**. The longest side is always opposite the right angle and is called the **hypotenuse**. It is standard practice to label the legs of a right triangle a and b, while the hypotenuse is labeled c. As discussed in previous lessons, the angles across from the sides are named with the capital letters of their opposite sides. Note in the next figure that the right angle is labeled C and the two acute angles are labeled A and B.

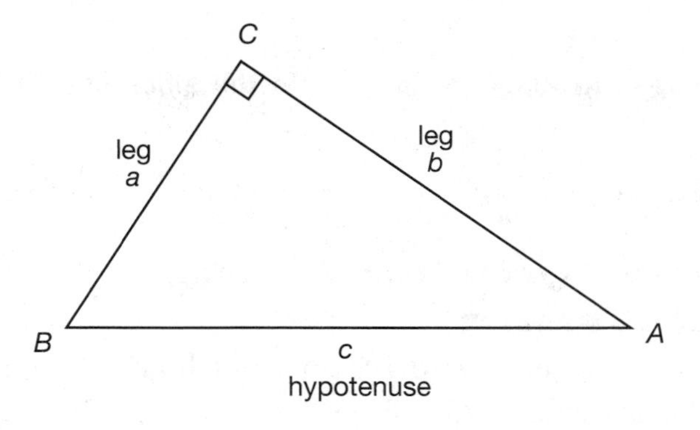

TIP: In a right triangle, the shorter sides are called the *legs* and are labeled *a* and *b*. The longest side opposite the right angle is called the *hypotenuse* and is labeled *c*.

The Pythagorean Theorem

Before the big reveal of how the Pythagorean theorem connects the side lengths in the triangle above, it should be mentioned that there is evidence that other mathematicians were using this theorem thousands of years before Pythagoras, but he is credited with writing the first formal proof of it.

The Pythagorean theorem is used to find the missing third side length in a right triangle when you are given any two sides. It is common to be given the lengths of the two legs, and to then use the Pythagorean theorem to solve for the hypotenuse. However, it is just as easy to use the Pythagorean theorem to solve for the length of one of the legs, when you're given the length of the other leg and the hypotenuse. The Pythagorean theorem is:

$$(\text{Leg \#1})^2 + (\text{Leg \#2})^2 = (\text{Hypotenuse})^2$$

Since the two legs are commonly labeled a and b, and the hypotenuse is commonly labeled c, the Pythagorean theorem is most commonly written as.

$$a^2 + b^2 = c^2$$

TIP: The Pythagorean theorem applying to the sides of right triangles is $a^2 + b^2 = c^2$.

Solving for the Hypotenuse Using the Pythagorean Theorem

Using the following figure, we will demonstrate how to solve for the missing hypotenuse using the Pythagorean theorem. (*Note:* It does not matter which leg you chose to label *a* and which leg you chose to label *b*.)

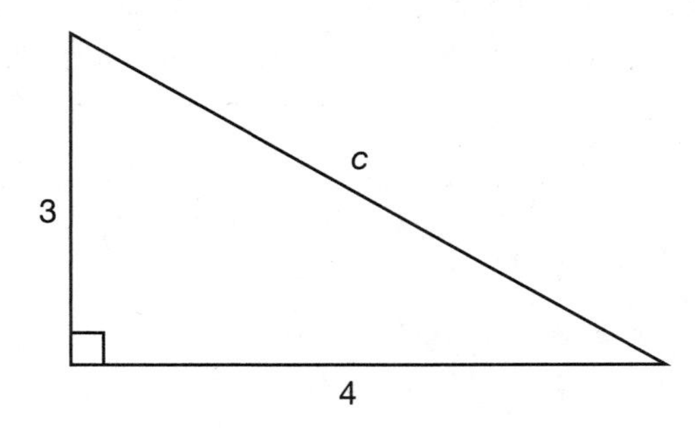

$a^2 + b^2 = c^2$
$3^2 + 4^2 = c^2$
$9 + 16 = c^2$ (To get rid of the c^2, take the square root of both sides.)
$25 = c^2$
$\sqrt{25} = \sqrt{c^2}$
$5 = c$, so the hypotenuse c in this example equals 5.

Solving for a Leg Using the Pythagorean Theorem

Using the following figure, we will demonstrate how to solve for a missing leg when the hypotenuse is one of the given pieces of information.

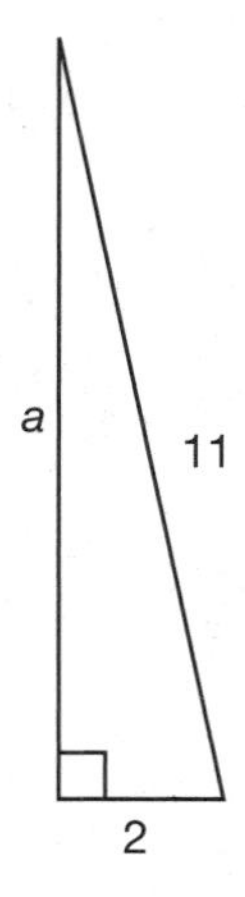

$a^2 + b^2 = c^2$

$a^2 + 2^2 = 11^2$. Notice that the hypotenuse length of 11 was subbed in for c, and not for *2* or b.

$a^2 + 4 = 121$

$a^2 = 117$

$\sqrt{a^2} = \sqrt{117}$

$a = \sqrt{117} \approx 10.8$. So the missing leg, a in this example, approximates to 10.8.

Be Careful of This Common Mistake!

The missing piece of information must be correctly identified and then shown in your equation when you are working with right triangles. A common mistake with the Pythagorean theorem is when students assume that whatever side is missing should be represented with c. You must be careful to notice whether you are solving for a missing leg (a or b) or a missing hypotenuse (c).

STANDARD ALERT!

Are you ready to practice Standard 8.G.B.7 by solving for the length of unknown sides of right triangles? There are two major pitfalls to avoid: (1) whether it's a variable or a number, make sure that you are always putting the hypotenuse of the triangle alone on one side of the equation; and (2) make sure that you are taking the square root as your final step. Ready, Set, Solve!

Practice 1

Use this figure to answer questions 1 through 4.

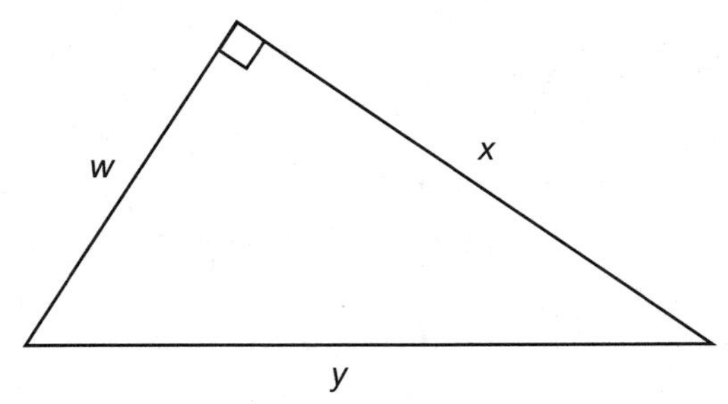

Note: Figure Not Drawn to Scale

1. Solve for y if $x = 8$ and $w = 6$.

2. Solve for x if $y = 13$ and $w = 5$.

3. Solve for w if $y = 10$ and $w = 5$.

4. Solve for y if $x = 4k$ and $w = 2k$. (Hint: Don't forget to square the variable k.)

5. Victoria has a shed that is 15 feet long and 10 feet wide. She wants to store the longest rolled-up carpet possible in the shed, diagonally on the floor. What is the largest rolled carpet that will fit in Victoria's shed?

6. Dorothy is standing directly 300 meters under a plane. She sees another plane flying straight behind the first. It is diagonally 500 meters away from her. How far apart are the planes from each other?

7. Eva and Carr pass at a corner. Eva turns 90° left and walks 5 paces; Carr continues straight and walks 6 paces. If a line segment connected them, how many paces would it measure?

Other Applications of the Pythagorean Theorem

Proving Right Triangles

Since the relationship between the legs and the hypotenuse in right triangles is $a^2 + b^2 = c^2$, we can work backwards to see if a given triangle is right or not. In the following figure, we can test to see if the lengths of the sides of the given triangle identify the Pythagorean theorem. If they result in a balanced equation, where the left side is equal to the right side, then ΔPIG is a right triangle. If the equation is not balanced, then you know that ΔPIG is not a right triangle.

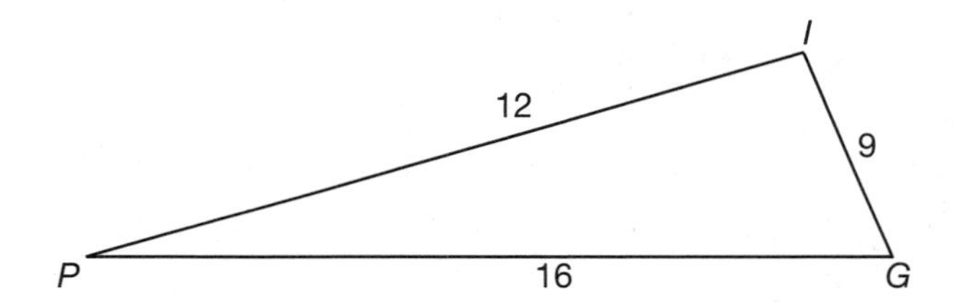

$a^2 + b^2 = c^2$

$9^2 + 12^2 = 16^2$ (Use the longest side as the hypotenuse.)

$81 + 144 = 256$

$225 \neq 256$

Since the result is a false statement, you can conclude that ΔPIG is not a right triangle.

Identifying Acute and Obtuse Triangles

Similarly, when you are given the length of all three sides of a triangle, you can use the Pythagorean theorem to identify whether it is an acute or an obtuse triangle. If the hypotenuse squared is larger than the sum of the squared legs, then the triangle is obtuse. See the next figure for an example that illustrates this.

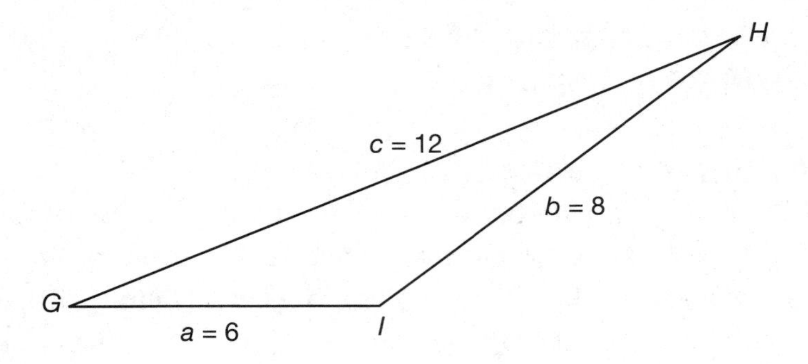

$a^2 + b^2 = c^2$

$6^2 + 8^2 = 12^2$

$36 + 64 = 144$

$100 < 144$

Therefore, ΔGHI is obtuse.

Conversely, if the hypotenuse squared is smaller than the sum of the squared legs, then the triangle is acute. See the next figure for an example that illustrates this.

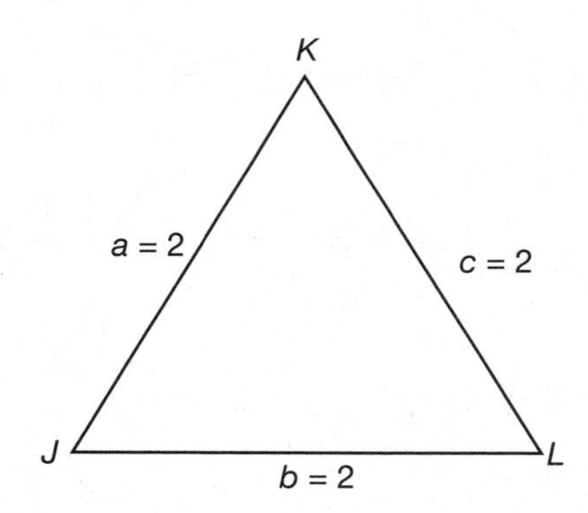

$a^2 + b^2 = c^2$

$2^2 + 2^2 = 2^2$

$4 + 4 = 4$

$8 > 4$

Therefore, ΔJKL is acute.

TIP: Allow c to be the longest side when using the following test to identify what type of triangle ABC is:

If $a^2 + b^2 = c^2$, then the triangle is a right triangle.

If $a^2 + b^2 < c^2$, then the triangle is an obtuse triangle.

If $a^2 + b^2 > c^2$, then the triangle is an acute triangle.

Practice 2

For questions 1 through 5, given the following side lengths, identify whether the triangle is right, acute, or obtuse:

1. 12, 16, 20

2. 8, 9, 12

3. 4.6, 7.8, 11.5

4. 11, 11, 16

5. $2\sqrt{3}$, 4, $2\sqrt{7}$

The 3-4-5 Right Triangle

One helpful ratio to memorize with triangles is the 3-4-5 triangle. A quick calculation using the Pythagorean theorem shows that a 3-4-5 triangle is right, since 9 + 16 = 25. Any triangle that has sides that can be reduced to a ratio of 3:4:5 will be a right triangle. For example, if you are given a triangle with side lengths of 30, 40, and 50, you should recognize that when you divide all the sides by 10, the ratio of the sides is 3:4:5. Seeing this, you can conclude that a triangle with sides 30, 40, and 50 is a right triangle.

The 3-4-5 property can also be used as a shortcut to the Pythagorean theorem. If you see that two legs can be reduced to fit the 3:4 ratio, then you can use this shortcut to determine the missing hypotenuse without going through the calculations of the Pythagorean theorem. For example, if you know that a right triangle has two legs that measure 6 and 8, you might notice that since $3 \times 2 = 6$ and $4 \times 2 = 8$, a multiple of 2 can be used to identify the hypotenuse by computing $5 \times 2 = 10$. Similarly, if you are given a leg and a hypotenuse that can be reduced to a 3:5 or 4:5 ratio, the same shortcut can be used. Consider a right triangle with a leg of 33 and a hypotenuse of 55. Since $3 \times 11 = 33$ and $5 \times 11 = 55$, a multiple of 11 can be used to determine that the second leg would be $4 \times 11 = 44$.

Practice 3

Given the two sides of right triangles, use the 3-4-5 ratio to find the missing side.

1. leg 1 = 12, leg 2 = 16; find the length of the hypotenuse.
2. leg 1 = 18, leg 2 = 24; find the length of the hypotenuse.
3. leg 1 = 21, hypotenuse = 35; find the length of leg 2.
4. leg 1 = 1.5, hypotenuse = 2.5; find the length of leg 2.
5. leg 1 = $3xy$, leg 2 = $4xy$; find the length of the hypotenuse.

Answers

Practice 1

1. $y = 10$
2. $x = 12$
3. $w = 5\sqrt{3}$ or 8.66 ($\sqrt{75} = \sqrt{25 \times 3} = \sqrt{25} \times \sqrt{3} = 5\sqrt{3} = 8.66$)
4. $y = 2k\sqrt{5}$ or $4.47k$ ($16k^2 + 4k^2 = 20k^2 = \sqrt{4 \times 5 \times k^2} = 2k\sqrt{5} = 4.47k$)
5. 18 feet ($\sqrt{325} \approx 18.03$)
6. 400 meters
7. $\sqrt{61} \approx 7.8$ paces

Practice 2

1. right
2. acute
3. obtuse
4. obtuse
5. right

Practice 3

1. 20
2. 30
3. 28
4. 2
5. $5xy$

10

45-45-90 and 30-60-90 Triangles

Geometry is the science of correct reasoning on incorrect figures.
—George Pólya

In this lesson you will apply what you learned about the Pythagorean theorem to the special side–angle relationships in 30°-60°-90° triangles and 45°-45°-90° triangles.

STANDARD SNEAK PREVIEW

This lesson is your bonus track on triangles! Although the two special triangles we investigate in the following pages are not specifically included in the Common Core State Standards for middle school, learning about these triangles is a great extension of the Pythagorean theorem. This lesson will allow you to see how the Pythagorean theorem is used to derive an algebraic equation that models the side relationships for 45-45-90 triangles. It also illustrates how the Pythagorean theorem confirms the given equation, which models the side relationships for 30-60-90 triangles. If you've have enough of triangles for now, you're free to move on to the next lesson since this is not a requirement for the middle school CCSS, but we encourage you to stay for a while, and have a little more fun with triangles! After all, you will come across these triangles in high school, so why not get familiar with them now that you're in a Pythagorean theorem state of mind!

45-45-90 Right Triangles

There are two common right triangles that have special angle–side relationships. The first noteworthy triangle is the **isosceles right triangle**, which is also referred to as the **45-45-90 right triangle**. As you remember, isosceles triangles have one unique vertex angle and two congruent base angles. If the vertex angle is 90°, the two base angles must sum to 90°, so they each measure 45°. That is why this special isosceles triangle is called the 45-45-90 right triangle. It is pictured in the next figure.

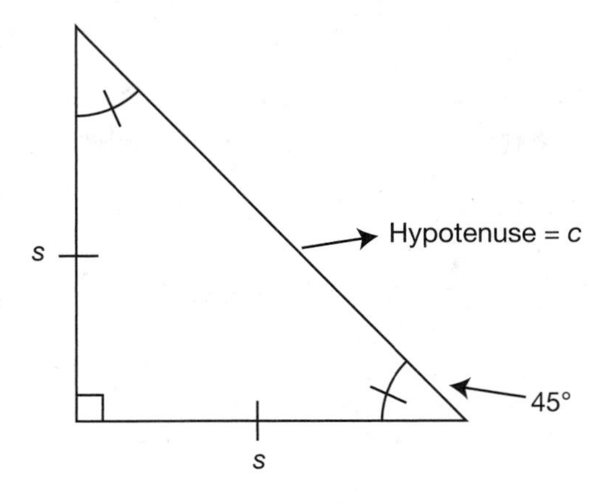

45–45–90 Right Triangle

The Pythagorean Theorem and the 45-45-90 Right Triangle

Since the legs of an isosceles triangle are congruent, they can each be labeled s instead of using a and b. The hypotenuse is labeled c. Then, applying the Pythagorean theorem, we can solve for c. This will reveal the special relationship between the legs and the hypotenuse in 45-45-90 right triangles:

$s^2 + s^2 = c^2$

$2s^2 = c^2$

$\sqrt{2s^2} = \sqrt{c^2}$

$s\sqrt{2} = c$, so the hypotenuse will always be s times $\sqrt{2}$ (which is the length of one leg times $\sqrt{2}$).

TIP: **In 45-45-90 right triangles, the length of the hypotenuse is equal to $\sqrt{2}$ times the length of one of the legs. This is written $s\sqrt{2}$.**

Working the other direction, it is possible to see that if you were given the length of the hypotenuse, you could divide it by $\sqrt{2}$ to get the length of one of the legs. However, because it is not acceptable to keep a radical sign in the denominator of a fraction, another way to write this leg is = $\frac{\sqrt{2}}{2}$ · hypotenuse.

TIP: In 45-45-90 right triangles, the length of one of the congruent legs is equal to the hypotenuse times ($\frac{\sqrt{2}}{2}$). This is written $\frac{\sqrt{2}}{2} \cdot c$.

The following illustration and formulas can be used when solving questions with 45-45-90 triangles:

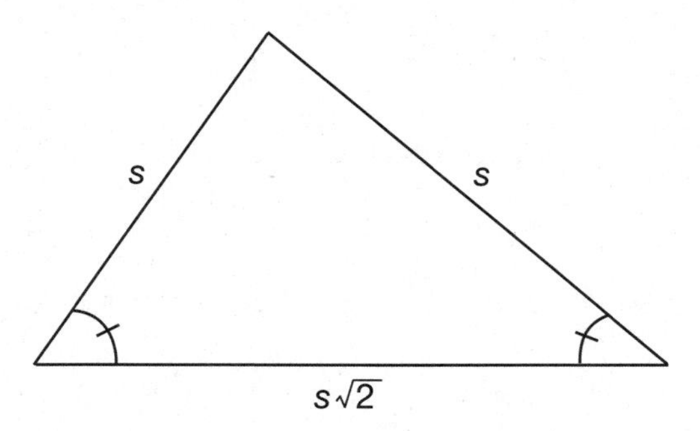

hypotenuse = leg $\cdot \sqrt{2}$

leg = $\frac{\sqrt{2}}{2} \cdot$ hypotenuse

Practice 1

Use the following figure to solve questions 1 through 6. (*Notice:* The sides are not labeled with the standard letters. This does not change how the rules for 45-45-90 right triangles apply.)

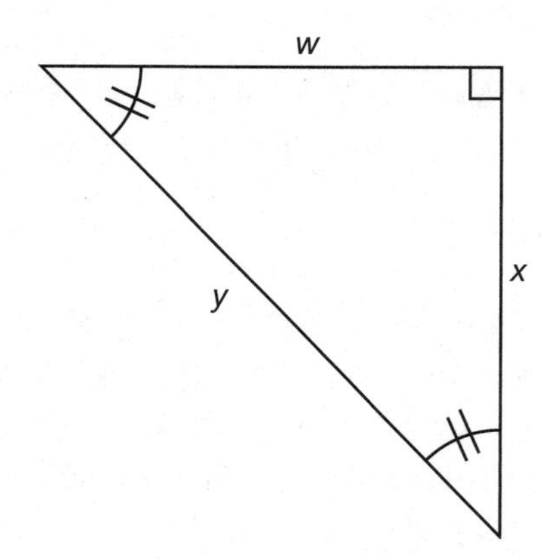

1. If $w = 12$, find the length of y.

2. If $y = 32$, find the length of x.

3. If $x = 5\sqrt{2}$, find the length of y.

4. If $y = 14\sqrt{6}$, find the length of x.

5. If $x = 7.5$, find the perimeter of the triangle to the nearest tenth.

6. If $y = 8\sqrt{2}$, find the perimeter of the triangle to the nearest tenth.

7. The perimeter of a 45-45-90 isosceles right triangle is 41 feet. Find the length of the triangle's legs and hypotenuse. (*Hint:* Assign variables to the legs and hypotenuse that reflect their relationship in 45-45-90 triangles, and then use those variable expressions in the perimeter formula.)

8. Izzy is making a gardening table from recycled wood to fit in the back corner of his yard. He wants the table to be an isosceles right triangle with the base edge facing out toward the garden. The longest piece of wood he has for the front edge of the table (the base of the triangle) is 5 feet 4 inches. How many inches long will each of the congruent sides of the table be?

9. The Rham family is restoring a large square farm table for their new backyard. Each edge of the table measures 6 feet 3 inches. They want to install a wire cable under the table to give it more stability. If they install this cable diagonally, from one corner to the opposite corner, how many inches long will it need to be?

10. Shelly is 40 miles directly north of Bend, Oregon. Adam is 40 miles directly east of Bend, Oregon. What is the shortest distance, diagonally, between Shelly and Adam? (Round to the nearest tenth of a mile.)

30-60-90 Right Triangles

The second common right triangle with special angle–side relationships is the **30-60-90 right triangle.** The 30-60-90 right triangle has a right angle and two acute angles that measure 30° and 60°. As you remember, the smallest side of a triangle is directly opposite the smallest angle. Therefore, we are going to call the length of the shortest leg opposite the 30° angle, n. The side opposite the 60° angle in 30-60-90 right triangles always has a length of $\sqrt{3}$ *times the shortest side* or $n\sqrt{3}$. The hypotenuse in 30-60-90 right triangles is always *twice the length of the shorter side*, or $2n$. Notice how the next figure is labeled.

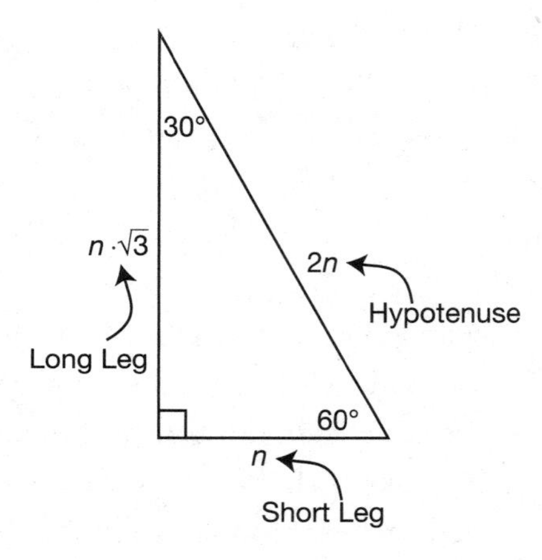

TIP: **In 30-60-90 right triangles, if the length of the smallest leg is *n*, then the length of the longer leg is $n\sqrt{3}$ and the length of the hypotenuse is $2n$.**

The Pythagorean Theorem and the 30-60-90 Right Triangle

The Pythagorean theorem can be used to confirm the special relationship between the sides of the 30-60-90 right triangle.

$a^2 + b^2 = c^2$
$(n)^2 + (n\sqrt{3})^2 = (2n)^2$
$n^2 + 3n^2 = 4n^2$
$4n^2 = 4n^2$ and since this is a true statement, the relationship between sides is confirmed to be correct.

TIP: When working with 30-60-90 triangles, use the following relationships to solve for missing sides:

Hypotenuse = (shorter leg) × (2)

Longer leg = (shorter leg) × ($\sqrt{3}$)

Working Backwards with 30-60-90 Triangles

If you are given the length of the hypotenuse in a 30-60-90 right triangle, then you can calculate the length of the shorter side by dividing it by 2. Similarly, if you're given the length of the longer leg, you can calculate the length of the shorter side by dividing it by $\sqrt{3}$. If you are not given the length of the shorter side, it is easiest first to solve for the shorter side and then to move forward from there.

Example

The length of the longer leg of a 30-60-90 right triangle is 10. Find the length of the shorter leg and of the hypotenuse.

We know that Longer leg = (Shorter leg) × ($\sqrt{3}$).
So 10 = (Shorter leg)($\sqrt{3}$).
Therefore, the shorter leg = $\frac{10}{\sqrt{3}} \approx 5.8$.
Since the hypotenuse = (shorter leg)(2), the hypotenuse = 5.8(2) = 11.6.

STANDARD ALERT!

Although solving for the side lengths in 45-45-90 and 30-60-90 triangles is an extension of the Pythagorean theorem, the work you are doing here is still helpful in reinforcing Standard 8.G.B.7 by solving for the length of unknown sides of right triangles.

Practice 2

Use the next figure to answer questions 1 through 5.

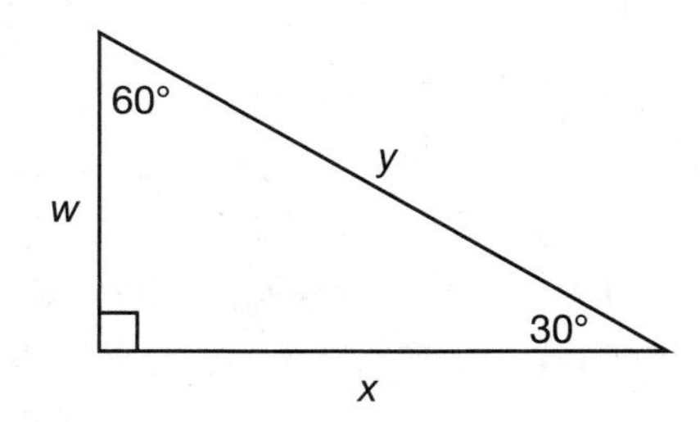

1. If $w = 7.5$, find the length of y.
2. If $y = 36$, find the length of w.
3. If $x = 6\sqrt{3}$, find the length of y.
4. If $y = 12\sqrt{10}$, find the length of x.
5. If $w = 22$, find the perimeter of the triangle to the nearest tenth.
6. ΔPAT is a 30-60-90 right triangle with the right angle at point A. If PA is the shorter leg, and $AT = 64$, find the length of the hypotenuse to the nearest tenth.
7. ΔSKY is a 30-60-90 right triangle with the right angle at point Y. If $SK = 38$ inches, find the perimeter to the nearest tenth.

Use the next figure to answer questions 8 through 10.

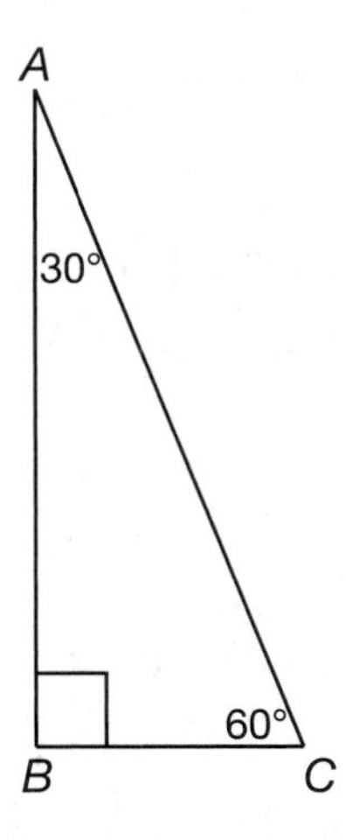

8. If the length of side AB is 9 cm, what is the length of side BC?

9. If the hypotenuse of triangle ABC is $6x + 2$ units long, what is the length of side AB?

10. If the sum of sides BC and AC is 12 units, what is the length of side AB? (*Hint:* Assign variables to show the relationship between the shortest leg and hypotenuse and use them with the given information.)

Answers

Practice 1

1. $y = 12\sqrt{2} \approx 16.9$ units
2. $x = 16\sqrt{2} \approx 22.6$ units
3. $y = 10$ units $(5\sqrt{2} \times \sqrt{2} = 5 \times 2 = 10)$
4. $y = 14\sqrt{6}$, and since this is the hypotenuse, multiply it by $\frac{\sqrt{2}}{2}$:
   $\frac{\sqrt{2}}{2} \times \frac{14\sqrt{6}}{1}$
   $\frac{14\sqrt{12}}{2}$
   $7\sqrt{12} = 7\sqrt{4}\sqrt{3} = 14\sqrt{3} \approx 24.2$
5. $P \approx 25.6$ units $(P = 7.5 + 7.5 + 7.5\sqrt{2} \approx 25.6)$
6. $P \approx 27.3$ units $(P = (8\sqrt{2}) + \frac{(8\sqrt{2}\sqrt{2})}{2} + \frac{(8\sqrt{2}\sqrt{2})}{2} = (8\sqrt{2}) + (16) \approx 27.3)$
7. Since it is a 45-45-90 isosceles right triangle, you can let each of the legs measure x units and, therefore, the hypotenuse will then measure $x\sqrt{2}$.
   $P = x + x + x\sqrt{2} = 41$
   $1x + 1x + 1.41x = 41$
   $3.41x = 41$, so $x \approx 12$
   Therefore, each of the legs is 12 and the hypotenuse is $12\sqrt{2}$.
8. The base is 5 feet 4 inches, which is $5(12) + 4 = 64$ inches. In order to find the length of each of the congruent sides, multiply 64 by $(\frac{\sqrt{2}}{2})$: $64(\frac{\sqrt{2}}{2}) = 32(\sqrt{2}) \approx 45.25$ inches
9. Each leg is 6 feet 3 inches or 75 inches long $(6' \times 12" = 72" + 3" = 75")$. The diagonal cable will be the hypotenuse of a 45-45-90 triangle, so it will be the leg $\times \sqrt{2}$, which is $75\sqrt{2} = 106$ inches long.
10. 56.6 miles

Practice 2

1. $y = 15$ units
2. $w = 18$ units
3. $y = 12$ units $(\frac{6\sqrt{3}}{\sqrt{3}} = 6$, which is the shorter side, then $6 \times 2 = 12)$
4. $x = 32.9$ units $(\frac{12\sqrt{10}}{\sqrt{3}} = 6\sqrt{10}$, which is the shorter leg. Then the longer leg is $6\sqrt{10} \times \sqrt{3} = 6\sqrt{30} \approx 32.9$ units)
5. $P \approx 104.1$ units $(P = 22 + 44 + 22\sqrt{3})$
6. Hypotenuse ≈ 74.0 units $(\frac{64}{\sqrt{3}} \approx 37.0$, which is the shorter side; then $37.0 \times 2 = 74)$

7. $P \approx 89.9$ inches ($P = 38 + 19 + 19\sqrt{3} \approx 89.9$)
8. $3\sqrt{3} \approx 5.2$ ($\frac{9}{\sqrt{3}} = \frac{9\sqrt{3}}{3} = 3\sqrt{3}$)
9. $(3x + 1)\sqrt{3}$ units. ($6x + \frac{6x + 2}{2} = (3x + 1)$ which is the shorter leg. Then multiplying that by $\sqrt{3}$ equals the length of the longer leg)
10. $4\sqrt{3}$ units. (12 = shorter leg (n) + hypotenuse ($2n$). So $12 = 3n$ and the shorter leg = 4.)

11 Polygons and Quadrilaterals

The value of a problem is not so much coming up with the answer as in the ideas and attempted ideas it forces on the would-be-solver.
—I.N. Herstein

In this lesson, you'll build upon your existing knowledge of polygons as you have an in-depth look at some of the shared characteristics of special polygons, including sides and angles.

STANDARD PREVIEW

The simple combination of onions, carrots, and celery is referred to as "the cooking Holy Trinity" in many kitchens across the world. Chefs combine this known trio of vegetables with their other knowledge of spices, vegetables, meats, and grains in order to create delicious food. Similarly, you are going to combine what you've learned in the previous lessons about parallel lines and angles to create new meaningful geometric relationships in polygons. In Lesson 5, *Standard 8.G.A.5* focused on the special pairs of congruent and supplementary angles that are made when a transversal intersects parallel lines, and those were important foundational concepts. Now, like skilled chefs, we will apply this information to parallelograms in order to see what special angle relationships we can cook up. You will then practice *Standard 7.G.B.5 by solving for unknown angles* in parallelograms. You will also be getting prepared to meet *Standard 6.G.A.1* in the next lesson, where you will find the area of polygons by breaking them down into composite shapes of triangles and rectangles.

Polygon Basics

Polygons are closed geometric figures that are made up of three or more straight line segments as sides. You've already spent a lot of time learning about triangles, which are the most basic types of polygons. But don't be fooled by the word *basic*: triangles are actually the building blocks of all polygons! This means that you can combine triangles of various shapes and sizes to create any other shape of polygon. No wonder they're so important! Polygons are named like triangles, which means you list their vertices in clockwise or counterclockwise order. It doesn't matter which vertex you start with, just make sure you don't leave any of the vertices out!

Classifying Polygons

The most basic way to classify a polygon is by the number of sides it has. Aside from triangles, there's a good chance you're already familiar with quadrilaterals (4 sides), pentagons (5 sides), hexagons (6 sides). Here's a list for you to look at for reference. It's helpful to be familiar with these polygons:

Number of Sides	Polygon Name
3	Triangle
4	Quadrilateral
5	Pentagon
6	Hexagon
7	Heptagon
8	Octagon
9	Nonagon
10	Decagon

Polygons are further classified by the congruence of their sides or angles. An **equilateral** polygon has *sides* that are all equal in length. An **equiangular** polygon has *angles* that are all equal in measure. The most specific polygon is the **regular** polygon, which is both equilateral *and* equiangular. An equilateral triangle is an example of a three-sided regular polygon. A rectangle is equiangular, but since it does not have to be equilateral, it is not a regular polygon. However, a square is regular since its sides and angles are *all* congruent. You can notice the difference in the next figure.

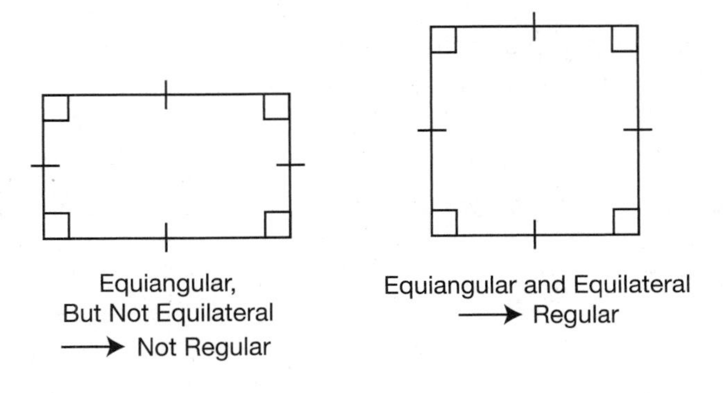

TIP: Polygons have the following classifications:

- **Equilateral** polygons have sides of equal length
- **Equiangular** polygons have angles of equal measure
- **Regular** polygons have equal sides *and* equal angles

Classifying Quadrilaterals

As recalled earlier, quadrilaterals are 4-sided polygons. In an earlier practice, Ryan had a rectangular garden plot that he was cutting diagonally to make a triangular succulent garden and a triangular vegetable garden:

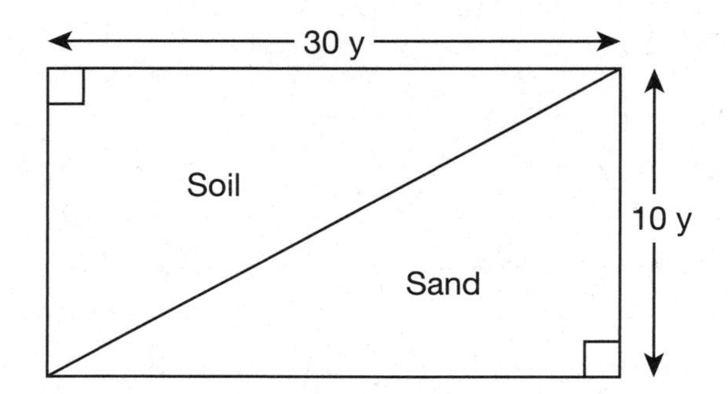

Not only can rectangles be cut into two triangles like this, but *all* quadrilaterals can be divided into two non-overlapping triangles. Since we know that the sum of a triangle's interior angles is 180°, we can therefore conclude that quadrilaterals have an interior angle sum of $2 \times 180° = 360°$.

TIP: All quadrilaterals have an interior angle sum of 360°.

Quadrilaterals are divided into three major categories according to how many pairs of parallel sides they have:

1. The first group has no pairs of parallel sides. It may or may not have sides and angles that are congruent, but it does not have any parallel sides.

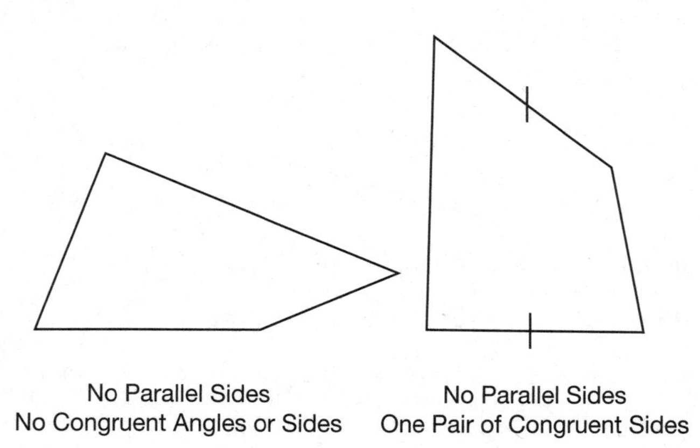

No Parallel Sides
No Congruent Angles or Sides

No Parallel Sides
One Pair of Congruent Sides

2. The second group has exactly one pair of parallel sides and is referred to as **trapezoids.** If a trapezoid has one pair of opposite congruent sides, then it is an **isosceles trapezoid**. See the next figure.

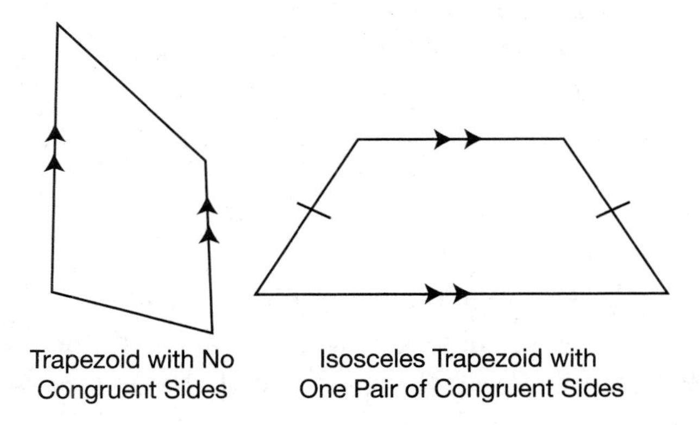

Trapezoid with No Congruent Sides

Isosceles Trapezoid with One Pair of Congruent Sides

3. The third classification of quadrilaterals is **parallelograms**. Parallelograms have two pairs of parallel sides.

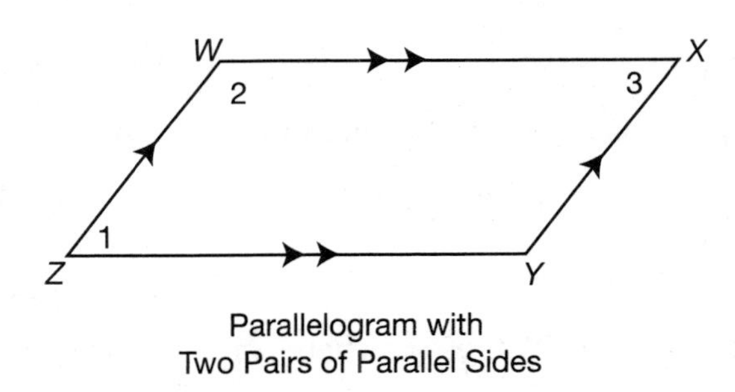

Parallelogram with Two Pairs of Parallel Sides

Properties of Parallelograms

Parallelograms are further classified by the number of congruent angles and sides they have. In this lesson we are going to begin by learning the basic properties that apply to all parallelograms.

STANDARD ALERT!

In Lesson 5 you learned how a transversal intersecting parallel lines creates special pairs of congruent and supplementary angles. Now, you'll apply this information from* Standard 8.G.A.5 *to parallelograms in order to discover the special angle relationships they have.

Interior Angles

In Lesson 5 we discussed the properties of congruent lines that are intersected by a transversal. In the parallelogram *WXZY* in the preceding figure, note that $\overline{WZ}$ is a transversal through parallel $\overline{WX}$ and $\overline{ZY}$. Angle 1 and angle 2 can then be considered *same-side interior* angles and one of our rules from Lesson 5 states that same-side interior angles are always supplementary. Therefore, we can conclude that in parallelograms, adjacent pairs of angles are supplementary.

TIP: Adjacent pairs of angles in parallelograms are supplementary.

Continuing our study of parallelogram *WXYZ*, we see that $\angle 1 + \angle 2 = 180°$ and that $\angle 3 + \angle 2 = 180°$. From there we can next identify that $\angle 1 + \angle 2 = \angle 3 + \angle 2$, since both of these sums equaled 180°. If we subtract $\angle 2$ from each side, we see that $\angle 1 = \angle 3$. This brings us to the next important property of parallelograms, which is that opposite pairs of angles are congruent. It is also true that if a quadrilateral has two pairs of opposite congruent angles, it must be a parallelogram.

TIP: Opposite pairs of angles in parallelograms are congruent.

These two properties can be seen in the next figure.

$$\angle A + \angle B = 180^\circ$$
$$\angle B + \angle C = 180^\circ \qquad \angle A = \angle C$$
$$\angle C + \angle D = 180^\circ \qquad \angle B = \angle D$$
$$\angle D + \angle A = 180^\circ$$

STANDARD ALERT!

Now that we've applied ***Standard 8.G.A.5*** *to uncover the special angle relationships formed by transversals in parallelograms, you can practice* ***Standard 7.G.B.5*** *to solve for unknown angles in parallelograms in Practice 1.*

Practice 1

1. A quadrilateral with one pair of parallel sides is called a/an ____________________.

2. True or false: A quadrilateral with one pair of parallel sides and one pair of congruent sides is called an equilateral trapezoid.

3. In quadrilateral *ASDF*, $m\angle A = 117^\circ$, $m\angle S = 39^\circ$, and $m\angle D = 86^\circ$. Find the $m\angle F$.

4. In parallelogram *RTYU*, $m\angle R = 74^\circ$. Find the measures of the other three angles.

5. In the next figure, $m\angle P$ is twice as large as $m\angle R$. Use the given information to determine the measures of angles *P* and *R*.

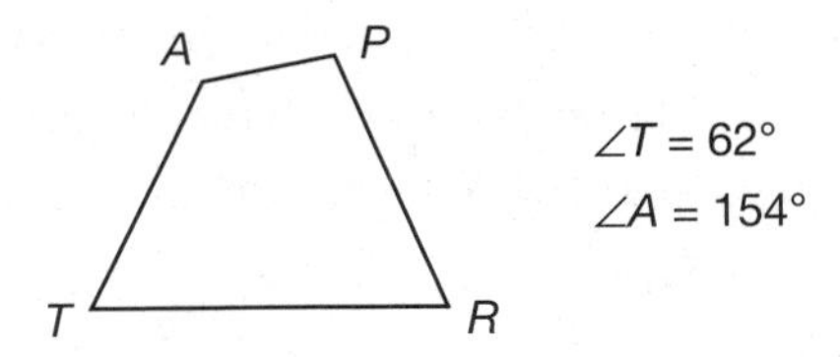

6. In quadrilateral $HJKL$, the measures of angles H, J, K, and L are in a ratio of 1:2:3:4. Find the measure of each of the angles in the quadrilateral.

7. True or false: In a parallelogram, there is one pair of congruent opposite angles.

Use the next figure to answer questions 8 through 11. The figures *are not drawn to scale*. Only the markings on the sides and angles should be used to identify the most specific name you can give to each four-sided polygon.

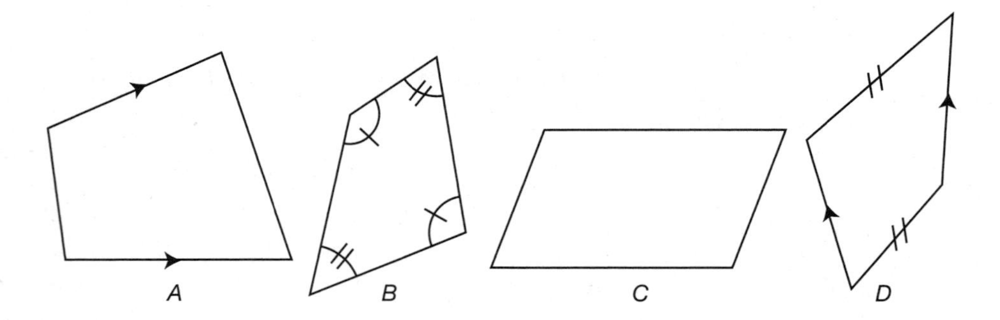

8. Shape A is a ____________________.

9. Shape B is a ____________________.

10. Shape C is a ____________________.

11. Shape D is a ____________________.

The Quadrilateral Family Tree

After learning all those neat things about parallelograms, you are well on your way to becoming a quadrilateral expert! You still have a few more quadrilaterals to learn before you can get that badge, though! The following illustration is a helpful visual that maps out how the different groups of quadrilaterals are related to each other. The Quadrilateral Family Tree starts with the most basic 4-sided polygon up top, and as you move down the family tree the quadrilaterals have more specific attributes and become more specialized. Notice how each quadrilateral contains all the characteristics of the shapes above it.

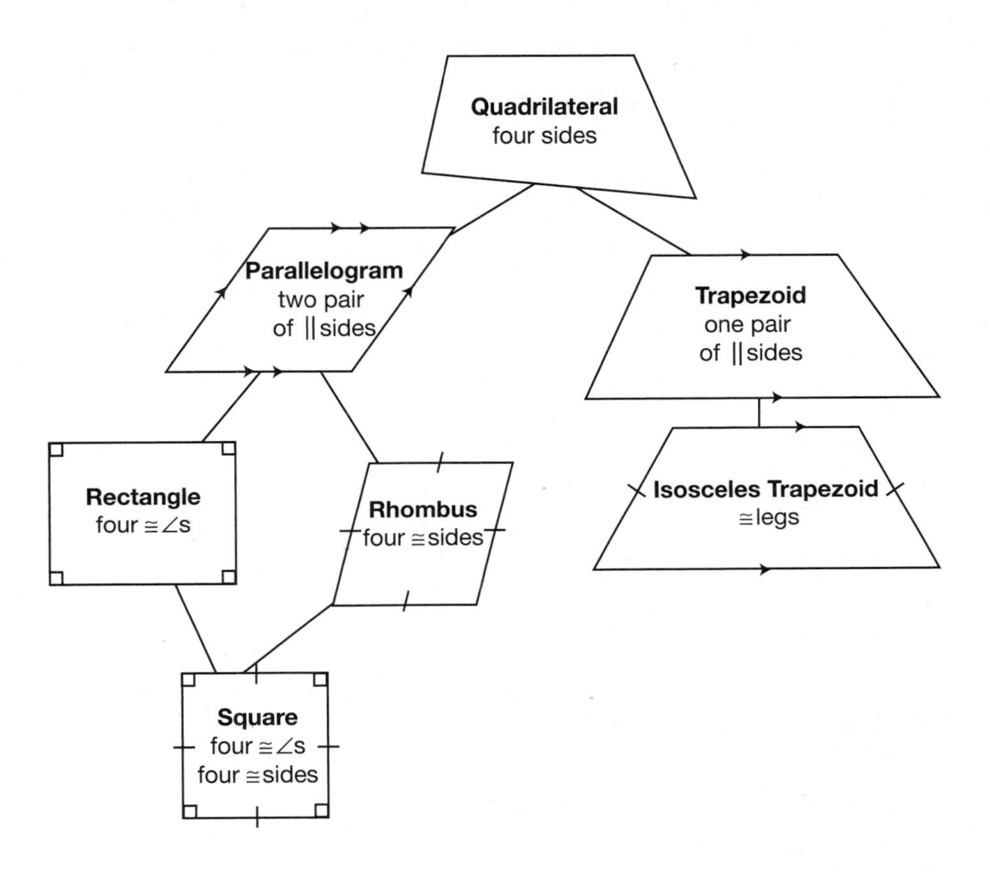

Trapezoids are on one-half of the quadrilateral family tree. Trapezoids are quadrilaterals with exactly one pair of parallel sides. A special type of trapezoid is the **isosceles trapezoid**, which has one pair of congruent sides.

The other half of the quadrilateral family tree represents parallelograms. Although we discussed the special properties of parallelograms, we did not investigate **rectangles**, **rhombuses**, or **squares**. These shapes are unique parallelograms with congruent angles, congruent sides, or congruent angles *and* sides. Be certain you are familiar with their attributes listed here:

- **Rectangle:** a parallelogram with four congruent *angles* that each measure 90°.

- **Rhombus:** a parallelogram with four congruent *sides* (the plural of *rhombus* is *rhombuses*).
- **Square:** a parallelogram with four congruent sides *and* four congruent angles that each measure 90°. Squares are both rectangles *and* rhombuses.

In the next lesson, you're going to learn how to calculate the area of lots of different polygons, but before you proceed, test your understanding of the attributes of quadrilaterals with the following practice problems.

Practice 2

Use the family tree diagram to identify whether each statement is true or false.

1. All trapezoids are quadrilaterals. ________________
2. All parallelograms are rectangles. ________________
3. All rectangles are quadrilaterals. ________________
4. All parallelograms are squares. ________________
5. All squares are rhombuses. ________________
6. All rectangles are parallelograms. ________________
7. All rhombuses are squares. ________________
8. All isosceles trapezoids are quadrilaterals. ________________
9. All squares are trapezoids. ________________
10. All quadrilaterals are squares. ________________

Answers

Practice 1

1. trapezoid
2. False. It is called an isosceles trapezoid.
3. $m\angle F = 118°$
4. $m\angle T = 106°$, $m\angle Y = 74°$, and $m\angle U = 106°$. (In a parallelogram, adjacent angles are supplementary, and opposite angles are congruent.)
5. $m\angle P = 96°$ and $m\angle R = 48°$. (Since $360° - 62° - 154° = 144°$, angles P and R must sum to 144°. Since $m\angle P$ is twice as big as $m\angle R$, set up and solve the equations $2x + x = 144°$, where $m\angle P = 2x$ and $m\angle R = x$.)
6. 36°, 72°, 108°, and 144°. (With ratio problems, multiply each number by the same variable x and set up an equation: $1x + 2x + 3x + 4x = 360°$, so $10x = 360°$ and $x = 36$.)
7. False. Parallelograms always have two pairs of congruent opposite angles.
8. Trapezoid—it has exactly one pair of parallel sides.
9. Parallelogram—it has two pairs of opposite, congruent angles.
10. Polygon—it looks like a parallelogram, but it has no markings to indicate congruence or parallel sides, so it can only be classified as a quadrilateral.
11. Isosceles trapezoid—it has one pair of parallel sides and one pair of opposite congruent sides.

Practice 2

1. True
2. False
3. True
4. False
5. True
6. True
7. False
8. True
9. False
10. False

12 Area of Composite Polygons

A man whose mind has gone astray should study mathematics.
—Francis Bacon

In this lesson you will build upon your knowledge of area as you learn and apply the area formulas for squares, rectangles, triangles, parallelograms, and trapezoids. You will also learn how to find the area of more complicated shapes by breaking them down into triangles and rectangles.

STANDARD PREVIEW

Life doesn't happen in neat squares and triangles. The CCSS acknowledge this in **Standard 6.G.A.1 by asking students to find the area of complicated polygons by breaking them down into triangles and rectangles.** A complex polygon—it doesn't fall into a neat category like *parallelogram* or *rectangle*—is called *composite*. Composite shapes can be divided into more manageable polygons. The standards for middle school geometry are pretty heavy on area, volume, and surface area, since these are some of the most useful concepts that people apply in their personal and work lives. Whether you need to buy expensive sun-resistant fabric to cover your pool, put a new roof on a shed, or cover a patio with Italian tiles, the ability to compute area is especially valuable. This lesson will give you lots of practice with **Standard 6.G.A.1 and Standard 7.G.B.6, including the chance to solve plenty of real-world problems.**

Basic Area Formulas

You probably remember from Lesson 8 that the **area** of a shape is the amount of space *inside* an object, which is measured in square units. For example, let's consider the front lawn of a home as represented by the next figure.

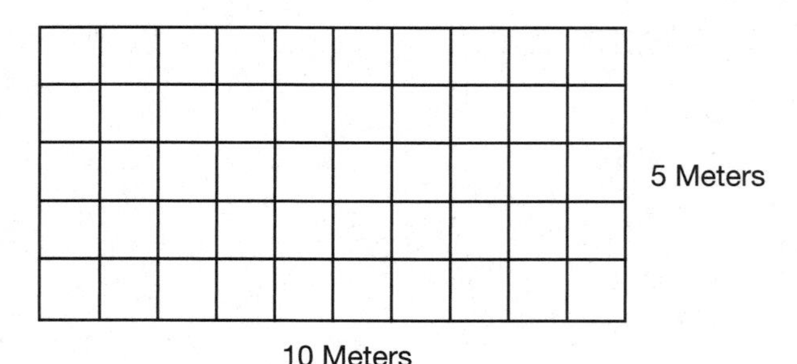

The perimeter would be used to help a homeowner figure out how much fencing to purchase to make a closed yard for his dog. The *area* of the lawn is the amount of space *inside* the fenced-off area and is measured in square meters. For every meter of length, there are five meters of depth to the lawn. This means that 10 *times* 5, or 50, square meters will fill the space of the lawn. This measurement would be helpful if the homeowner was purchasing square meters of sod to plant in the front lawn. When representing area, remember to use the squared exponent to show that area is a two-dimensional measurement. In this case, it would be incorrect to write that the area is 50 meters, since that implies that the area is a straight line. It would be *correct* to write that the area is equal to 50 m^2 or 50 $meters^2$.

Area Formulas for Rectangles, Squares, and Triangles

We already saw in the preceding example that the area of the rectangular lawn was calculated by multiplying the 10 meters of length by the 5 meters of width. The general formula for the area of a rectangle is length *times* width.

TIP: The *area of a rectangle* = (length) × (width), or $A = lw$

Since the length and width in squares are the same size, the formula for the area of a square is the length of one of its sides, times itself. In squares it is typical to refer to the side length as *s* for *side*, instead of using *length* or *width* as is used in rectangles.

TIP: The *area of a square* = (side) × (side), or $A = s^2$

STANDARD ALERT!

One goal of the CCSS is to encourage students to understand the origins and reasoning behind mathematic theorems, formulas, and procedures. One of the formulas that is important to understand thoroughly is the area of triangles, since using this formula is a skill associated with many standards in middle school math. Follow along the informal proof offered here to see where $A = \frac{1}{2}$(base)(height) comes from.

The formula for the area of a triangle is a concept that was presented in Lesson 8. Now that you understand where the formula for the area of a rectangle comes from, it is helpful to see how that relates to the area of a triangle.

Looking at the following figure, you can see that when a diagonal is drawn in rectangle *ABCD*, two congruent triangles are formed. Since the area of the rectangle is length *times* width, you can identify that the area of one of the triangles would be half of length *times* width. Replacing "base" for the "width," and "height" for the "length," this brings us to the formula for area of a triangle presented in Lesson 10: Area = $\frac{1}{2} \times$ (base) $\times$ (height).

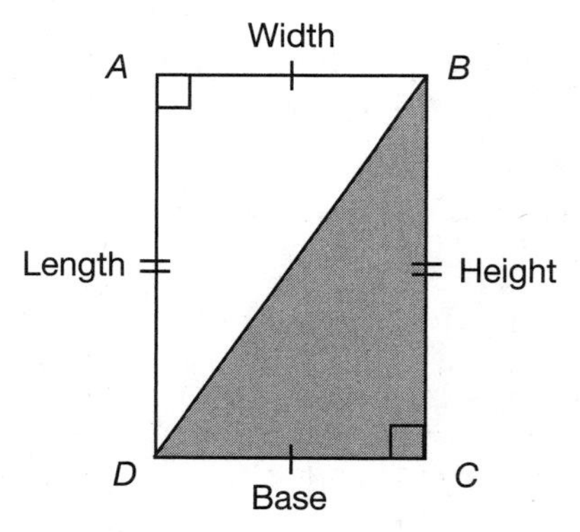

Area of Rectangle = Length × Width
Area of Triangle = $\frac{1}{2}$ Base · Height

TIP: The *area of a triangle* = $\frac{1}{2} \times$ **(base)** $\times$ **(height) or** $A = \frac{1}{2}bh$

STANDARD ALERT!

Our lives are filled with complicated shapes—shapes that we can break down into "friendlier" polygons to use in calculations. This process is an essential part of standard 6.G.A.1. In this lesson, you will learn two different methods for finding the area of complicated polygons.

Area of Piecewise Figures

Sometimes, a figure must be broken up into separate polygons in order to identify its area. Consider the next figure, which represents the lawn around a square patio. In order to find the area of the lawn there are two methods.

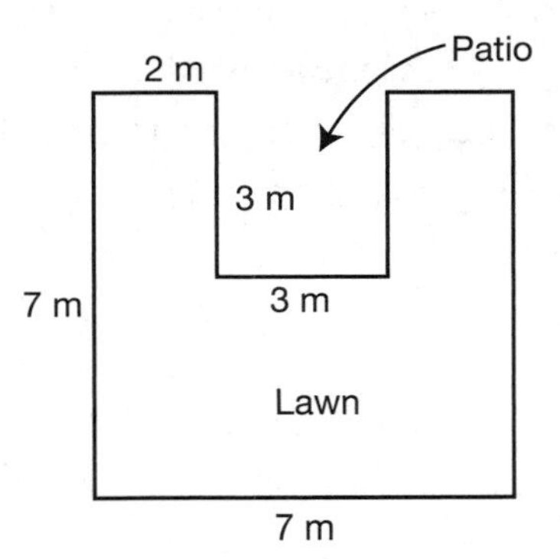

Method 1: Find the area of the entire figure and then subtract the smaller portion that is not included. Here, the larger shape is a square that is 7 meters long by 7 meters wide. The area of that square is therefore 49 m^2. The smaller patio section that is not included in the lawn is 3 meters by 3 meters. The area of the patio is 9 m^2. This must be subtracted from the larger area: 49 m^2 – 9 m^2 = 40 m^2. The area of the lawn is 40 m^2.

Method 2: The second method involves breaking the figure up into separate polygons as illustrated in this next figure. Sections A and C will both have areas of 6 m² and section B will have an area of 28 m². Adding these three separate sections together, the combined area will be 40 m².

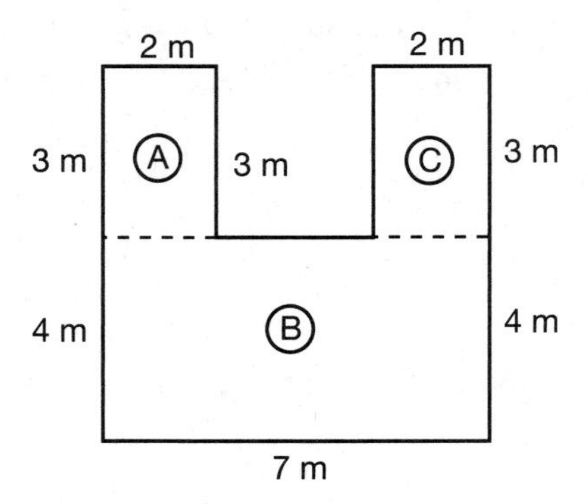

STANDARD ALERT!

In the first set of practice problems, you'll put your new knowledge to work as you help a landscape architect determine how much concrete she needs to order for her upcoming projects. This is going to make standard 8.G.B.6 very happy, since it is all about real-world applications of area!

Practice 1

Erin is a landscape architect who designs garden and patio spaces in Ojai, California. The following composite shapes represent the new cement surfaces she's planning for her upcoming projects.

Find the area of each shape so that she can order the correct amount of concrete to complete these jobs. All dimensions are in feet. (Note: Since Erin is an architect, her drawings are exact—all lines that appear to be parallel *are* parallel, and all lines that appear to be perpendicular *are* perpendicular.)

1.

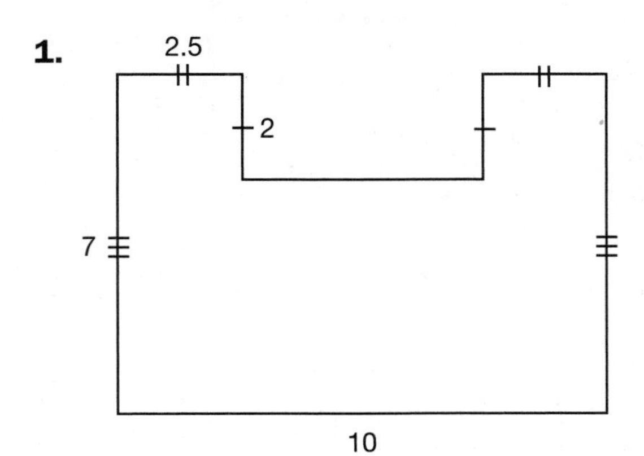

2.

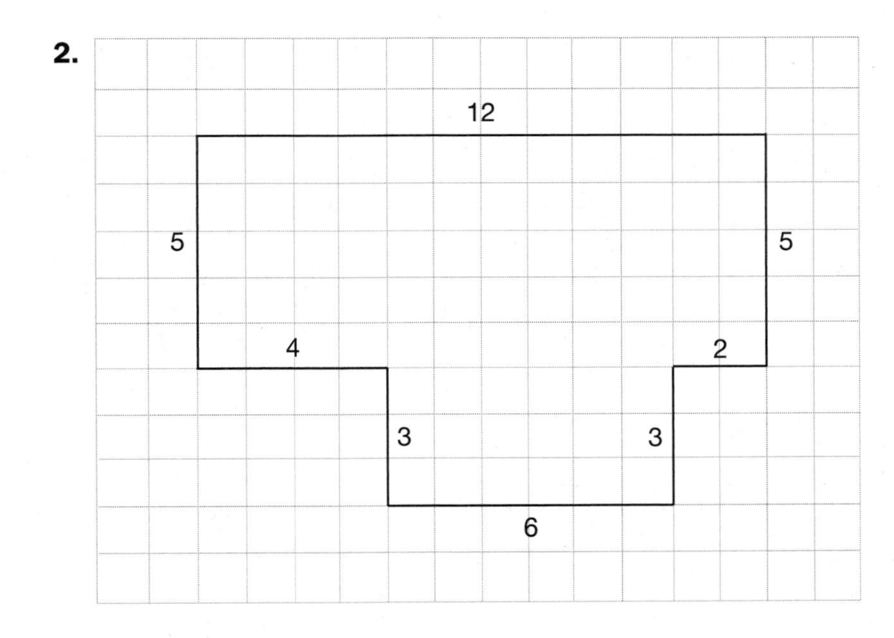

3.

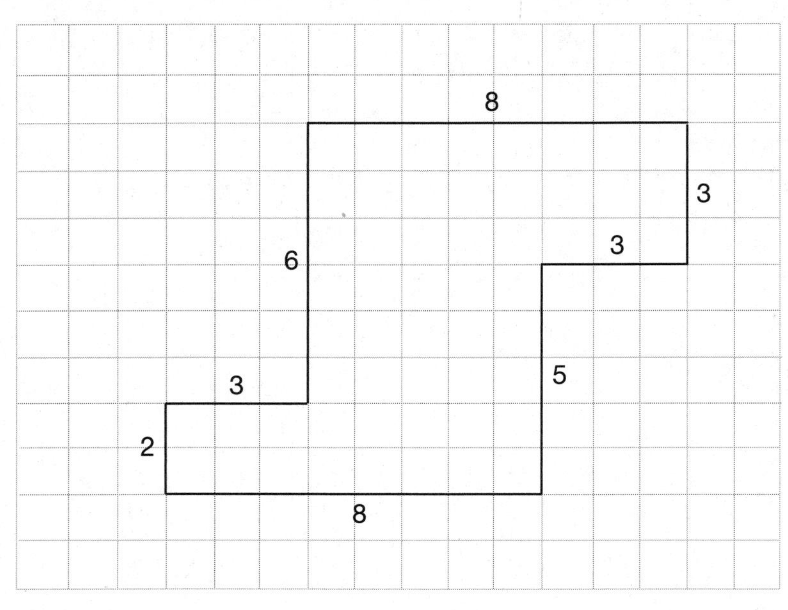

4.

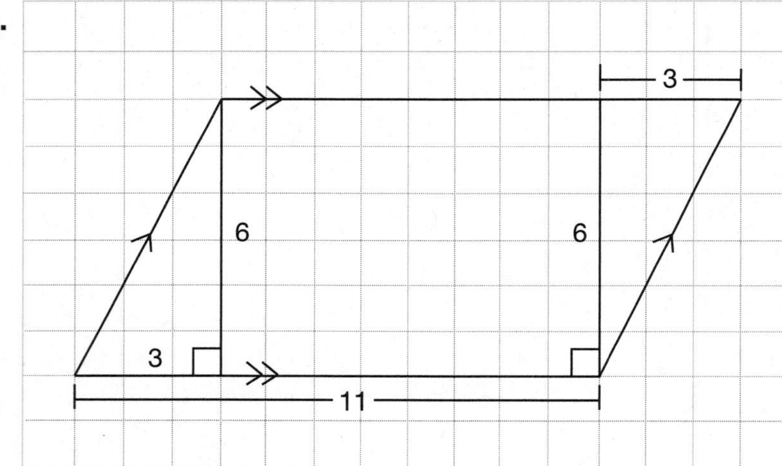

5.

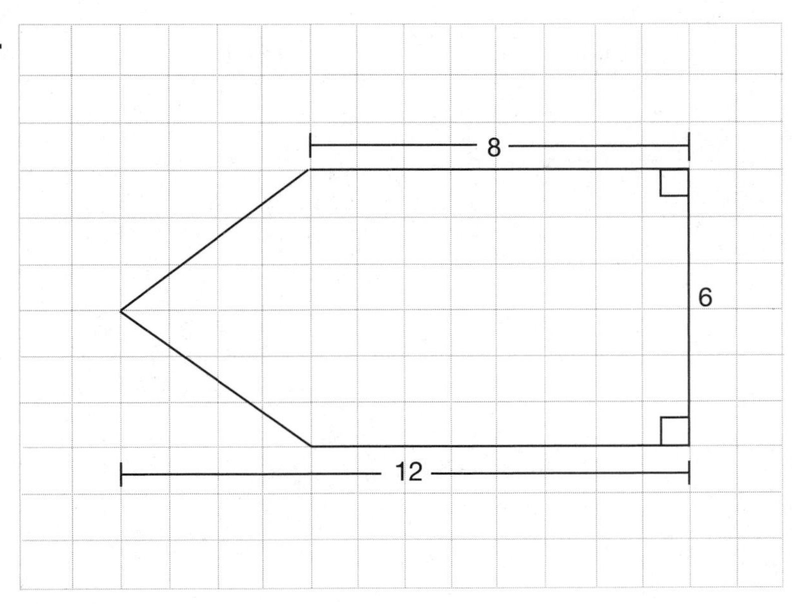

6.

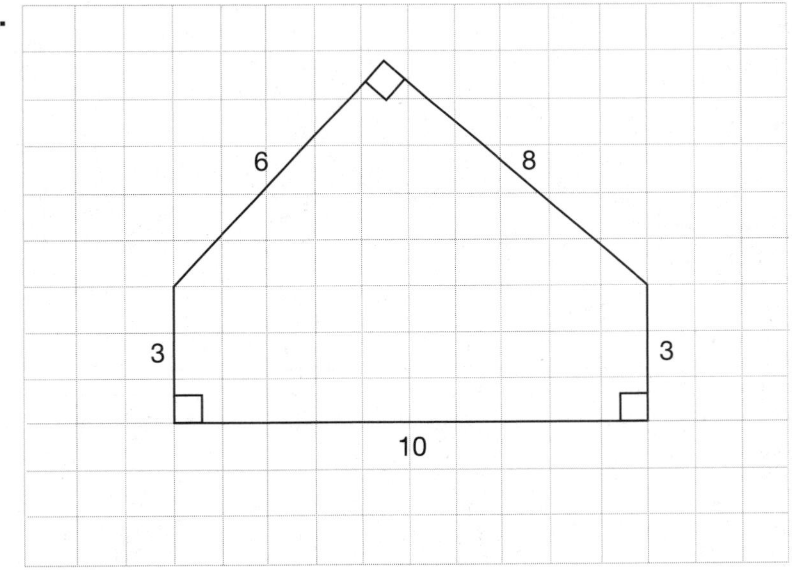

7.

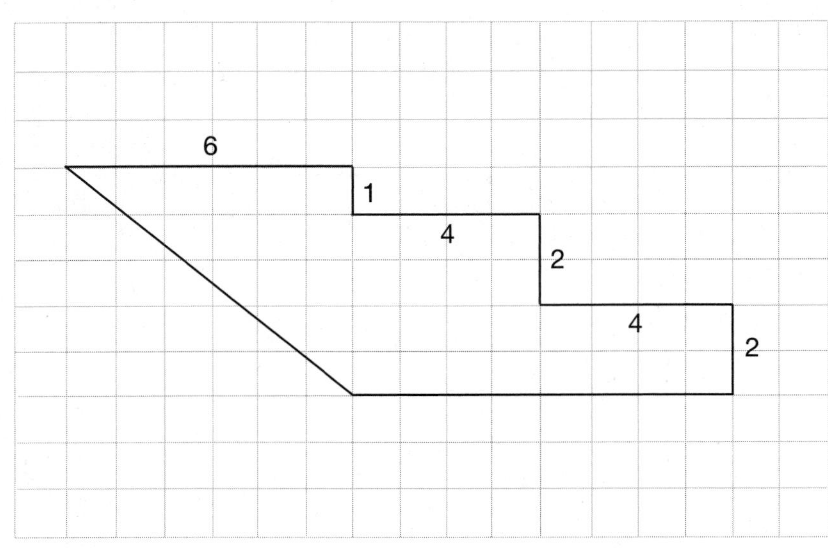

8.

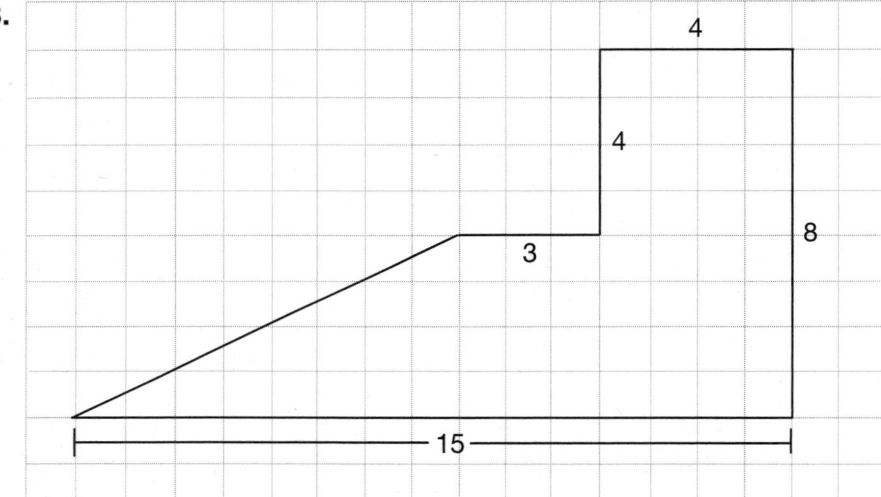

Area of Parallelograms and Trapezoids

STANDARD ALERT!

You are about to see the formula for the area of a parallelogram appear magically before your eyes! Well, it's not really magic if you can see that this area formula is actually the result of breaking down a parallelogram into a triangle and a trapezoid and then rearranging it so it's just a rectangle. Standard 6.G.A.1 is encouraging us to slice, dice, and formulate!

The relationship between the **base** and the **height** of a parallelogram is the same that is found in triangles. Any side of a parallelogram can be called the base. The corresponding altitude must be drawn from that base to the opposite vertex so that it forms a right angle. This altitude is called the height and is illustrated in the next figure.

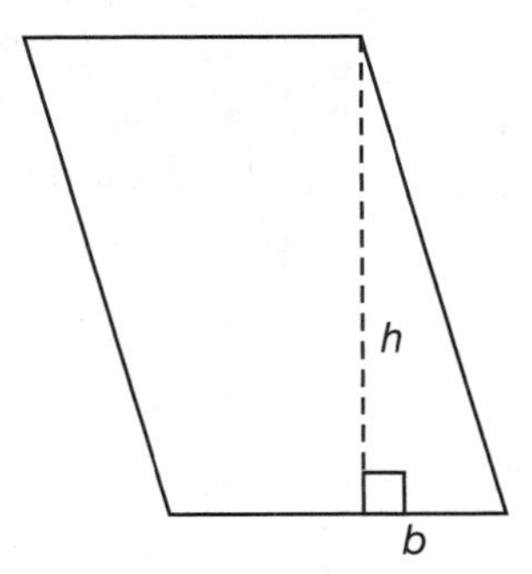

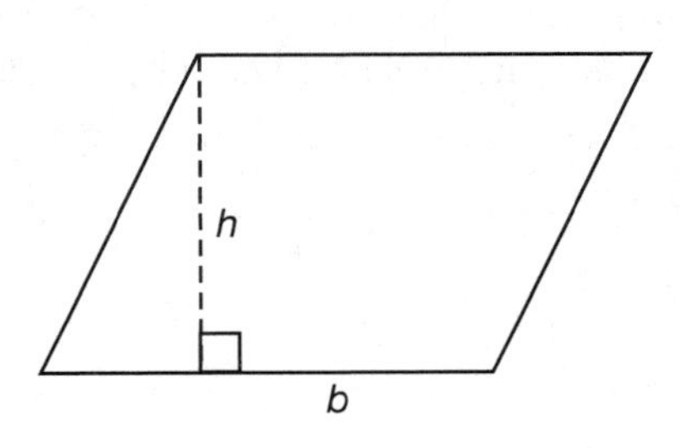

To understand the area formula for parallelograms, notice that once the height has been drawn in, any parallelogram can be dissected into two pieces and rearranged to form a rectangle as shown in the next figure.

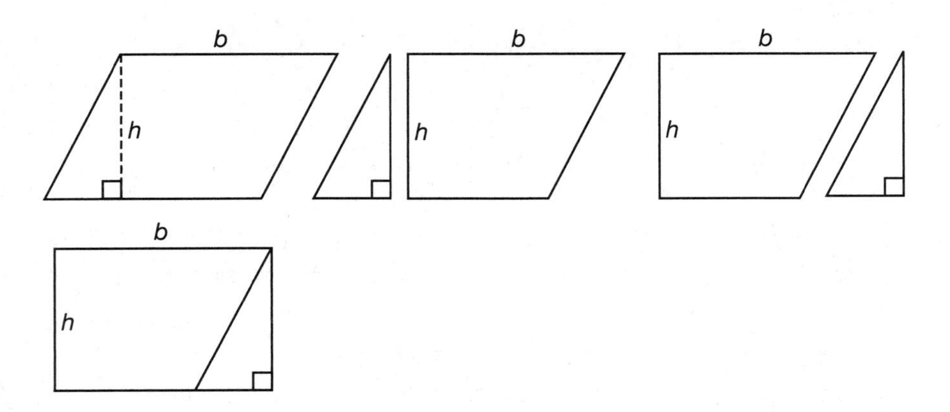

Looking at this figure, it is probably clear to you that the area of a parallelogram can be determined by multiplying the base times the height.

TIP: The *area of a parallelogram* = (base) × (height) or $A = bh$

In a trapezoid, the bases are parallel, but not congruent, as in a parallelogram. In order to find the area of a trapezoid, the average of the two bases is multiplied by the height. This is more commonly written as one-half the product of the height times the sum of the bases. This is shown in the next figure.

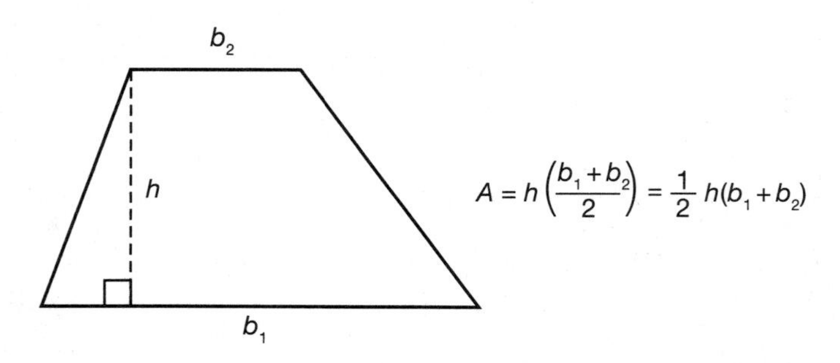

TIP: The *area of a trapezoid* = $\frac{1}{2}$(height)(base$_1$ + base$_2$), or $A = \frac{1}{2}(h)(b_1 + b_2)$

Practice 2

The following drawings are all examples of what architects, engineers, and designers may see in scale drawings for future projects. Notice that the units change from illustration to illustration, so make sure you properly label the area of these parallelograms and trapezoids.

1.

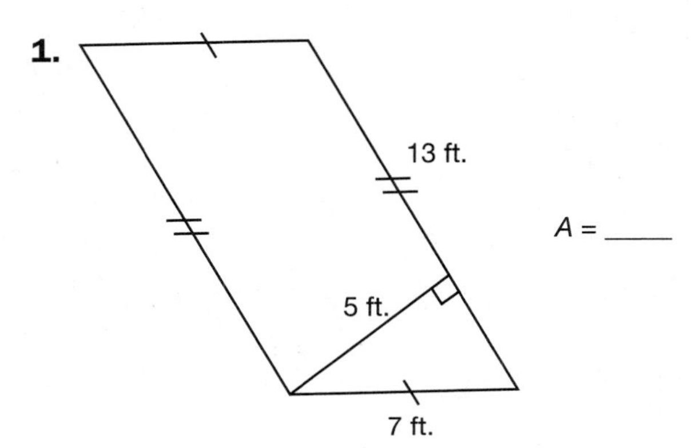

$A =$ _____

2.

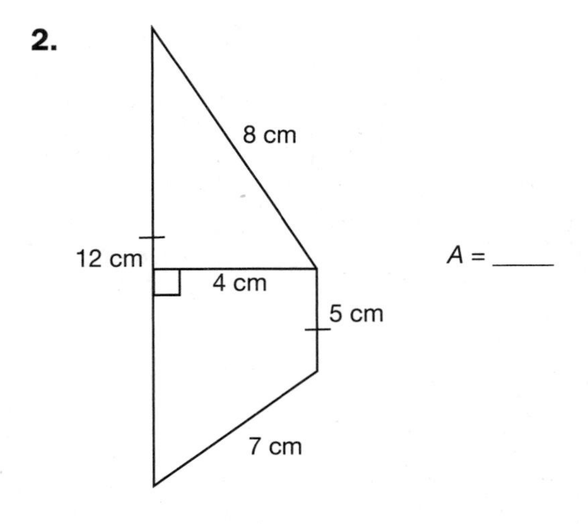

$A =$ _____

3.

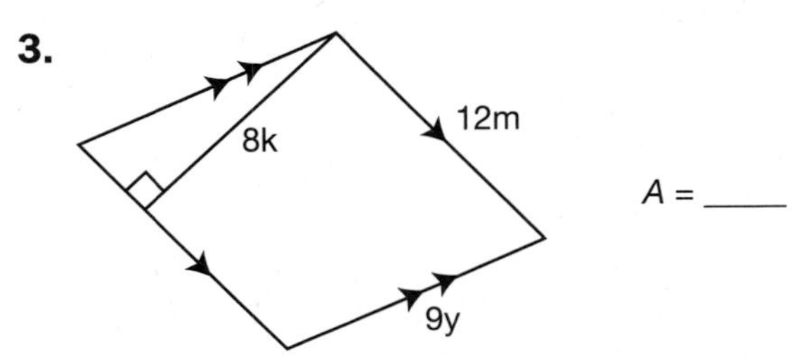

$A =$ _____

4.

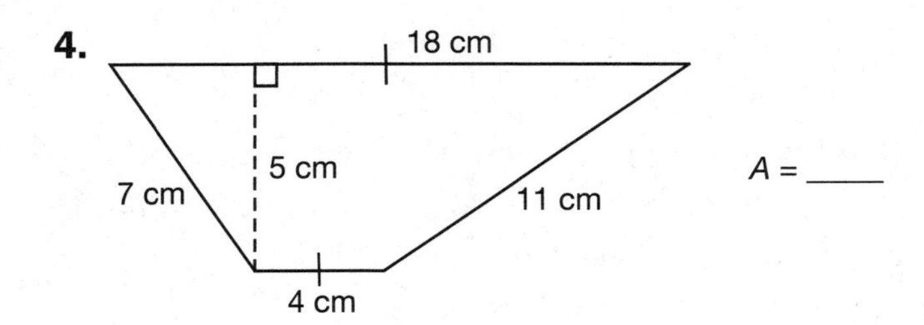

$A =$ _____

Find the area of each figure below.

5. Find the area of quadrilateral *ABCD*.

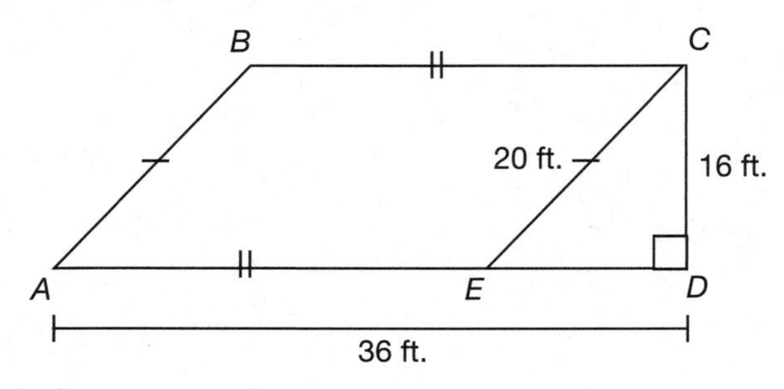

6. Find the area of polygon *RSTUV*.

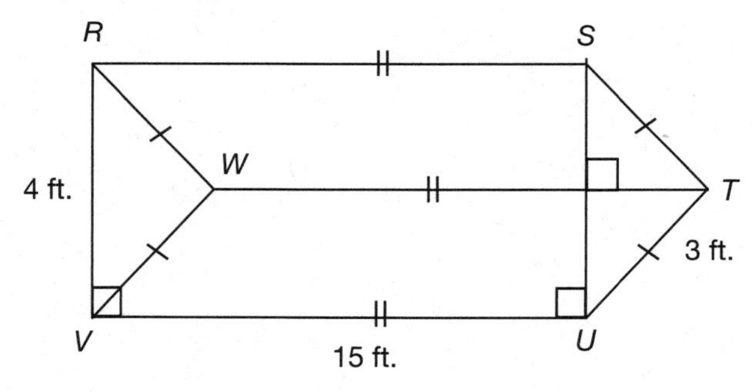

Answers

Practice 1

1.

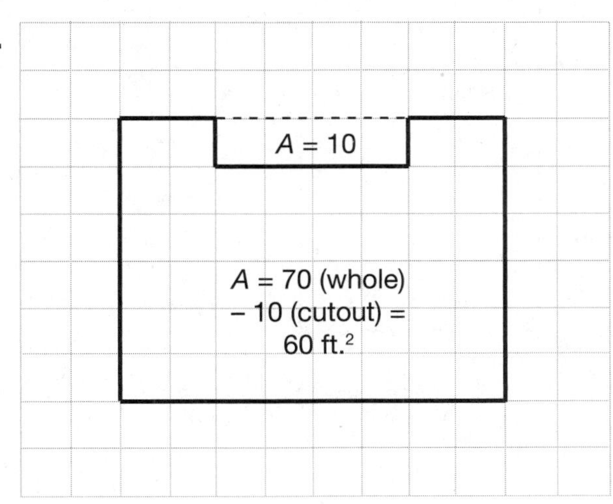

2.

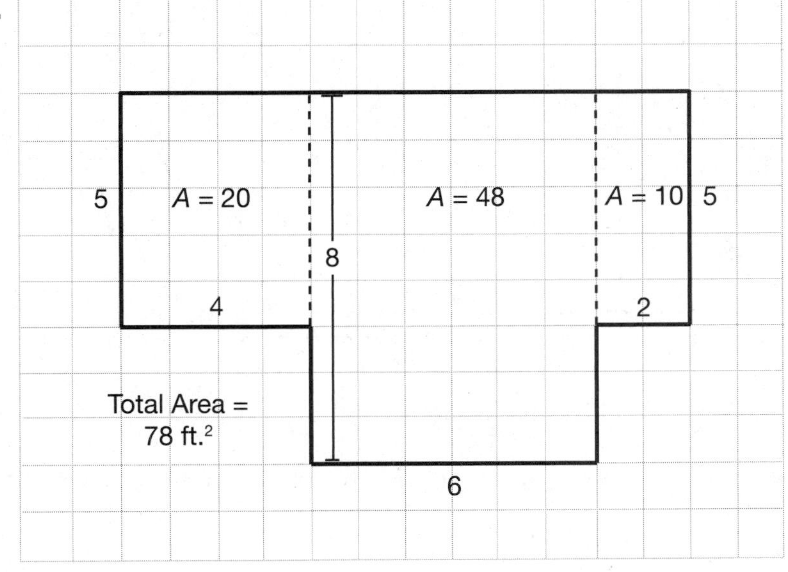

3.

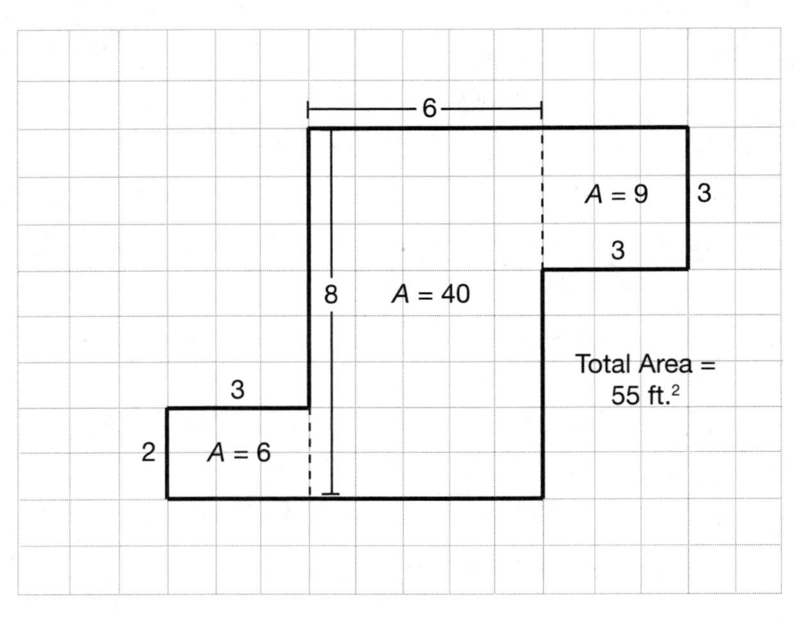
6
A = 9
3
3
8
A = 40
Total Area =
55 ft.²
3
2
A = 6

4.

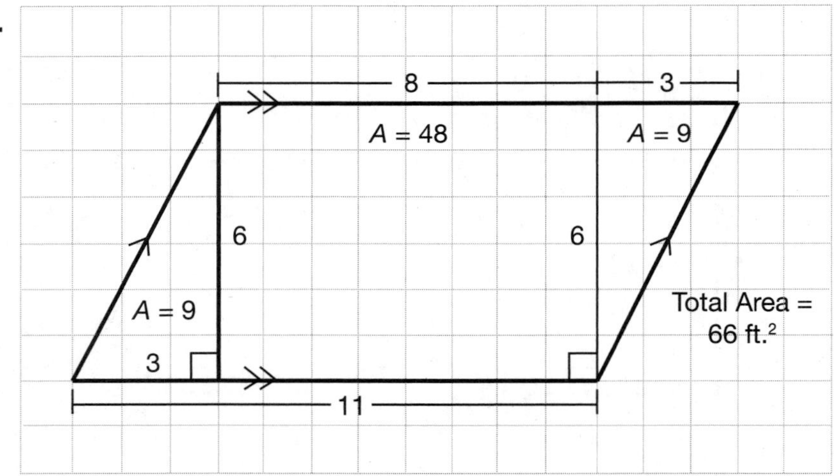
8
3
A = 48
A = 9
6
6
A = 9
Total Area =
66 ft.²
3
11

5.

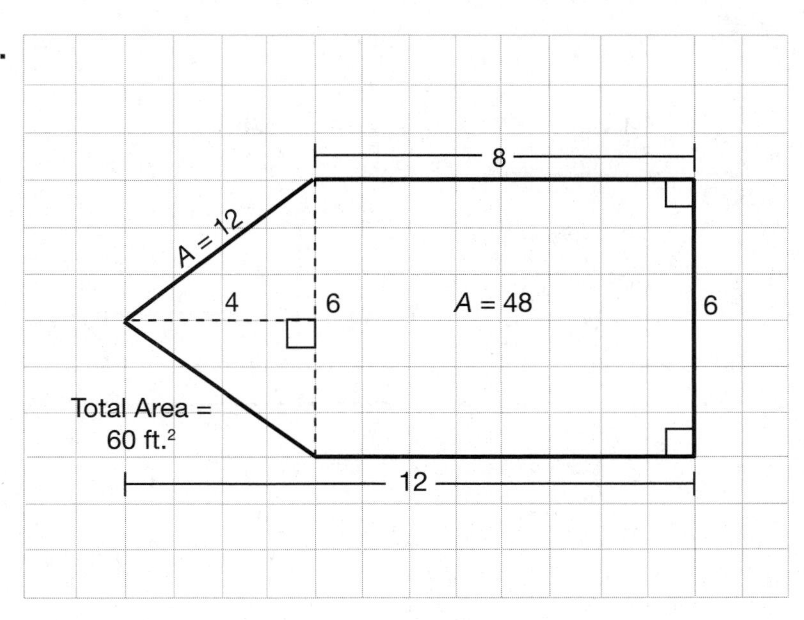

6.

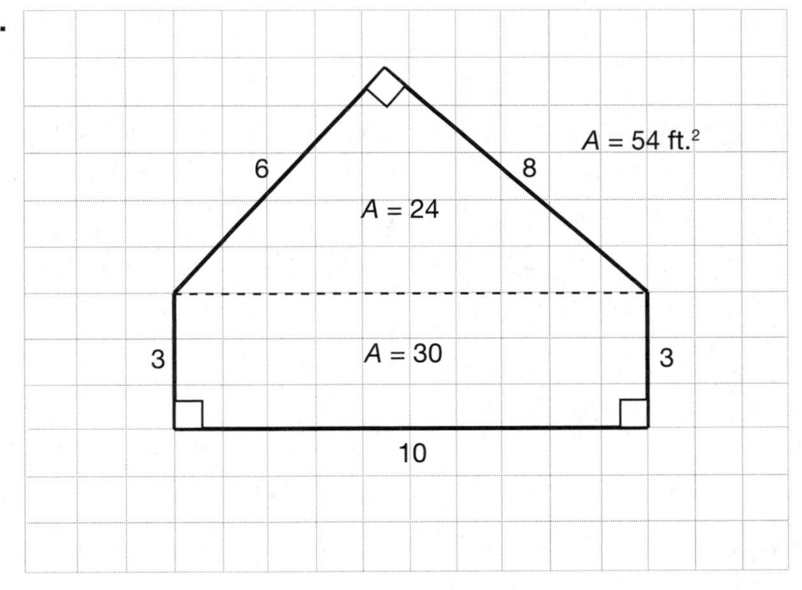

7.

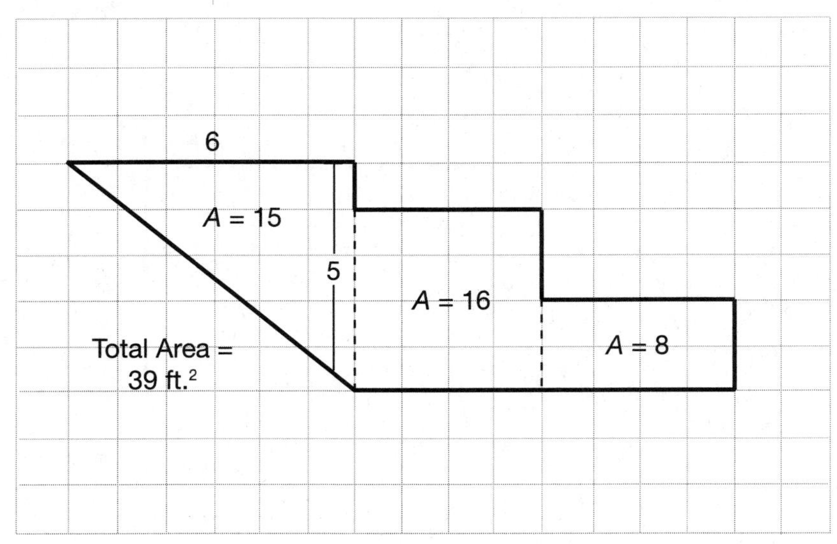

8.

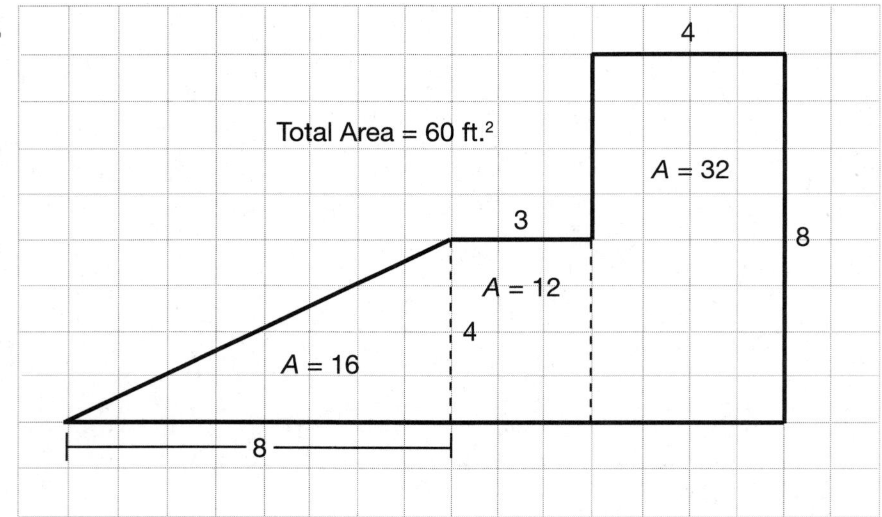

Practice 2

1. $A = 5(13) = 65 \text{ ft.}^2$
2. $A = \frac{1}{2}(4)(12 + 5) = 34 \text{ cm}^2$
3. $A = 8(12) = 96 \text{ m}^2$
4. $A = \frac{1}{2}(5)(18 + 4) = 55 \text{ cm}^2$

5. $(\overline{ED})^2 = 20^2 - 16^2$

$(\overline{ED})^2 = 144$

$(\overline{ED}) = 12$

$A(\Delta EDC) = \frac{1}{2}(12)(16) = 96 \text{ ft.}^2$

$\overline{AE} = 36 \text{ ft.} - 12 \text{ ft.} = 24 \text{ ft.}$

$A(AECB) = 24(16) = 384 \text{ ft.}^2$

$A(ABCD) = 384 + 96 = 480 \text{ ft.}^2$

6.

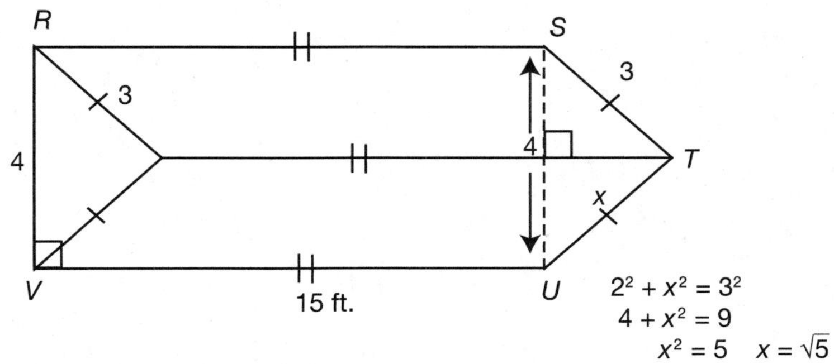

$2^2 + x^2 = 3^2$

$4 + x^2 = 9$

$x^2 = 5 \quad x = \sqrt{5}$

A (Rectangle *RSUV*) = (15 ft.)(4 ft.) = **60 ft.²**

Next calculate area of triangle:

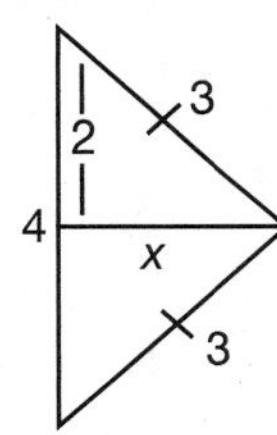

$A = \frac{1}{2}(4)(\sqrt{5}) = \mathbf{2\sqrt{5}}$

Total Area = 60 ft.² + $2\sqrt{5}$ ft.²

13

Ratios and Proportions

There is no excellent beauty that has not some strangeness in the proportion.
—Sir Francis Bacon

In this lesson you will begin by exploring ratios. You will learn how to use ratios to set up proportions and how to apply proportions to problem solving with scale drawings of geometric figures.

STANDARD PREVIEW

You might not realize that you come across ratios and proportions on a daily basis. One example of a ratio is when a protein shake calls for 5 scoops of powder for every 2 cups of water. Another example is a map of the United States where 1 inch represents 300 miles. Ratios are the building blocks for proportions, and proportions are used to create scale models of real-life scenarios that affect us every day—in city planning, graphic illustration, architecture, and interior design. Ratios and proportions are such important concepts in the real world that they actually make up their own subcategory in the middle school Common Core State Standards! In this lesson you will tackle *Standard 7.G.A.1, which asks students to apply proportions and scale to geometric figures*. You will find missing side lengths, compute areas from scale drawings, and reduce or enlarge scale drawings. You will incorporate what you learn here in Lesson 22, which takes a closer look at similar polygons.

Ratios and proportions are widely used concepts in the everyday world. A common example of a ratio that you might see in a supermarket is "3 boxes of cereal for $7.00." A common use of proportions is estimating distances between cities when working with the scale on a map. Before we get into problem solving with proportions, you first need to understand what ratios are and how they are used to construct proportions.

Ratios: The Building Block of Proportions

A **ratio** is a comparison of two different quantities or numbers. Ratios can be written with colons or as fractions. They have the same meaning, regardless of which style is used to present them. The following two ratios are said "dogs to cats": dogs:cats and $\frac{\text{dogs}}{\text{cats}}$. Notice that the first word said must be the first term when you're using a colon or it must be the numerator of the fraction.

TIP: A *ratio* is a comparison of two numbers, written as the quotient of two quantities in fraction form, or as two terms separated by a colon.

Ratios are always reduced into simplest terms. For example, if an animal shelter has 20 dogs and 30 cats, the ratio of dogs to cats will be expressed in simplest terms by reducing $\frac{20}{30}$ to $\frac{2}{3}$. This can also be written as 2:3. Ratios sometimes represent the "part" of a quantity to the "whole" of the quantity. In the preceding case of the animal shelter, there are 50 animals total (20 dogs and 30 cats), so the ratio of dogs to animals would be $\frac{20}{50}$ or $\frac{2}{5}$. This ratio means that there are 2 dogs for every 5 animals at the shelter. One important thing to keep in mind when you're reducing ratios to simplest terms is that they should always reflect two quantities and not be written as just a whole number. For example, if the number of apples to oranges in a fruit basket is 12 to 4, this is written as $\frac{12}{4} = \frac{3}{1}$. It is not written as just "3."

Practice 1

1. What is the ratio of weekdays to weekend days in any given week?

2. What is the ratio of weekdays to the total number of days in a week?

Consider the set of whole numbers 1 through 10 for questions 3 through 5.

3. What is the ratio of odd numbers to even numbers in this set?

4. What is the ratio of prime numbers to composite numbers in this set?

5. What could the ratio 1:2 represent, given the set of whole numbers 1–10?

Using the following figure for questions 6 through 9, represent each ratio in simplest form.

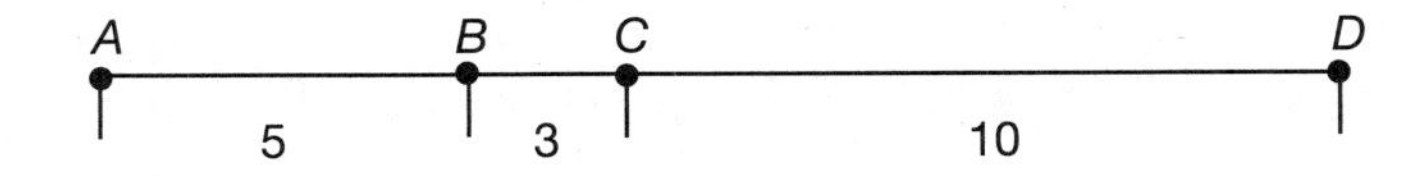

6. $\frac{AB}{BC}$

7. $\frac{AB}{CD}$

8. $\frac{CD}{AB}$

9. $\frac{AC}{AD}$

Proportions: The Relationship between Two Ratios

A **proportion** is an equation where two ratios equal each other. For example, consider the following lemonade instructions. The directions on a jar of lemonade concentrate suggest that one gallon of lemonade can be made by mixing 1 quart of lemonade concentrate with 3 quarts of water. It also suggests that for two gallons of lemonade, 2 quarts of lemonade concentrate should be mixed with 6 quarts of water. A proportion modeling the two different recipes will look like this:

$$\frac{\text{concentrate}}{\text{water}} = \frac{1}{3} = \frac{2}{6}$$

or, it would also be correct to model the directions in this manner:

$$\frac{\text{water}}{\text{concentrate}} = \frac{3}{1} = \frac{6}{2}$$

Notice that it does not matter if the lemonade concentrate is in the numerator or the denominator of the ratios. As long as a single proportion has the quantities of lemonade concentrate in the same position in both of the ratios, it is correct.

TIP: A *proportion* is an equation with two ratios that are equal to each other.

The Means–Extremes Property of Proportions

The first and last numbers in a proportion are called the *extremes*. The middle numbers are called the *means*.

4:6 = 2:3
means
extremes

$\frac{4}{6} = \frac{2}{3}$
means extremes

There is a special relationship between the means and the extremes of proportions. Consider the following proportion:

$\frac{3}{6} = \frac{1}{2}$

You can see that these fractions are equal to each other since $\frac{3}{6}$ reduces to $\frac{1}{2}$. Take a closer look at what is true when the means are multiplied together ($2 \times 3 = 6$) and when the extremes are multiplied together ($1 \times 6 = 6$). You can see that in this proportion, the *product of the means equals the product of the extremes*. This rule actually holds true for *all* proportions and is an important postulate when working with proportions. Rather than memorize that *the product of the means equals the product of the* extremes, it's easier to remember that proportions have *equivalent cross products*. A *cross product* is the answer when multiplying diagonally:

The *equivalent cross products* of $\frac{a}{b} = \frac{x}{y}$ are represented as $ay = bx$.

TIP: In a proportion, the cross products are equal to each other. If $\frac{a}{b} = \frac{x}{y}$, then $ay = bx$.

Since you now know that the cross products of proportions are equal, it shouldn't surprise you to learn that the converse is also true: If the cross products of two ratios are not equal, then they are not proportional. (Another way to say this would be to say that the proportion is not true.) For example, let's test the cross products to see if $\frac{5}{7}$ and $\frac{17}{20}$ are proportional:

$\frac{5}{7} \stackrel{?}{=} \frac{17}{20}$

$7 \times 17 = 119$

$5 \times 20 = 100$

Since the two cross products are not equal, we can determine that $\frac{5}{7}$ and $\frac{17}{20}$ are not proportional and that $\frac{5}{7} = \frac{17}{20}$ is a false statement.

TIP: If the cross products of two ratios are *not* equal, then the ratios are not proportions and the proportion is false. Given ratios $\frac{a}{b}$ and $\frac{x}{y}$, if $ay \neq bx$, then $\frac{a}{b} \neq \frac{x}{y}$.

Practice 2

For questions 1 through 3, identify whether the given proportion is true or false.

1. $\frac{8}{9} = \frac{11}{12}$
2. $\frac{2}{5} = \frac{16}{40}$
3. $\frac{1.25}{0.5} = \frac{25}{10}$
4. $\frac{3}{2w + 1} = \frac{6}{4w + 1}$

Use the following information for questions 5 through 7: A map of New York City uses a scale that represents 500 meters with $\frac{1}{2}$ inch.

5. Write a ratio that shows the relationship between meters and inches.
6. Two subway stations on the New York City map are $3\frac{1}{2}$ inches apart. Write a proportion that sets your ratio from question 5 equal to the distance in *m* meters between the two subway stations.
7. Two parks in New York City are 3,200 meters away from each other. What proportion would be used to calculate how many inches, *i*, the parks should be apart on the park map? (Use your ratio from question 5 to begin this proportion.)

Solving Proportions

STANDARD ALERT!

Now that you understand how to write a ratio and how to test to see if two ratios are proportional, you are ready to lay the foundation for Standard 7.G.A.1, which is working with scale. You may have recognized that problems 5 through 7 above got you started with this skill, so read on and learn more about it.

Let's take a closer look at how to translate a word problem into a proportion that can be solved. The last three problems in the practice set above should have given you an idea of how to translate a word problem into a proportion, but let's map it out clearly here:

Step 1: Begin by writing a ratio in words to represent the two different units in the problem. For the New York City map problem, we wrote $\frac{\text{meters}}{\text{inches}}$ to remind us that all of the meter measurements would go in the numerator and their corresponding inch measurements would go in the denominator.

Step 2: Use the first pair of corresponding data (500 meters and $\frac{1}{2}$ inch) to fill in the first ratio, making sure that the units correctly correspond with the words:

$$\frac{\text{meters}}{\text{inches}} = \frac{500}{\frac{1}{2}}$$

Step 3: Set that ratio equal to a second ratio that uses a variable to represent the unknown information you are solving for. In this case 3,200 represents meters and i represents the number of inches:

$$\frac{\text{meters}}{\text{inches}} = \frac{500}{\frac{1}{2}} = \frac{3{,}200}{i}$$

Step 4: Find the cross products of proportion, set them equal to each other, and solve for the variable, i:

$$\frac{1}{2}(3{,}200) = 500i$$
$$1{,}600 = 500i$$
$$i = 3.2$$

Therefore, the two parks should be 3.2 inches apart on the map.

Comparing "Part to Part" vs. "Part to Whole"

One factor that is important to pay close attention to when working with proportions is to be aware of whether you are comparing two different *parts* to each other, or a *part* of something to its *whole*. Look at the two similar examples below to see the difference:

Example 1:
Wendy and Dan are making punch for a summer dinner party. The recipe calls for 6 cups of passion fruit juice, 8 cups of orange juice, and 7 cups of guava juice. If Wendy plans to use 14 cups of orange juice, how many cups of passion fruit juice does Dan need to buy at the store? (Even though there are 3 parts to this recipe, this is still an example of comparing part to part. We will use the steps given above and the two ingredients that this question involves.)

Step 1: $\frac{\text{orange juice}}{\text{passion fruit juice}}$

Step 2: $\frac{\text{orange juice}}{\text{passion fruit juice}} = \frac{8}{6}$

Step 3: $\frac{\text{orange juice}}{\text{passion fruit juice}} = \frac{8}{6} = \frac{14}{p}$

Step 4: $8(p) = 6(14)$
$$8p = 84$$
$$p = 10.5$$

So Dan needs to buy 10.5 cups of passion fruit juice.

Example 2:
Use the same recipe above to determine how many cups of guava juice are needed if Dan and Wendy decide they want to make 60 cups of juice for their party.

This is an example in which you need to compare one part of a ratio to the whole. Since the recipe calls for 6 cups of passion fruit juice, 8 cups of orange juice, and 7 cups of guava juice, that means a single batch of the recipe will yield 21 cups of juice. Compare the 7 cups of guava juice to the 21-cup recipe in the following solution:

Step 1: $\frac{\text{guava juice in recipe}}{\text{total cups in juice recipe}}$

Step 2: $\frac{\text{guava juice in recipe}}{\text{total cups in juice recipe}} = \frac{7}{21}$

Step 3: $\frac{\text{guava juice in recipe}}{\text{total cups in juice recipe}} = \frac{7}{21} = \frac{g}{60}$

Step 4: $7(60) = 21(g)$
$420 = 21g$
$g = 20$

So Dan and Wendy will need 20 cups of guava juice.

Practice 3

Use the following information to solve questions 1 through 4. In Mr. Mallory's class the ratio of girls to boys is 3:4.

1. If there are 12 boys in the class, what proportion would be used to solve for how many girls, g, are in Mr. Mallory's class?

2. If there are 12 girls in Mr. Mallory's class, how many boys are in his class?

3. There are 21 students in Mr. Mallory's class. What proportion would be used to solve for how many boys, b, are in the class? (*Hint:* You will be comparing the number of boys to the *total number* of students.)

4. How many girls are in Mr. Mallory's class, if there are 28 students in his class?

Using Proportion to Work with Scale

Whether you're comparing guava juice to orange juice in a recipe or inches to meters on a map, decoding word problems and setting up proportions is just a matter of following the steps laid out in the previous section. Let's take a look at how proportions are applied to working with scale drawings and geometric figures.

STANDARD ALERT!

Standard 7.G.A.1 wants student to be able to solve problems involving scale drawings of geometric figures. In this section you will learn to compute actual lengths and areas from scale drawings, and you'll learn how to reproduce a scale drawing using a different sale.

What Is Scale?

You have already seen a few examples of scale in this lesson, but let's formally define it. **Scale** is the ratio between the measurements of a model and the measurements of the real-life equivalent it is representing. Scale is used to make models that enlarge or reduce the size of their subject matter. For example, a map discussed above used a scale where $\frac{1}{2}$ inch represented 500 meters. A graphic designer could make an enlarged poster of a proposed magazine advertisement with a scale of 1 foot : 3 inches. Scale is normally written with a colon, but it can be changed into its fractional form in order to set up and solve a proportion. Study the following examples to learn how scale is used to compute actual lengths and areas from a scale drawing.

Example 1: Using scale to compute actual lengths from a scale drawing.

The figure below shows the open floor plan of a kitchen and living room in an apartment in Silverthorne. If the scale used is 2 centimeters : 5 feet, find the dimensions of the living room.

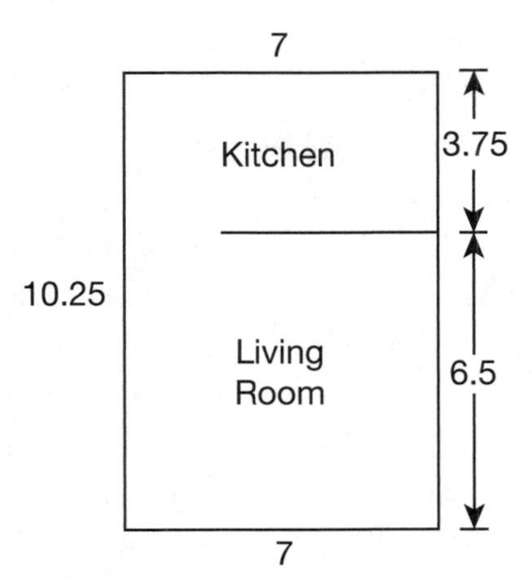

A scale of "*2 centimeters : 5 feet*" is stating that every 2 centimeters on the drawing represents 5 feet of the actual home. Use the steps you learned above to set up and solve a proportion for the width of the living room. First, write a ratio in words, and then set this equal to the given scale of 2 to 5. Finally, fill in the given and missing information before solving the problem algebraically:

$$\frac{\text{model (cm)}}{\text{real world (ft.)}} = \frac{2}{5} = \frac{6.5}{w}$$
$$2w = 32.5$$
$$w = 16.25$$

So the width of the living room is 16.25 feet.

Next solve for the length of the living room:

$$\frac{\text{model (cm)}}{\text{real world (ft.)}} = \frac{2}{5} = \frac{7}{l}$$
$$2l = 35$$
$$l = 17.5$$

So the length of the living room is 17.5 feet. The dimensions of the room are 16.25' × 17.5'.

Example 2: Using scale to compute area from a scale drawing. Find the area of the living room.

In order to find the area from a scale drawing, find the required dimensions of the real-world figure and then apply them to the area formula. In this case, the living room is a rectangle that is 16.25' × 17.5', so put these dimensions into the area formula:

$Area = length \times width = 16.25 \times 17.5 = 284.375 \text{ ft.}^2$

Standard 7.G.A.1 wants you to be able to blow up or shrink a figure's dimensions to reproduce it at a different scale. This type of task is commonly given with a scale factor. While scale factor is technically the ratio between two corresponding parts of similar figures, it will often be given as a single number, like 2 or 3.5. If you need to enlarge a figure using a scale factor, *multiply* the dimensions of the original by the scale factor. If you want to shrink a figure using a scale factor, *divide* the dimensions of the original by the scale factor.

TIP: Enlarge a scale drawing by ***multiplying*** **all the dimensions by the scale factor. Shrink a scale drawing by** ***dividing*** **all the dimensions by the scale factor.**

Example 3: Using scale to enlarge a drawing using a different scale.

Enlarge the following illustration using a scale factor of 2.

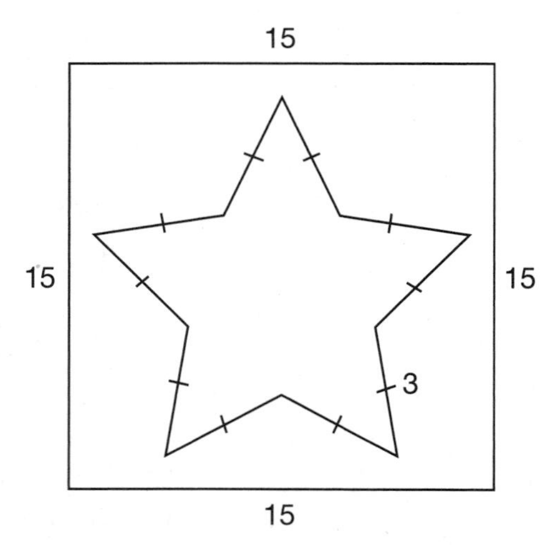

Since all the side lengths of the star are 3 units and congruent, compute the side lengths of the new star by multiplying 3 units by the scale factor of 2 to get 6. Doing the same for the exterior box around the star gives a new square side length of 15(2) = 30 units. Enlarge the star using the found dimensions:

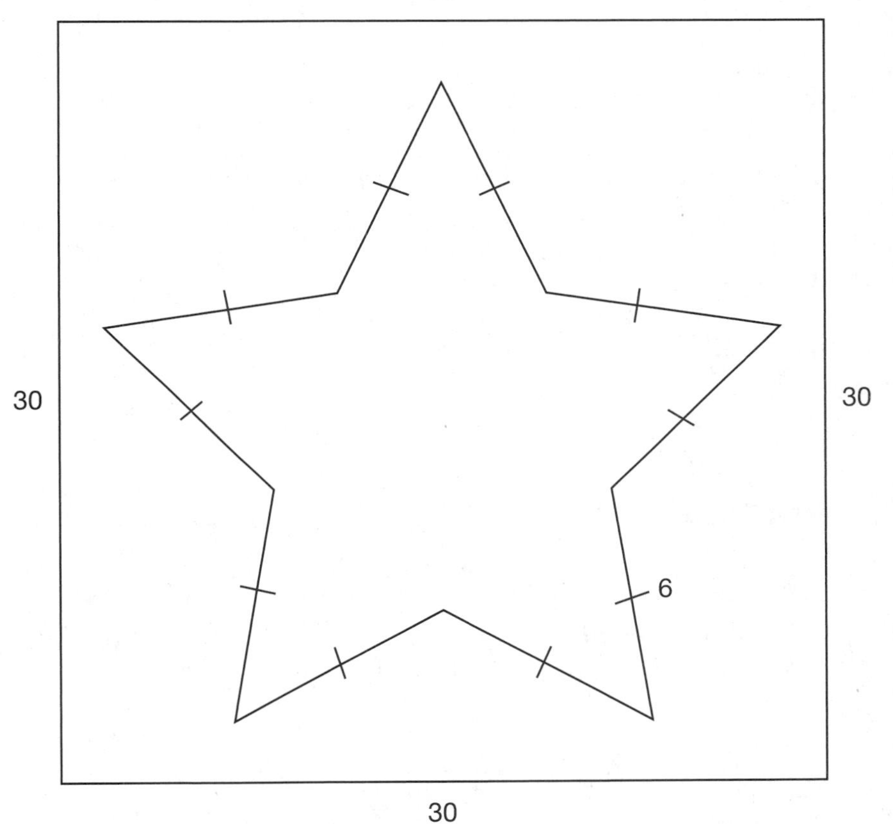

Example 4: Using scale to reduce a drawing using a different scale.

Reduce the following triangle using a scale factor of 3.

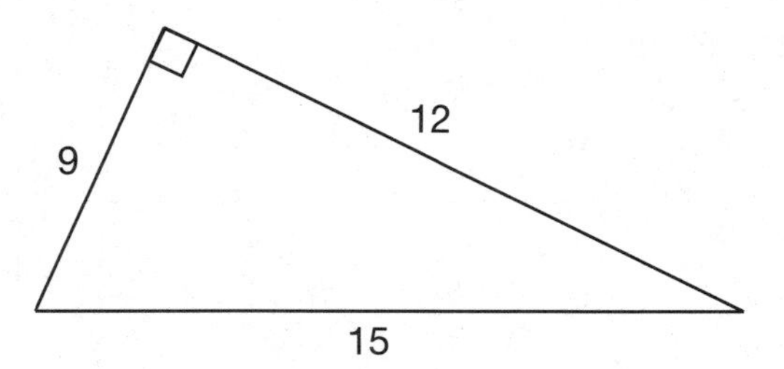

Since the triangle is being reduced, divide all of the side lengths by 3 to get the new dimensions: $\frac{9}{3} = 3$; $\frac{12}{3} = 4$; $\frac{15}{3} = 5$. Draw the reduced triangle using these dimensions:

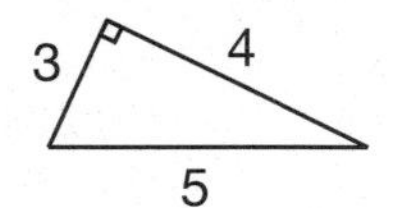

Practice 4

1. A jewelry firm uses the following illustration as part of the letterhead on its business stationery. If the firm wants to enlarge this illustration to be 2 feet wide on a banner, what will be the height of the illustration?

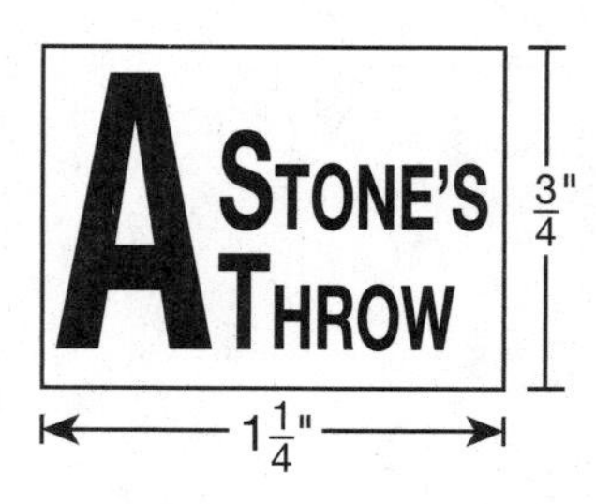

2. Using the illustration from question 1, how long would the illustration be if it were reduced for a business card with a width of $\frac{3}{4}$ of an inch?

3. Use the floor plan for the apartment in Silverthorne in Example 1 (pages 170–171) to find the area of the kitchen.

The following scale drawing represents a proposal for how a playground will be divided into cement for picnic tables and turf for various slides, swings, monkey bars, etc. Use this illustration for questions 4 through 6.

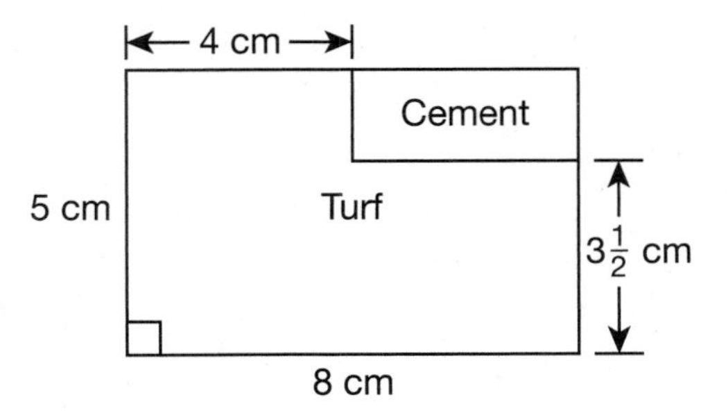

4. The scale factor for the drawing is 1 cm : 7 feet. Find the dimensions in feet of the cement portion of this playground.

5. Find the area of the cement portion of the playground in order to determine how many square feet of concrete will need to be poured.

6. The designer wants to blow up a copy of this model using a scale factor of 3 so she can include it in a packet she is preparing for her client. Determine all the new dimensions, and then enlarge this model using a ruler that has centimeters.

Answers

Practice 1

1. $\frac{\text{weekdays}}{\text{weekend}} = \frac{5}{2}$
2. $\frac{\text{weekdays}}{\text{total}} = \frac{5}{7}$
3. $\frac{5}{5} = \frac{1}{1}$ (*Remember:* This cannot be written as just *1*—keep it in fraction or colon form.)
4. $\frac{\text{prime}}{\text{composite}} = \frac{4}{5}$ (*Remember:* 1 is neither prime nor composite. 2, 3, 5, and 7 are prime. 4, 6, 8, 9, and 10 are composite.)
5. The ratio 1:2 could be representing the number of even (or odd) numbers in the set to the total amount of numbers in the set, since 5:10 reduces to 1:2.
6. $\frac{AB}{AC} = \frac{5}{3}$
7. $\frac{AB}{CD} = \frac{5}{10} = \frac{1}{2}$
8. $\frac{CD}{AB} = \frac{10}{5} = \frac{2}{1}$
9. $\frac{AC}{AD} = \frac{8}{18} = \frac{4}{9}$

Practice 2

1. False. Since the cross products are not equal, $\frac{8}{9}$ and $\frac{11}{12}$ are not proportional. ($9 \times 11 = 99$ and $8 \times 12 = 96$)
2. True. Since the cross products are equal, $\frac{2}{5}$ and $\frac{16}{40}$ are proportional. ($2 \times 40 = 80$ and $5 \times 16 = 80$)
3. True. Since the cross products are equal, $\frac{1.25}{0.5}$ and $\frac{25}{10}$ are proportional. ($1.25 \times 10 = 12.5$ and $0.5 \times 25 = 12.5$)
4. False. Since the cross products are not equal, $\frac{3}{2w+1}$ and $\frac{6}{4w+1}$ are not proportional. [$3(4w + 1) = 12w + 3$ and $6(2w + 1) = 12w + 6$. With a little further investigation, you should determine that $12w + 3$ cannot equal $12w + 6$, because $12w + 6$ will always be 3 bigger than $12w + 3$.]
5. $\frac{\text{meters}}{\text{inches}} = \frac{500}{\frac{1}{2}} = \frac{1{,}000}{1}$
6. $\frac{\text{meters}}{\text{inches}} = \frac{500}{\frac{1}{2}} = \frac{m}{3\frac{1}{2}}$ or $\frac{1{,}000}{1} = \frac{m}{3\frac{1}{2}}$
7. $\frac{\text{meters}}{\text{inches}} = \frac{500}{\frac{1}{2}} = \frac{3{,}200}{i}$

Practice 3

1. $\frac{\text{girls}}{\text{boys}} = \frac{3}{4} = \frac{g}{23}$
2. 16 boys

   $\frac{\text{girls}}{\text{boys}} = \frac{3}{4} = \frac{12}{b}$

   $4(12) = 3(b)$

   $48 = 3b$

   $b = 16$
3. $\frac{\text{boys}}{\text{total}} = \frac{4}{3+4} = \frac{b}{21}$ which is $\frac{4}{7} = \frac{b}{21}$

   $\frac{\text{boys}}{\text{total}} = \frac{4}{7} = \frac{b}{21}$
4. There are 12 girls.

   $\frac{\text{girls}}{\text{total}} = \frac{3}{3+4} = \frac{g}{28}$

   $\frac{\text{girls}}{\text{total}} = \frac{3}{7} = \frac{g}{28}$

   $7(g) = 3(28)$

   $7g = 84$

   $g = 12$

Practice 4

1. Set up a proportion that compares the width and the height of the model with the width and height of the reproduced image. (In this case, the first ratio will be in inches, and the second ratio will be in feet.)

   $\frac{width}{height} = \frac{1.25}{0.75} = \frac{2}{w}$

   $1.25(w) = (2)(0.75)$

   $1.25w = 1.50$

   $w = 1.2$ or $1\frac{2}{10}$

   The width of the illustration on the banner will be $3\frac{1}{3}$ feet.
2. $\frac{width}{height} = \frac{1.25}{0.75} = \frac{0.75}{h}$

   $0.75(0.75) = 1.25h$

   $0.45 = h$

   The height of the illustration on the business card will be 0.45 inch, or almost half an inch.

3. In the scale drawing, the kitchen is 3.75 cm by 7 cm, so apply the scale of 2 cm : 5 ft. to calculate the dimensions of the kitchen in feet, and then put these dimensions into the area formula:

$\frac{\text{model (cm)}}{\text{real world (ft.)}} = \frac{2}{5} = \frac{3.75}{w}$

$2w = 18.75$

$w = 9.375$

The kitchen is 7.5 feet deep, and it was determined in the example that the length of the room was 17.5 feet, so multiply the two dimensions together to find the area: 9.375' × 17.5' = 160 square feet.

4. Since the scale factor for the drawing is 1 cm : 7 feet, you can multiply each centimeter in the drawing by 7 feet to find the actual dimensions. Using subtraction, you'll see that the dimensions of the cement portion are 4 cm by 1.5 cm. Therefore, the dimensions of the cement portion will be 28 feet by 10.5 feet.

5. $28 \times 10.5 = 294$ ft^2

6. Your new model should be 24 centimeters wide by 15 centimeters tall, with a cement portion that is 12 centimeters wide by 4.5 centimeters tall.

14

Pi and the Circumference of Circles

A circle is a round straight line with a hole in the middle.
—Mark Twain

In this lesson you will learn about circles, the irrational number π, and how it connects with the circumference of circles.

STANDARD PREVIEW

Of course you know what a circle is—it's pizza, it's apple pie, it's the Ferris wheel at the summer state fair. Circles make up the gears of important machinery and are the wheels on bikes and cars that take us from here to there. Just like other figures, circles have their own formulas for calculating area and perimeter, and Standard 7.G.B.4 wants you to be proficient at using these formulas. In this lesson we are going to learn about pi (not the delicious one you enjoy on Thanksgiving, but the Greek one that look like this: π) and how it relates to the circumference of a circle. In fact, circles are so fancy that they have their own name for perimeter, *circumference*, so let's not delay our circle investigation any longer!

Basic Terms of Circles

The formal definition of a circle is *a collection of points that are all equal distance from one singular point*. That one special point is the **center** of a circle, and it is what defines the name of any given circle. So, if the center point of a circle is labeled F, it will be referred to as *circle F*.

TIP: Circles are named by their center point.

The distance from the center of the circle to any point on the edge of the circle is called the **radius** of the circle. A circle has an infinite number of radii, which are all the same length ("**radii**" is the plural of "radius"). In the next figure, $\overline{FA}$, $\overline{FB}$, and $\overline{FC}$ are three congruent radii in circle F.

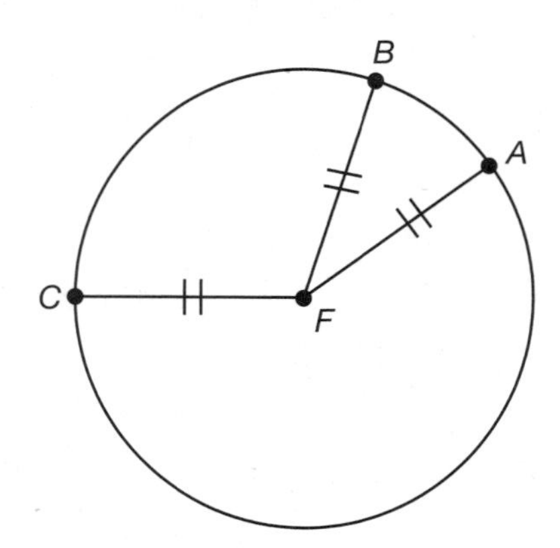

TIP: The distance from the center of the circle to a point on the circle is called the radius.

Another important term when you are dealing with circles is **chord**. A chord is a line segment that extends from one point on the circle to another point on the circle. Middle school geometry doesn't study chords, but we wanted to point them out to illustrate that chords are *not* radii. In the next figure, line segments $\overline{PQ}$ and $\overline{VW}$ are both chords in circle G.

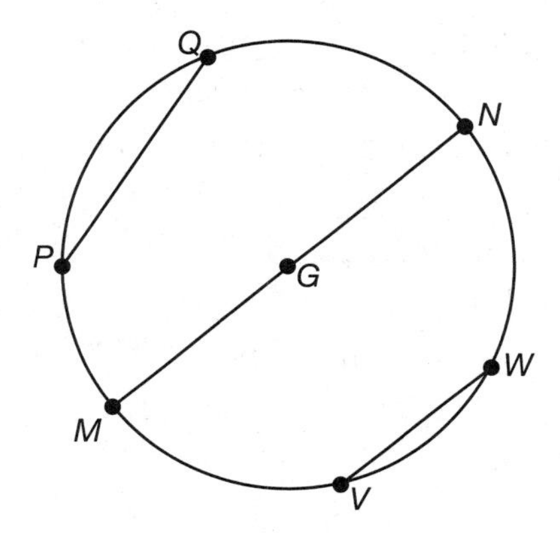

Look at chord $\overline{MN}$ in the preceding figure. When a chord passes through the center of a circle, it is called a **diameter**. Since the diameter is essentially made up of two radii extending from the center in opposite directions, the diameter will always be *twice the length* of a radius. Therefore the radius will always be half the diameter.

TIP: A *diameter* is a special chord that passes from side to side of and through the center of a circle. The length of the diameter is twice the radius. The radius is half the diameter.

What Is π?

The irrational number π (which is spelled pi and pronounced "pie"), is a key concept when working with circles. It relates to the **circumference** of a circle, which is the perimeter around the outside of a circle.

TIP: The distance around a circle is called its *circumference*.

More than 2,000 years ago, mathematicians worked to find a shortcut to calculate a circle's circumference, by using the length of the circle's diameter. This may seem like a nearly impossible task, but you do not have to be an expert mathematician to investigate this relationship. You could do this yourself with just a ruler, a piece of string, and several different-sized cylinders, such as a toilet paper roll, a can of soup, and a salad bowl. If you measured the distance around each cylinder and divided it by the cylinder's diameter, you would find that $\frac{\text{circumference}}{\text{diameter}}$ is always just a little bit larger than 3. This would be a great start to estimating π, which is defined as the ratio of a circle's circumference to its diameter. This relationship is the same in all circles, and the most commonly used approximation of the ratio between a circle's circumference and its diameter is 3.14. Pi does not stop at 3.14 though; π is an irrational number that carries on infinitely and does not end.

TIP: Pi, or π, is an irrational number equal to the ratio $\frac{\text{circumference}}{\text{diameter}}$ which is the same number for all circles. π is commonly approximated as 3.14 (or $\frac{22}{7}$).

Circumference of a Circle

Now that the relationship between a circle's circumference and diameter has been determined, it is possible for you to come up with a formula used to calculate circumference:

$$\frac{\text{circumference}}{\text{diameter}} = \pi$$

$$\frac{\text{circumference} \times (\text{diameter})}{\text{diameter}} = \pi \times (\text{diameter})$$

$$\text{circumference} = \pi \times (\text{diameter})$$

$$C = \pi d$$

Since the diameter is always twice the length of the radius, r, another way to express circumference is $C = 2\pi r$.

TIP: Circumference = πd or $2\pi r$, where π = 3.14 or $\frac{22}{7}$.

Since π is an irrational number, answers relating to circumference are often left in terms of "π," instead of being evaluated with π as 3.14. In general, questions state whether to leave your answer in terms of π, or to round it to the nearest tenth or hundredth. When you're rounding your answer, it is good practice to use the symbol for approximation, which is ≈, instead of an equals sign. Refer to the next figure for the following examples.

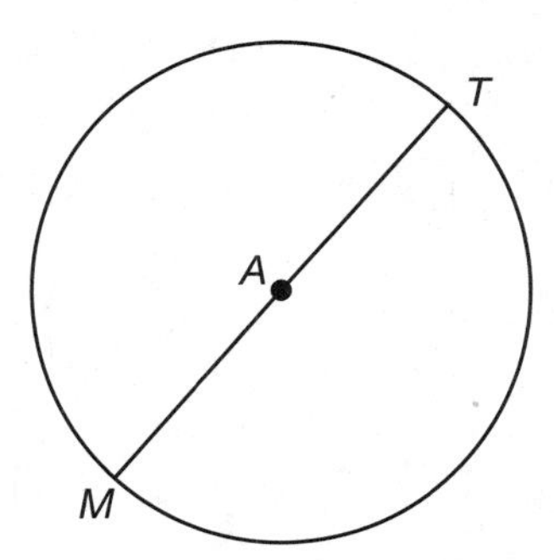

Example 1: Find the circumference of circle A, if $\overline{MT} = 6$ (leave your answer in terms of π).
Solution: Since Circumference = πd, $C = 6\pi$.

Example 2: Find the circumference of circle A, if $\overline{AT} = 5$ (round your answer to the nearest tenth).
Solution: Since Circumference = $2\pi r$, $C \approx 2(3.14)(5) \approx 31.4$.

Example 3: If the circumference of circle $A = 36\pi$, find the length of $\overline{MA}$.
Solution: Since **C** $= 2\pi r$, $36\pi = 2\pi r$ and $\frac{36\pi}{2\pi} = r$, so $\overline{MA} = 18$.

STANDARD ALERT!

*Now that you know about π and have seen how the circumference formula is calculated, work through the following problems to practice the requirement presented in **7.G.B.4.***

Practice 1

Use the first two figures to answer questions 1 and 2.

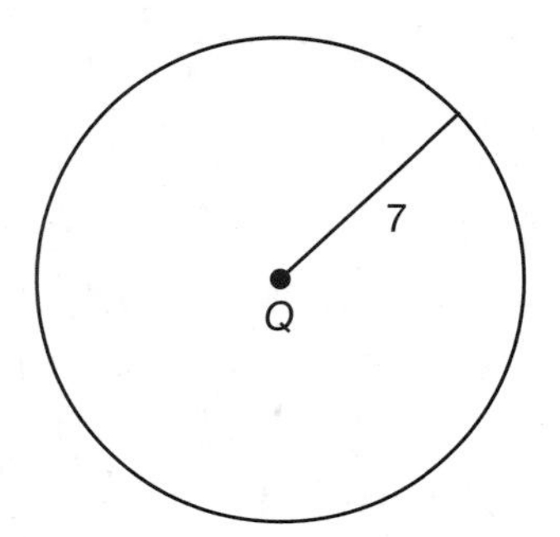

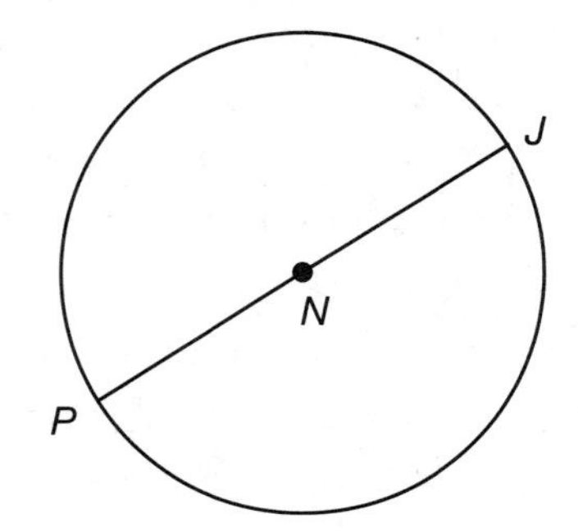

1. Find the circumference of circle Q to the nearest hundredth.

2. If $\overline{PJ} = 19$, find the circumference of circle N in terms of π.

Use the following figure to answer questions 3 through 5.

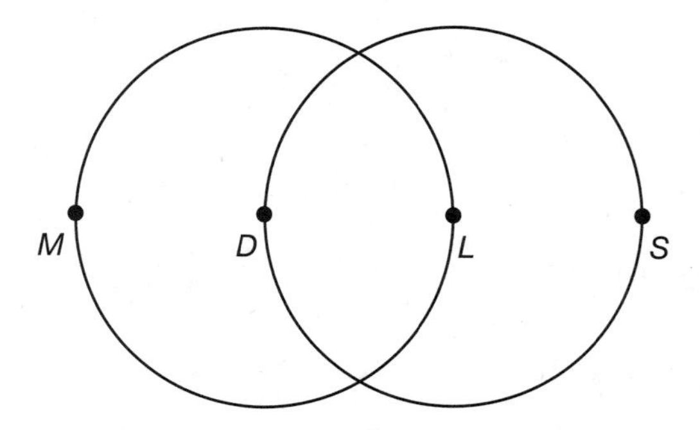

3. $\overline{DL}$ is the radius of both circle D and circle L. If $\overline{MS} = 15$, find the circumference of circle D to the nearest hundredth.

4. If $\overline{MD} = 2x$, what is the circumference of circle L in terms of π and x?

5. If the circumference of circle L is $(14)(v)(\pi)$, what is the length of $\overline{MS}$ in terms of v?

6. Your circular rose garden has a diameter of 25 feet. Fencing is sold by the yard at Francine's Fencing Surplus. Given that a yard equals three feet, how many yards of fencing must you purchase in order to encircle your rose garden?

7. If the radius of a circular hatbox is 8 inches, approximately how many inches of ribbon will you need to circle your hatbox one time?

8. The circumference of a circular plastic container is approximately $39\frac{1}{4}$ inches. What is the radius of the container?

Combining Circumference with Perimeter

In the real world, many objects are a combination of two or more shapes put together. A corner office desk may be made up of two different-sized rectangles, or a coffee table might be a square with circular ends. It is important to be able to do calculations when you're working with these more complex shapes. Often, you will need to identify which portion of the circle you're using—using half of the circle or one-quarter of the circle are common. In addition, you will need to decide if there are parts of the polygon that should be omitted when you're calculating the perimeter. These problems are best done by breaking them up into smaller, separate parts. Consider the following problem that accompanies the next figure.

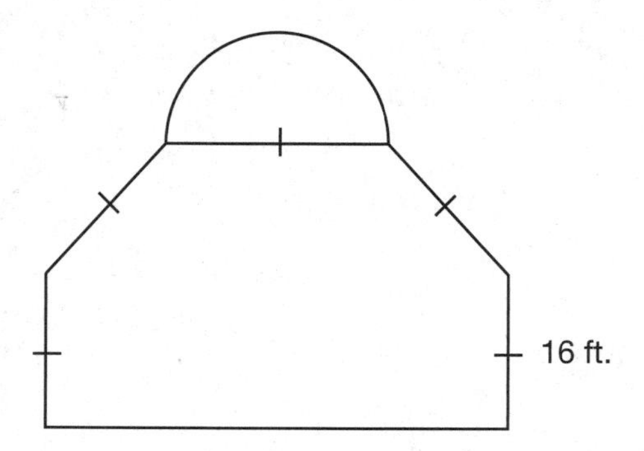

Example: The main hall of the building is the top part of a regular octagon. Its vertical walls, slanted walls, and ceiling are all congruent. The top of the building is a semicircle. If some students were trying to calculate how many feet of twinkle lights were needed to line the outside of this building, not including the floor, how would this be done?

Solution: Since there are 4 edges that are 16 feet each, the straight edges will need 4(16) = 64 feet of lights. Last, the semicircle on the top will need $\frac{1}{2}$ the circumference of lights. $C = \pi d = 16\pi$, so one-half of this is 8π. 8π estimates to just over 25 feet, so if the students bought 64 + 26 = 90 feet of lights, they would be able to line the outside edges of this building.

STANDARD ALERT!

Circles, semi-circles, and quarter-circles are often used in the design of buildings, gardens, and home furniture. Therefore, it is important to be able to incorporate what you learned for standard ***7.G.B.4*** *with other shapes in order to find the perimeters and areas of more complex figures. We'll get to practice area in the next lesson, but for now, try your hand at finding the perimeter of these composite shapes.*

Practice 2

1. Melissa is giving a party and has designed invitations that will look like the petals of a flower when the semicircular edges are folded over the square invitation. She wants to line the outside edge of the four semicircular parts with silk ribbon. If the inner square part of the invitation measures 5 inches by 5 inches, approximately how many inches of ribbon will she need to line each invitation?

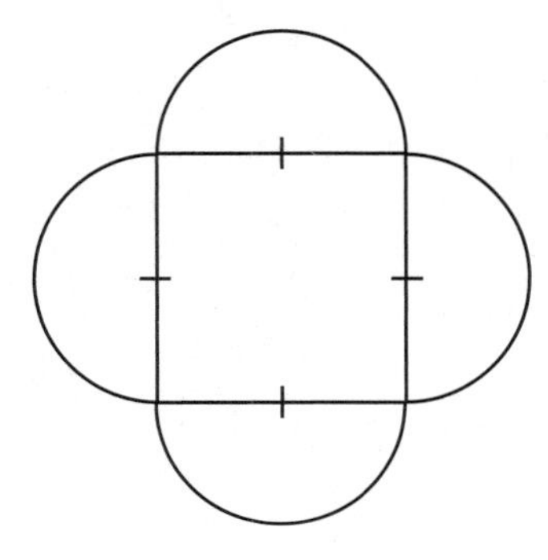

2. Fred is hosting a party for his two-year-old daughter. He needs to buy foam to line the outside edge of a coffee table to protect the heads of any kids who might fall against the table. The table has a rectangular section that is 3 feet wide by 5 feet long, and at each edge of the table there is a semicircular section. Approximately how many feet of foam does Fred need to buy?

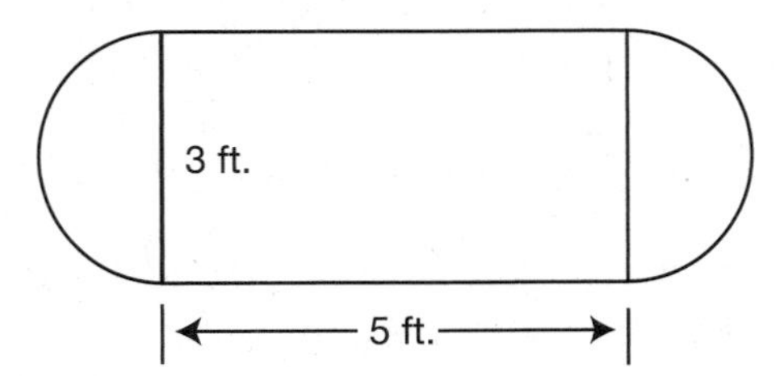

3. A door is 4 feet wide by 8 feet tall and it is symmetrically centered in an archway that is 10 feet wide by 11 feet tall. If Juan wants to line the outside edge of the archway with tiles, how many feet of tiles will he need to complete the job? (The edge to be lined with tiles has been darkened.)

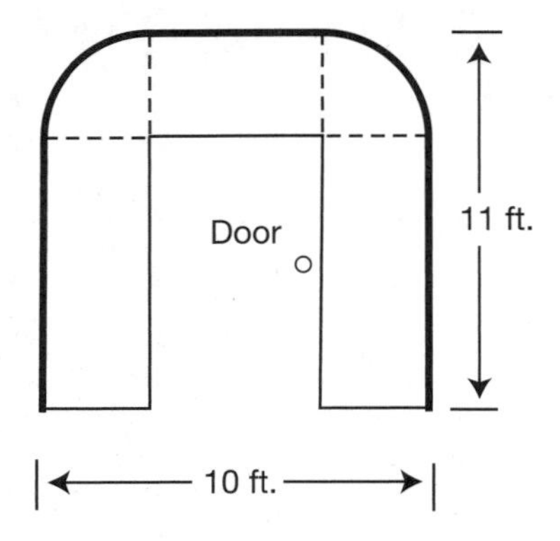

Answers

Practice 1

1. $C = 2\pi r = 2\pi 7 \approx 14(3.14) \approx 43.96$
2. $C = \pi d = 19\pi$
3. Since $\overline{MS} = 15$, then $\overline{DL} = 5$ since $\overline{MS}$ is made up of three congruent radii. Therefore, the circumference of circle $D = 2\pi r \approx 10(3.14) \approx 31.4$.
4. Since $\overline{MD} = 2x$, then $\overline{DS} = 4x$ and the circumference of circle $L = 4x\pi$.
5. Since $C = 2\pi r$ then $2\pi r = (14)(v)(\pi)$. Divide both sides by 2π to get $r = 7v$. $\overline{MS}$ is three radii in length, so $\overline{MS} = 3(7v) = 21v$.
6. $C = \pi d = 25(3.14) = 78.5$ feet. Now change this into yards by dividing it by 3 feet: $\frac{78.5}{3} = 26.166$. Therefore, you need to purchase 27 yards of fencing to encircle the garden.
7. $C = 2\pi r$, so $C = 2\pi 8 \approx 16(3.14) \approx 50.24$ inches
8. $C = 2\pi r$, so $39.25 = 2\pi r$
   $2\pi r \approx 2(3.14)r \approx 39.25$, so $r \approx \frac{39.25}{6.28} \approx 6.25$ inches

Practice 2

1. Melissa's invitation has four semicircles, each with a diameter of 5 inches. So the distance around the outside is (4 parts)($\frac{1}{2}$ of each circle's circumference) = $(4)(\frac{1}{2})\pi d = 2\pi 5 = 10\pi \approx 31.4$ inches.
2. The foam needed to cover the two straight edges of the table is $5 \times 2 =$ 10 feet. The circumference of the two outer semicircles is (2 parts)($\frac{1}{2}$ of each circle's circumference) = $(2)(\frac{1}{2})\pi d = 1\pi 3 \approx 9.42$ feet. So Fred should buy 19.5 or 20 feet of foam to surround the table.
3.

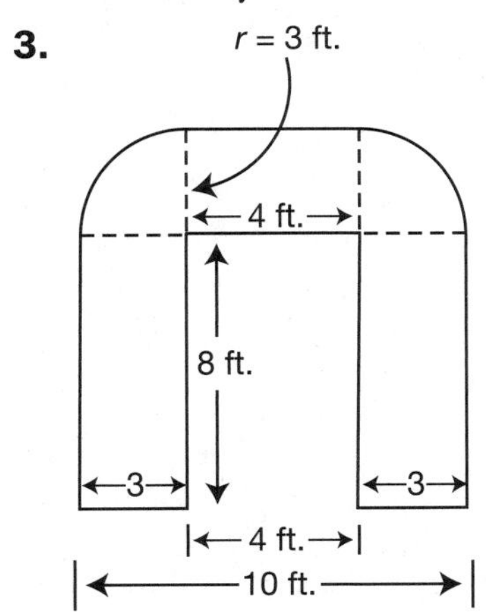

The outside of the archway can be broken up into three rectangles and two quarter-circles. These two quarter-circles can be combined to form a semicircle with a radius of 6. The circumference of this semicircle is ($\frac{1}{2}$ of the circumference) = $(\frac{1}{2})2\pi r = 1\pi 3 \approx 9.42$ feet. The vertical edges of the archway are each 8 feet long, so 16 feet of tiles will be needed there. Last, the top edge of the archway is 4 feet. The total outside border of the archway is $9.42 + 16 + 4 \approx 29.42$ feet.

15

The Area of Circles

In mathematics you don't understand things.
You just get used to them.
—Johann van Neumann

In this lesson, you will learn how to apply π to the area of circles. You will also learn how to determine the area and perimeter of sectors of circles when you're given the central angle measurement.

STANDARD PREVIEW

Now that you've gotten familiar with calculating the circumference of circles, you're ready to step up to the big leagues and learn how to calculate area. ***Not only does Standard 7.G.B.4 want you to know the area formula, but it asks you to understand how the area formula relates to the circumference formula.*** A pretty neat and informal derivation of the area formula can be found by breaking down a circle into lots of triangles, which makes ***Standard 6.G.A.1*** happy, so we'll start with that. We'll end this lesson with a non-essential but advanced extension of ***Standard 7.G.B.4*** as you learn to calculate the area of sectors and length of intercepted arcs in circles.

Area of a Circle

Deriving the Area Formula for a Circle

In a previous lesson, we showed visually that the area of a triangle is $\frac{1}{2}$(base)(height) by demonstrating how a rectangle can be broken down into two congruent triangles. In a different lesson you learned why the area of a parallelogram is *base* × *height*: We broke up a parallelogram into a triangle and a trapezoid and rearranged these two shapes to form a rectangle. It is also possible to informally derive the formula for the area of a circle by breaking it up into triangles. Consider the circle that follows, which has been broken up into congruent gray and white triangular forms. Notice that the gray triangular forms constitute one half of the area and one half of the circumference as well. The white triangular forms constitute the other half of the area and circumference. Try to connect how the *base* × *height* area formula for rectangles is being applied to the reconstructed circle in the following illustration:

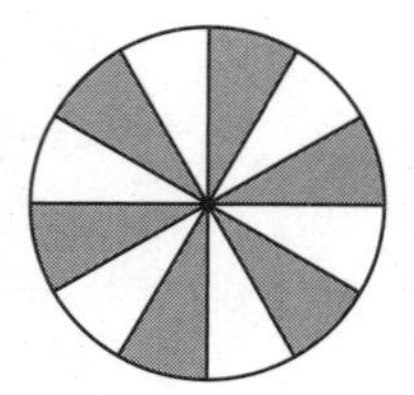

height = *r*

base = $\frac{1}{2}C = \frac{1}{2}2\pi r = \pi r$

In the reconstructed figure on the right, the base of the parallelogram is equal to the circumference of the gray triangular forms. You learned in the previous lesson that $C = 2\pi r$, and since the gray triangular forms constitute half the circumference, the length of the base is πr. The height of the parallelogram is the radius of the circle, r. Follow the substitution used here to derive the formula for the area of a circle:

$A = base \times height$, where base = πr and height = r
$A = (\pi r) \times (r)$
$A = \pi r^2$

STANDARD ALERT!

Did the previous section make your head hurt? Part of 7.G.B.4 is being able to give an informal derivation of the relationship between the circumference and area of a circle, so it's important that you make sense of it. It may be helpful to redraw the figures on your own and try to rework the formulas independently, looking only at this text to check your work. Do that until you have a thorough grasp of the derivation above!

TIP: The area of a circle is equal to πr^2.

It is important to notice that the formula for the area of circles uses radius and not diameter. Therefore, if you are given only the diameter of a circle, you must first divide it in half before using the area formula. (Remember that the radius *squared* is not the same thing as the radius times 2. You *cannot* use diameter in place of r^2.)

Use the circles in this figure for examples 1 and 2 that follow.

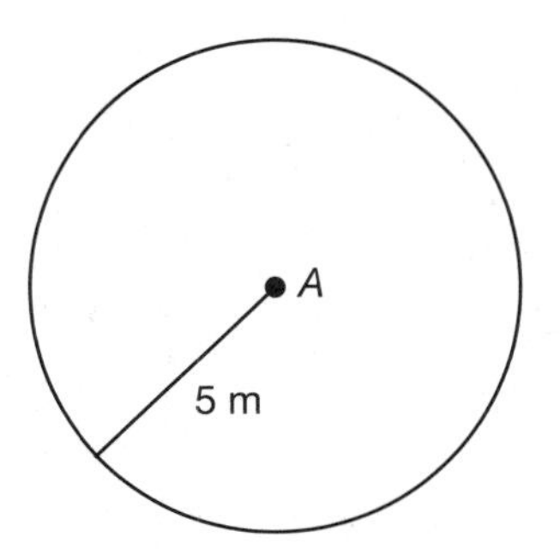

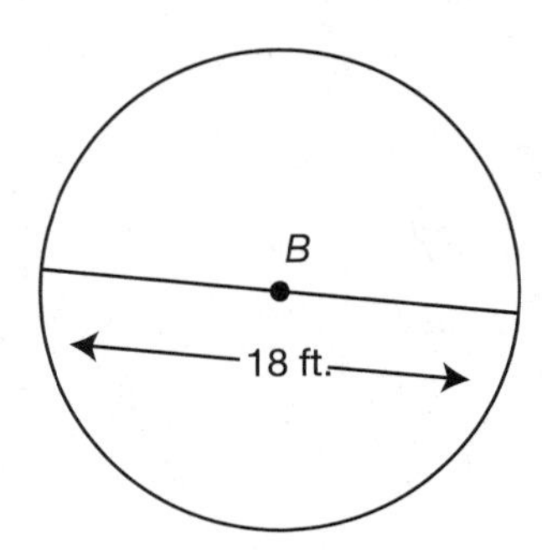

Example 1: Find the area of circle A to the nearest tenth of a meter.
Solution: Area = $\pi r^2 = \pi(5^2) \approx 3.14(25) \approx 78.5$ m^2.

Example 2: Find the area of circle B. Keep your answer in terms of π.
Solution: The radius of circle B is $\frac{18}{2} = 9$ feet. Area = $\pi r^2 = \pi(9^2) = 81\pi$ ft.^{2}.

STANDARD ALERT!

Now that you know how to derive the area formula for circles and how to apply it, work though the mathematical and real-world problems below to test your 7.G.B.4 skills.

Practice 1

1. In the next figure, what is the area of circle H? Keep your answer in terms of π.

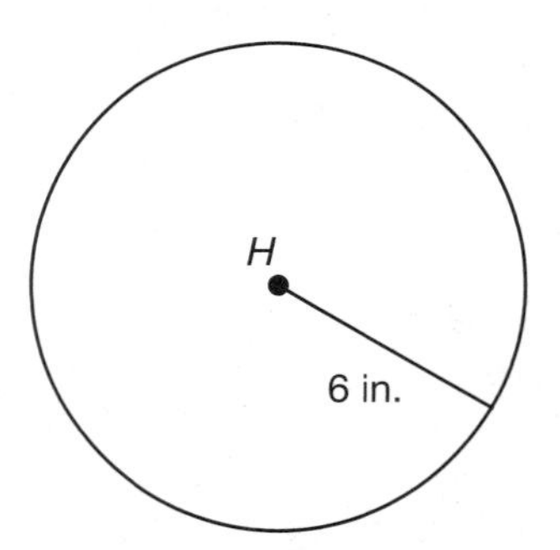

2. In the next figure, what is the area of circle I? Keep your answer in terms of π.

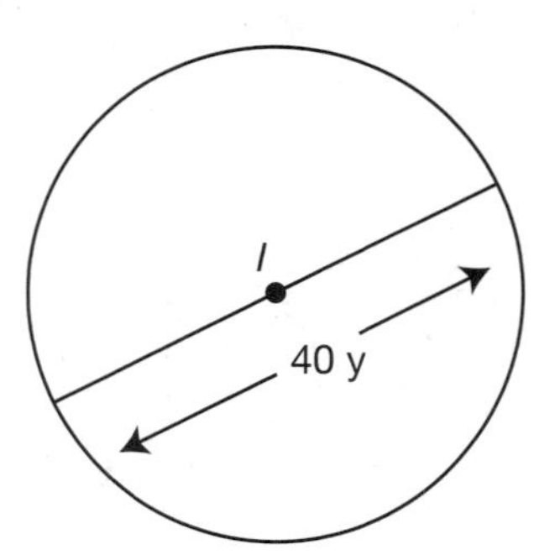

3. In the following figure, what is the area of the shaded half of the circle? Round your answer to the nearest tenth.

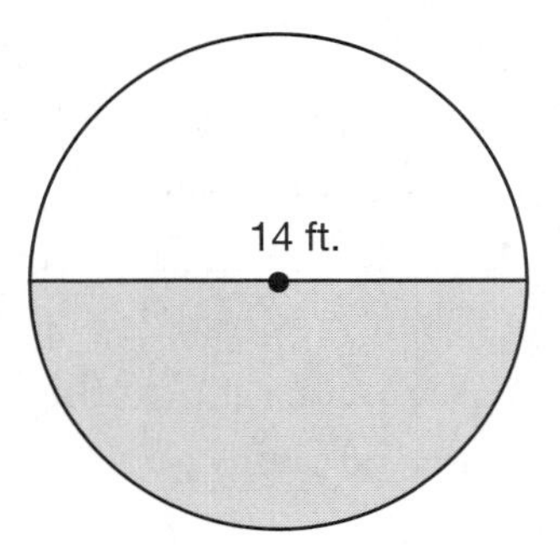

4. Victoria makes Parker Elizabeth a round birthday cake that had a diameter of 16 inches. What is the area of the top of the cake? (Round your answer to the nearest tenth.)

5. Michael is helping Maddie Winn paint a round sun on her bedroom wall. Maddie Winn wants the sun to go from the ceiling to halfway down the wall toward the floor. If her ceiling is 9 feet high, how many square feet of paint will be used to paint the sun? (Round your answer to the nearest tenth.)

6. In the previous lesson we calculated the number of feet of fencing needed for a circular rose garden with a diameter of 25 feet. How many square feet of mulch will be needed to cover the garden? (Round your answer to the nearest tenth.)

7. If the area of cover for a circular children's pool is 38.5 ft.^{2}, what is the radius of the pool?

8. The area of a circular stage is 452.2 m^2. What is the diameter of the stage?

STANDARD ALERT!

The following material is not part of 7.G.B.4 skills but is an extension for students who would like to challenge themselves. These topics are part of the high school CCSS but are a natural extension of 7.G.B.4, so they have been included for the enjoyment of more adventurous students!

Arcs, Degrees, and Angles in Circles

Arcs in Circles

When you hear the word *arc*, you probably think of something curved. In circles, an **arc** is a curved section of the outside of the circle that is defined by two points on the circle. In this section we are going to focus on arcs that are formed by two radii, but arcs can also be defined by a chord. Arcs are written similarly to line segments, except the line over them is curved. In the following figure, the radii $\overline{FC}$ and $\overline{FB}$ define arc $\overset{\frown}{CB}$.

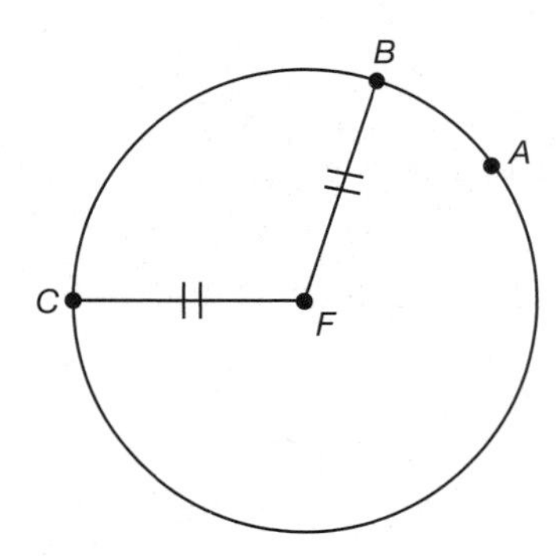

When you're reading $\overset{\frown}{CB}$, you cannot be certain if what is being referred to is the smaller clockwise distance from C to B or the longer counterclockwise distance from B to C. To clarify this, the smaller arc, or **minor arc**, is referred to as $\overset{\frown}{CB}$. The larger arc, or **major arc**, would need a third letter along the circumference to define it, such as $\overset{\frown}{CAB}$.

TIP: An **intercepted arc** is the curved section of the outside of the circle that is defined by two points on the circle.

Degrees in Circles

The length of an arc can be represented in units such as centimeters or inches, but more commonly, the length is represented as a measure of degrees. A complete revolution around a circle contains 360°. A semicircle contains 180° since it is half the distance around a circle. A *minor arc* is any arc that measures less than 180° and a *major arc* is any arc that measures more than 180°.

TIP: A circle contains 360°. A semicircle contains 180°.

Central Angles in Circles

A central angle is an angle whose vertex is the center of the circle with two endpoints on the circle. Central angles are formed by two radii of a circle. $\angle CFB$ in the previous figure is a central angle of circle F.

TIP: A *central angle* is an angle formed by two radii. Its vertex is the center of the circle and whose endpoints sit on the circle.

The minor arc formed by a central angle is referred to as the ***intercepted arc***. There is a special relationship between the measure of a central angle and its intercepted arc. Although you will study central angles more in depth in your future math classes, it is sufficient for now for you to know that a central angle's measure is equal to the measure of its intercepted arc. Therefore, a 90°central angle will make an intercepted arc that also measures 90°. (This should not be surprising, since a circle contains 360° and a 90° central angle will define one-quarter of the outside of the circle.) See the following figure:

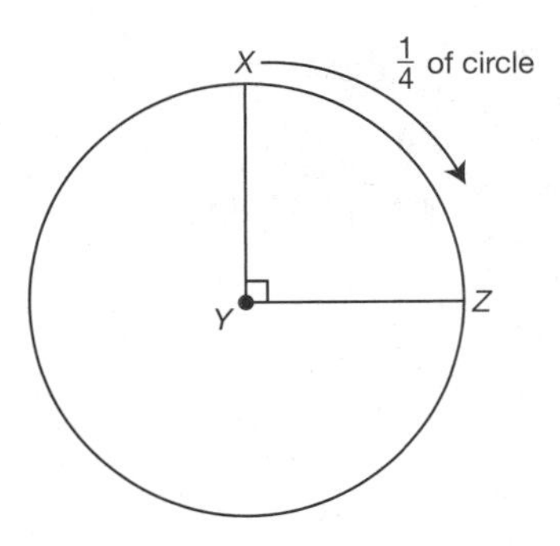

TIP: A **central angle** is an angle whose vertex is the center of a circle with two endpoints on the circle. A central angle's measure is equal to the measure of its intercepted arc.

Finding the Area of a Sector of a Circle

A **sector** is a part of a circle bound by two radii and the arc made along the circumference. A sector looks like a piece of pie, or an ice cream cone.

Sometimes, you will to need calculate the area or perimeter of just a sector of a circle. Remember that a central angle has the same degree measurement as its intercepted arc. Let us consider a circle that has a central angle of θ. The sector enclosed by that angle will occupy a portion of that circle that is equal to the *measure of angle* divided by the *total number of degrees* in the circle: $\frac{\theta}{360}$ ($\frac{\theta}{360}$ represents the fractional part of the area you will be solving for). Therefore, you calculate the area of the sector of a circle by multiplying the *fractional part* of the circle by the *total area* of the circle: $(\frac{\theta}{360})(\pi r^2)$.

TIP: Area of a sector of a circle = $(\frac{\theta}{360})(\pi r^2)$

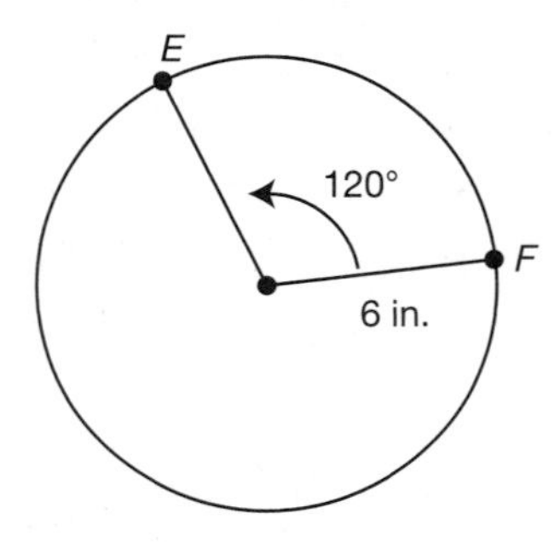

Example: Find the area of the sector in this figure.
Solution:

$A(\text{sector}) = (\frac{\theta}{360})(\pi r^2)$

$A(\text{sector}) = (\frac{120}{360})(\pi 6^2)$

$A(\text{sector}) = (\frac{1}{3})(36\pi)$

$A(\text{sector}) = 12\pi \text{ in.}^2$

Finding the Length of a Sector's Arc

The same principle can be used to calculate the length of the intercepted arc formed by a central angle. Multiply the fractional part of the circle by the total circumference of the circle: $(\frac{\theta}{360})(\pi d)$.

TIP: Length of a sector's arc = $(\frac{\theta}{360})(\pi d)$

Example: Find the length of the intercepted arc $\widehat{EF}$ in the preceding figure.

Solution:

$C(\text{arc}) = (\frac{\theta}{360})(\pi d)$

$C(\text{arc}) = (\frac{120}{360})(\pi 12)$

$C(\text{arc}) = (\frac{1}{3})(12\pi)$

$C(\text{arc}) = 4\pi$ inches

Practice 2

1.

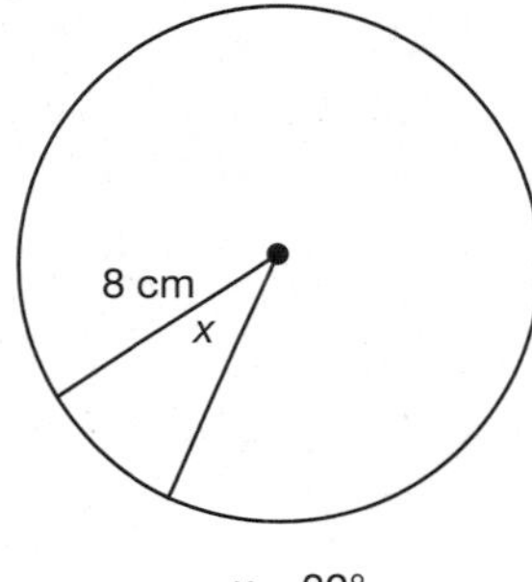

$x = 32°$

Find the area to the nearest hundredth of the sector defined by angle x.

2. Using the figure from question 1, find the length of the intercepted arc defined by angle x. Round it to the nearest tenth.

3.

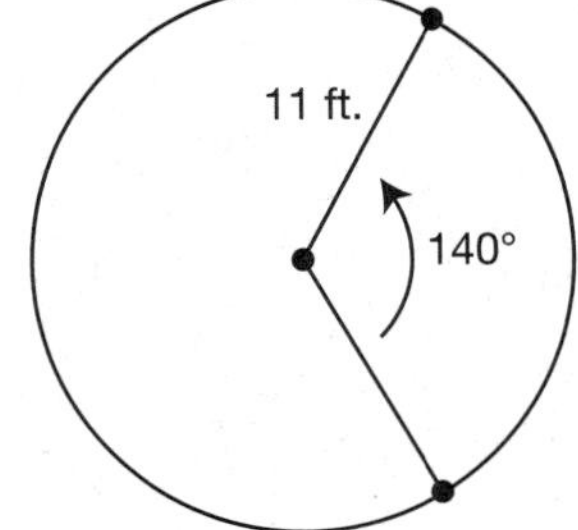

This figure shows a circular garden space. The sector created by the 140-degree angle will have woodchips and benches for people to sit. The remainder of the garden space will have grass. How many square feet of this area will be covered in grass? Round to the nearest whole foot.

Answers

Practice 1

1. Area = $\pi r^2 = \pi(6^2) = 36\pi$ in.^{2}.
2. The radius of circle *B* is $\frac{40}{2} = 20\ y$. Area = $\pi r^2 = \pi(20^2) = 400\pi\ y^2$.
3. The radius of the circle is $\frac{14}{2} = 7$ ft. Area = $\pi r^2 = \pi(72) = 49\pi$ ft.$^2 \approx 153.86$ ft.^{2} The area of the shaded half is therefore 76.9 ft.^{2}.
4. The radius of the cake is $\frac{16}{2} = 8$ inches. Area = $\pi r^2 = \pi(8^2) = 64\pi$ in.$^2 \approx$ 200.96 in.$^2 \approx$ 201.0 in.^{2} when rounded to the nearest tenth.
5. Since the ceilings are 9 feet tall, and the sun will go halfway down the wall, it will have a diameter of 4.5 feet. The sun will therefore have a radius 2.25 ft. Area = $\pi r^2 = \pi(2.25^2) = 5.0625\pi$ ft.$^2 \approx 15.9$ ft.^{2}. Michael will need to buy enough paint to cover approximately 16 square feet.
6. The rose garden has a radius of 12.5 feet. Area = $\pi r^2 = \pi(12.5^2) \approx$ 490.6 ft.^{2}
7. Area = πr^2; $38.5 = \pi r^2$; $r^2 = \frac{38.5}{3.14} = 12.26$, so $r \approx 3.5$ feet.
8. Area = πr^2; $452.2 = \pi r^2$; $r^2 = \frac{452.5}{3.14} = 144.01$, so $r \approx 12$ feet and $d \approx 24$ feet.

Practice 2

1. $A(\text{sector}) = (\frac{\theta}{360})(\pi 8^2)$

   $A(\text{sector}) = (\frac{32}{360})(64\pi)$

   $A(\text{sector}) = 17.86\ \text{cm}^2$

2. $C(\text{arc}) = (\frac{\theta}{360})(\pi d)$

   $C(\text{arc}) = (\frac{32}{360})(\pi 16)$

   $C(\text{arc}) = 4.5$ inches

3. To calculate, find the area of the entire garden and then subtract the area of the sector:

   $A(\text{whole}) - A(\text{sector}) = (\pi r^2) - (\frac{\theta}{360})(\pi r^2)$

   $A(\text{whole}) - A(\text{sector}) = (\pi 11^2) - (\frac{140}{360})(\pi 11^2)$

   $A(\text{whole}) - A(\text{sector}) = (380) - (148)$

   $A(\text{whole}) - A(\text{sector}) = 232\ \text{ft.}^2$

16

Area and Perimeter in the Real World

We only think when confronted with a problem.
—John Dewey

In this lesson you will learn how to calculate the missing side lengths in polygons when you know the area or perimeter. You will also calculate the areas of more complex figures.

STANDARD PREVIEW

Mario is a costume designer and he needs to buy enough fabric to make a dozen dresses. Anne installs hardwood floors and she needs to order just the right number of feet of an expensive custom molding for her next job. Bill is a farmer and he needs to know how many horses he can graze in his small pasture. Mario, Anne, and Bill all need to be skilled in calculating perimeter and area to do their jobs. Even if you end up having a job that doesn't demand these skills, it's likely that they'll be useful to you in your personal life. ***Now that you are equipped with the formulas for calculating area, circumference, and perimeter for many types of shapes, you are ready to test your skills in some real-world applications, and that is what Standard 7.G.B.6 is about!***

Using Area to Find Missing Information

In some situations, the area of a polygon will be given, and you will need to solve for its width or length. To do so, you will need to choose the correct formula, plug in the given information, and then solve for the missing dimension. Sometimes, you will use a formula for perimeter, and other times it will be appropriate to use a formula for area.

Example: Marni is painting the walls of her house. One gallon of paint covers 400 $ft.^2$ of wall space with a single coat. If Marni's walls are 8 feet tall, how many horizontal feet of wall space can she cover with a single coat of paint?
Solution: Painting walls in a house involves knowing about the space inside rectangles, so plug the given information into the formula for the area of the rectangles. Then solve for the missing dimension:

A = length × width
400 = length × (8)
Length = 50

Marni can cover 50 horizontal feet of wall space with one gallon of paint.

Practice 1

1. Daphne boards her horse, Lightning, at a stable located in the hills outside her home in the city. She has requested that when the stable needs to tether Lightning, that Lightning be able to walk at least a 250-foot circle around his tethering post. What is the shortest length of rope that the stable hand can use to tether Lightning?

2. Ryan has a package of zinnia seeds. The instructions say that there are enough seeds to cover 50 $ft.^2$ of ground. He wants to make a flower box in front of his living room window that is 12.5 feet long. How wide should the box be?

3. Jocelyn purchased 16 feet of decorative trim to put around the edge of a rectangular painting she will make over the weekend. What are three different possible combinations for the length and width of the painting, if she wants to use all of the trim she purchased and not have any fractional edge lengths?

4. Using the information in problem 3, Jocelyn decides to make a circular painting and still wants to use all 16 feet of the decorative trim around the outside edge. What will the diameter of this circular painting be?

5. Ivanna has 7 yards of sequined trim that she plans to stitch around the border of a square tablecloth. She doesn't want to waste any of the trim and wants to make the largest square tablecloth possible. What is the side length in feet of the largest tablecloth Ivanna can decorate with her 7 yards of sequined trim?

Finding the Area between Shapes

Another common type of problem relating to area is when one shape is contained within another shape. You will encounter this situation often in the real world, especially when it comes to landscaping a yard or outdoor area. For example, a patio might be in the middle of a yard, which will reduce the number of square feet of sod needed for the lawn. In order to solve this type of problem, it is necessary to find the area of the larger figure, and then subtract from that measure the area of the smaller figure. Look at the following figure as you read through the following example.

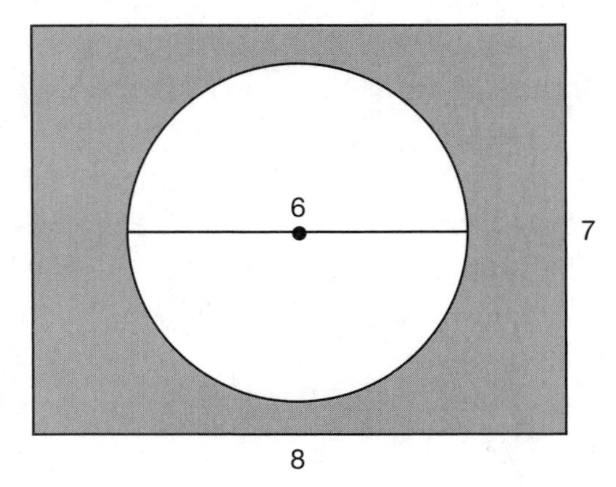

Example: A circular fountain sits in a rectangular garden space. The garden space is 8 feet long by 7 feet wide and the fountain has a diameter of 6 feet. What is the area, to the nearest tenth, of the garden space that will not be taken up by the fountain?
Solution: The area of the garden space will be $A = lw = (7)(8) = 56$ ft.^{2}. The area of the circular fountain will be $A = \pi r^2 = \pi(3^2) \approx 28.26$ ft.^{2}. Therefore, the area of the garden space that is not covered by the fountain is the difference of the two shapes: $56 - 28.3 = 27.7$ ft.^{2}.

TIP: To find the area between two different figures, subtract the area of the smaller figure from the area of the larger figure.

One thing to be aware of when you're working with real-world problems is that the units may not always be the same. For example, if you are given one dimension in inches and one in feet, it is necessary to convert both dimensions into feet, or both into inches, before attempting to solve the problem.

Practice 2

1. In their rectangular backyard, the Spieth family has a fire pit as shown in the following figure. The fire pit is 8 feet long by 3 feet wide. They want to fill a 14-foot by 11-foot area surrounding the fire pit with concrete. How many square feet will the concrete section surrounding the fire pit contain?

2. Matt and Jen just had a wood deck built around their pool. The length of the deck is 38 feet and the width is 20 feet, and their pool sits within it. Their pool has a semi-circular staircase leading into it, which has a six-foot radius. The straight length of the pool is 24 feet and the width is 12 feet. They need to buy sealant to protect their deck, and one can of sealant is enough to cover 100 square feet. How many cans will they need to cover their exposed deck?

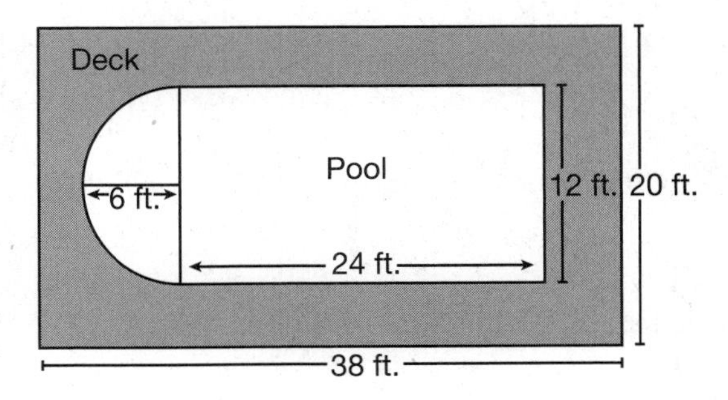

3. A bathroom counter has two circular sinks as shown in the following figure. The counter is 5 feet by 2 feet in size and each sink is $1\frac{1}{4}$ feet in diameter. How many square feet of tiles will Kerry need to cover the counter?

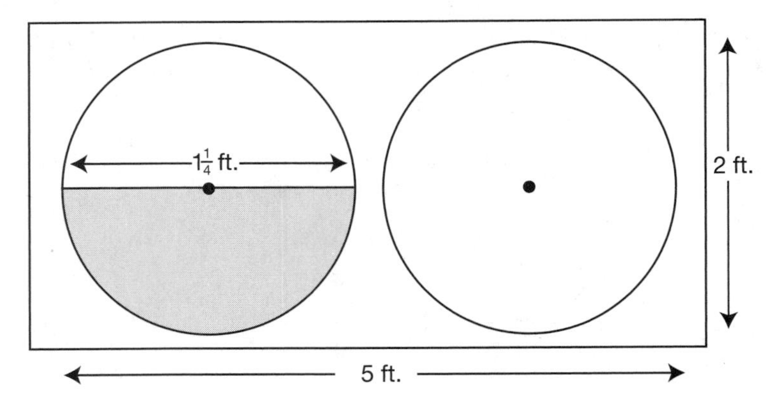

4. Tivoli makes a flag to bring to the beach so that her friends can find her. She first uses a piece of teal fabric that is 5 feet long by 3 feet wide. She then adds to it three pieces of printed material: two triangles and one semicircle with the dimensions given in the next figure. What is the area in square feet of the original teal fabric that will be visible in her flag?

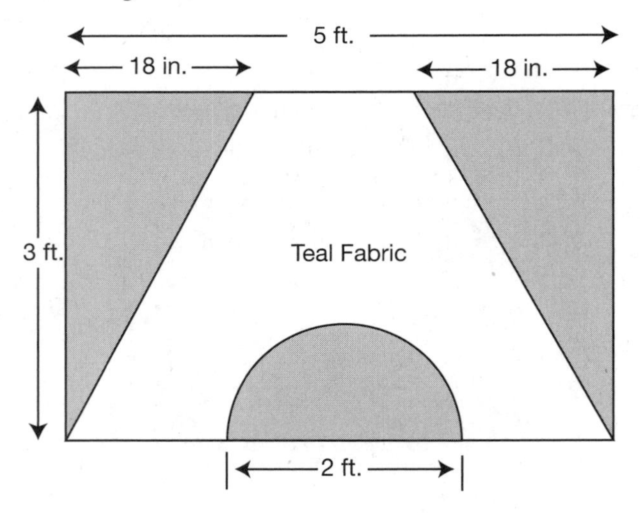

5. Jago is using tiles that are 3 inches long by 2 inches wide to cover a kitchen counter. If the counter is 6 feet long by 2.5 feet wide, how many tiles will he need to cover the counter completely? (*Hint:* Find the area of the counter *in inches*, and divide it by the area in square inches of each tile.)

Answers

Practice 1

1. Daphne wants Lightning to be able to walk a 250-foot circumference, so you must solve for the radius of a circle with a diameter of 250 feet.
   $C = \pi d$ gives
   $250 \approx 3.14(d)$
   $d \approx 79.6$ feet
   Since the radius is one-half of the diameter, Lightning should have a tether of 40 feet or longer.
2. This problem involves the area of a rectangle, since it is discussing the space within a rectangular flower box.
   $A = lw = 12.5w = 50$, so $w = 4$. The box should be 4 feet wide.
3. This problem involves the perimeter of a rectangle, which uses the formula $P = 2l + 2w$ or $P = 2(l + w)$. Since the perimeter is 16 feet, $16 = 2(l + w)$ or $8 = l + w$. Therefore, any pairing of length and width that sum to 8 would work as a correct answer. Some possibilities are l = 7 ft., $w = 1$ ft.; $l = 6$ ft., $w = 2$ ft.; and $l = 5$ ft., $w = 3$ ft. Even $l = 4$ ft., $w = 4$ ft. would work since although this is a square, by definition, a square is a rectangle.
4. This problem involves the circumference, since it is investigating how large a circle 16 feet of ribbon can go around.
   $C = 2\pi r$, so $16 = 2\pi r$
   $2\pi r \approx 2(3.14)r \approx 16$, so $r \approx \frac{16}{6.28} \approx 2.5$ feet. The diameter of the painting could be ≈ 5 feet.
5. This problem is giving information in yards and asking for information in feet. Begin by changing 7 yards into feet: $7 \times 3 = 21$ feet. Since the perimeter of a square is $P = 4s$, replace the perimeter with 21, and get $21 = 4s$, so $s = 5.25$ feet, or $5\frac{1}{4}$ feet. Since $\frac{1}{4}$ of a foot is $\frac{1}{4}(12) = 3$ inches, Ivanna's tablecloth can have a side length of 5 feet and 3 inches.

Practice 2

1. Since all given dimensions are in feet and both figures are rectangles, just use the formula $A = LW$ for each shape and find the difference:
   $A(\text{whole}) - A(\text{fire pit}) = (L)(W) - (l)(w)$
   $A(\text{whole}) - A(\text{fire pit}) = (14)(11) - (8)(3)$
   $A(\text{whole}) - A(\text{fire pit}) = (154) - (24)$
   $A(\text{whole}) - A(\text{fire pit}) = 130 \text{ ft.}^2$

2. First, find the area of the entire 38-by-20-foot footprint of the deck, since that is the larger space. $A = lw = 38$ ft. × 20 ft. = 760 square feet. Next, find the area of the pool by breaking it into the semi-circular entrance to the pool plus the rectangular section of the pool. The semi-circular area is found by halving the area of a circle with a radius of 6: $\frac{A}{2} = \frac{\pi r^2}{2} = \frac{\pi 6^2}{2} = \frac{36\pi}{2} \approx 56.5$ square feet.
The rectangular section of the pool will be $A = lw = 24$ ft. × 12 ft. = 288 square feet.
So the total area of the pool is 288 ft.^{2} + 56.5 ft.^{2} = 344.5 ft.^{2}.
The area of the footprint minus the area of the pool gives the area of the exposed deck: 760 ft.^{2} – 344.5 ft.^{2} = 415.5 ft.^{2}. Since each can of sealant covers 100 square feet, Matt and Jen will need to buy five cans to protect their deck.
3. The area of the entire counter will be 5 × 2 = 10 ft.^{2}. The radius of the sinks is $\frac{1.25}{2} = 0.625$ feet. The combined area of the two sinks will be $2(A) = 2(\pi r^2) = 2(\pi)(0.6252) \approx 2.5$ ft.^{2}. The difference of these two areas is 7.5 ft.^{2}.
4. The area of the teal fabric that Tivoli begins with is 5 × 3 = 15 ft.^{2}. The base of each of the triangles is 3 feet and their heights are each 1.5 feet ($\frac{18 \text{ inches}}{12 \text{ inches}}$ = 1.5 feet). Therefore, the area of these two triangles is $2(A) = 2(\frac{1}{2})$(base)(height) = (1)(3)(1.5) = 4.5 ft.^{2}. The area of the semi-circle with a radius of 1 is $(\frac{1}{2})(A) = (\frac{1}{2})(\pi r^2) = (\frac{1}{2})(\pi 1^2) = 1.57$ ft.^{2}. Subtracting the circular and triangular material from the teal gives 15 – 4.5 – 1.57 = 8.93 ft.^{2}.
5. The area of the counter is 6 × 2.5 = 15 ft.^{2}. Every square foot has an area of 144 square inches, since a square foot is made up of 12 inches by 12 inches. Therefore, for each of the 15 square feet of counter, there are 144 square inches, so we can calculate that there are 15 × 144 or 2,160 square inches of area within the counter. Each tile is 2 × 3 = 6 in.^{2}. Last, to see how many 6-square-inch tiles are needed to cover 2,160 square inches, divide the counter area by the tile area: $\frac{2{,}160}{6} = 360$ tiles.

17

Prisms, Nets, and Surface Area

Since the mathematicians have invaded the theory of relativity, I do not understand it myself anymore.
—Albert Einstein

In this lesson you will learn how to identify 3-dimensional objects, how to calculate their surface area, and how to represent 3-dimensional figures with 2-dimensional nets. You will also gain an understanding of how nets are useful models for deriving the surface area formulas for 3-dimensional figures.

STANDARD PREVIEW

You already know that area is the space within a 2-dimensional space. This is useful for calculating how much sod to buy to make a grassy patch in your yard, or how much fabric will be needed to create a canopy over your patio, or how much sealant you need to order to waterproof a deck. This *area* concept sure finds its way around town! **Surface area** is the combined area of 3-dimensional spaces. Surface area will tell you how much paint is needed to color an entire box or the paper needed to wrap a gift. ***Standard 6.G.A.4 wants students to understand the relationship between 3D figures and 2D composite shapes called "nets."*** When folded along certain edges, these 2D nets become 3D objects, so nets are a great way to model and understand surface area. After you gain a context for understanding surface area with nets, ***you will give Standard 7.G.B.6 some more attention by applying your surface area problem-solving skills to some real-world applications***.

Identifying Prisms

Although the word "prism" is something that you may not have heard until today, you have been looking at prisms all day. A **prism** is a three-dimensional figure that has two congruent polygon ends and parallelograms as the remaining sides. A pizza box is a prism that you are probably familiar with. The sides of prisms can all be called **faces**. The congruent ends are more specifically referred to as **bases**. The remaining sides, which are parallelograms, are the **lateral faces**. The *shape of the polygon base* is what determines the name of a prism as shown in the next figure.

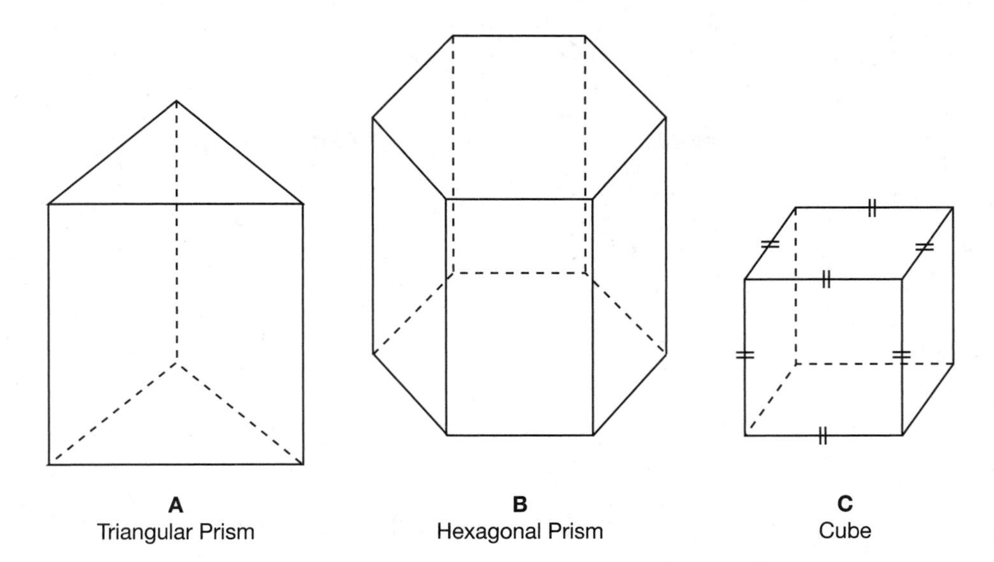

Prism A has two parallel triangular bases, so it is a **triangular prism**. Prism B has parallel hexagons as bases, so it is a **hexagonal prism**. Prism C has a square base and the lateral faces are also all squares. Instead of being called a "square prism," this figure is called a **cube**. Notice that in all three prisms, the lateral faces are rectangles. **Edges** in prisms are the segments formed where two faces come together. Edges that are parallel always have the same length. **Vertices** are the corners where three edges come together.

TIP: A *prism* is a three-dimensional figure with two congruent and parallel polygon bases and parallelogram sides. Prisms are named after their polygon bases.

Practice 1

1. How many faces in total does a rectangular prism have?

2. How many faces in total does an octagonal prism have?

3. How many edges does a triangular prism have?

4. How many vertices does a triangular prism have?

For questions 5 through 8, name the following prisms.

5.

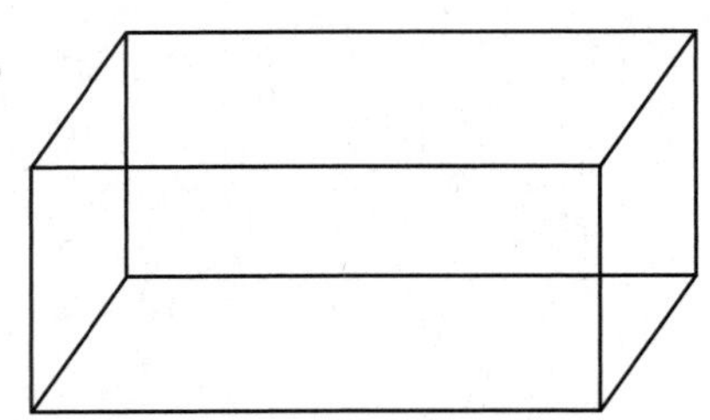

6.

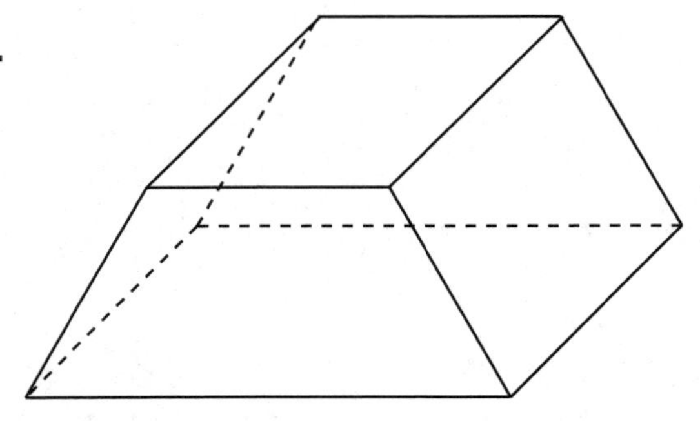

7.

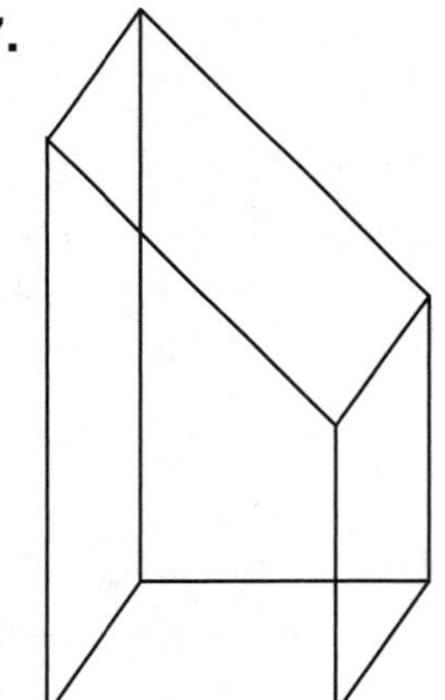

8.

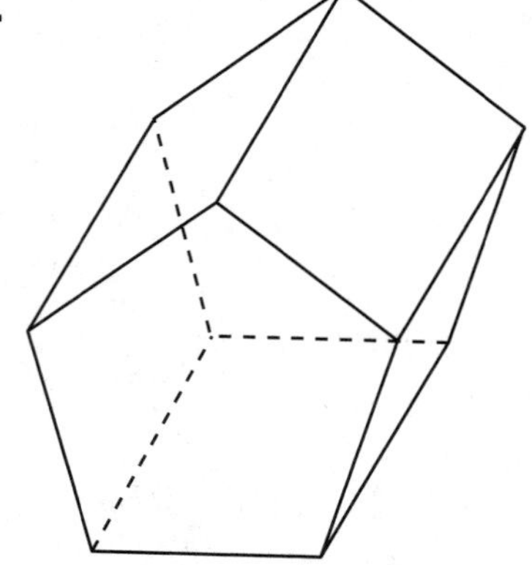

What Is Surface Area?

In previous lessons you learned that *perimeter* and *circumference* are one-dimensional measurements of the distance around a two-dimensional figure, such as a square or rectangle. You also learned that *area* is a two-dimensional measurement of the space *inside* a two-dimensional figure, such as a square or parallelogram. **Surface area** is a third type of measurement that you apply to three-dimensional figures, such as prisms and cylinders. The surface area of a 3D figure is the combined area of all the figure's faces. For example, if you needed to wrap a birthday gift in a box, you would need to cover the top, bottom, front, back, and two sides of the box. The combined area of these six sides is called the *surface area* of the box.

TIP: The *surface area* of a three-dimensional figure is the combined area of all the figure's outward-facing surfaces.

Surface Area of Rectangular Prisms

A rectangular prism has six faces. Its three dimensions are referred to as length, width, and height. In the next figure, you can see that the length of the front face is 9 and its height is 4. The width of the prism is 3 units.

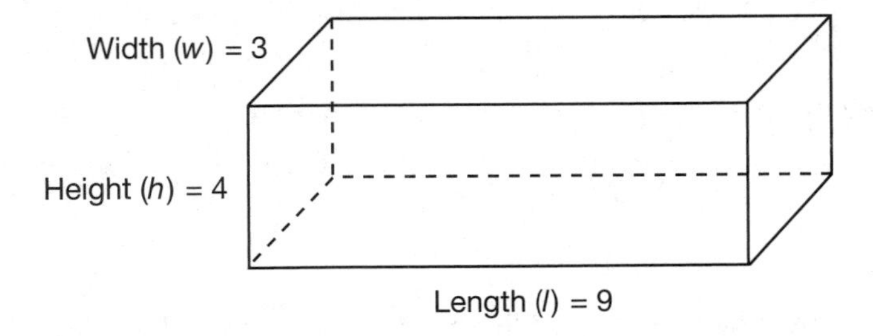

The six faces of rectangular prisms include three pairs of opposite congruent faces that have the same dimensions. In this figure, the front and back faces have dimensions of 9 units by 4 units; the left and right faces have dimensions of 4 units by 3 units; and the top and bottom faces have dimensions of 9 units by 3 units. To find the surface area of this prism, we could set up a table as in the next figure.

Faces	Formula for Area	Area of One Face
Front and back	$l \times h$	$9 \times 4 = 36$
Left and right	$h \times w$	$4 \times 3 = 12$
Top and bottom	$l \times w$	$9 \times 3 = 27$

Subtotal: 36 + 12 + 27 = 75 units
Multiplied by 2 since there are 2 faces with each dimension: (75)(2) = 150 units = Surface Area

The area of each pair of faces is the product of two different dimensions. The three distinct products have a sum of 75 units. You then double this subtotal, since there are two identical faces with each given area. The result is a total surface area of 150 units^2. This would be a lot of work to go through every time you wanted to find the surface area (SA) of a rectangular prism.

Therefore, consider the formula, $SA = 2lh + 2hw + 2lw$. In this formula, the three different products, lh, hw, and lw, represent the three areas of one of the distinct front, top, and left faces. Each of these areas is multiplied by 2, and this sum is the total surface area. This formula can be simplified by factoring a 2 out of the terms and writing it as $SA = 2(lh + hw + lw)$.

TIP: The surface area of a rectangular prism is: $SA = 2(lh + hw + lw)$.

Representing 3D Prisms as 2D Nets

STANDARD ALERT!

Are you feeling a little dizzy after seeing that area formula? Let's take a step back and look at how the rectangular prism in the past section can be deconstructed to create a 2D net. From there, we will reinvestigate the surface area formula! This is the heart and soul of Standard 6.4.G.A.

It might surprise you to learn that all 3-dimensional shapes can be broken into 2-dimensional **nets**. A net is a 2-dimensional figure that can be folded to make a 3-dimensional figure. Similar to how you can cut a cardboard box at a few seams and lay it out flat on the floor, you can unfold the $9 \times 4 \times 3$ prism we used earlier to look like this 2-dimensional net:

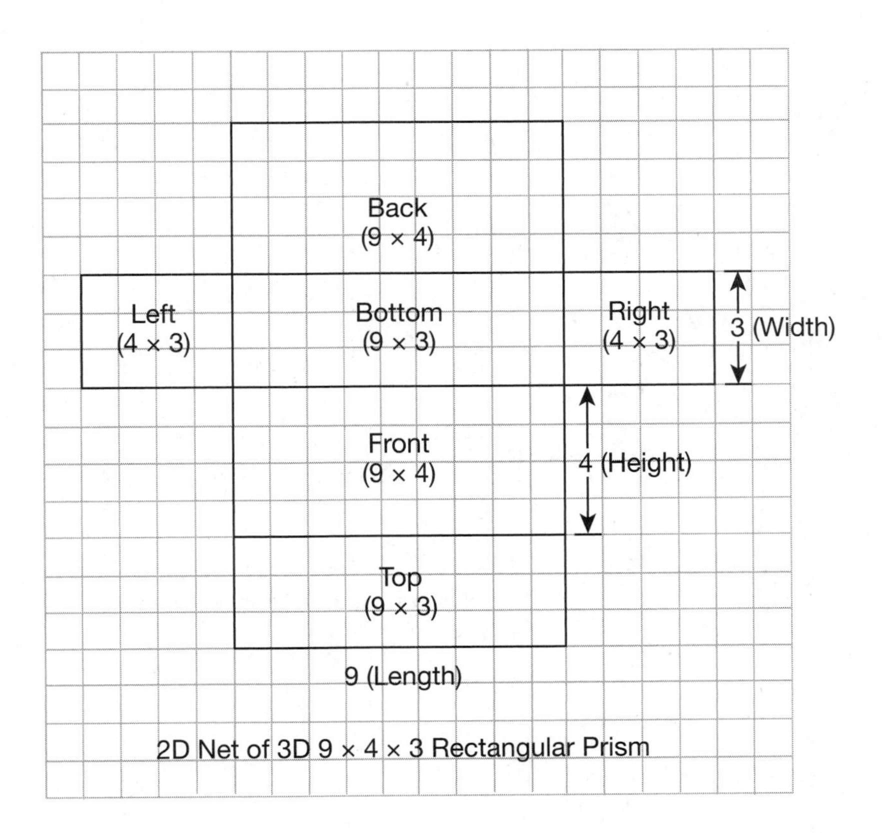

2D Net of 3D 9 × 4 × 3 Rectangular Prism

Notice how the different sides of the net correspond to the dimensions of the 3D prism and also to the faces that are listed in the chart before it. Looking at this net should help illustrate why the surface area formula for rectangular prisms is two times the sum of all the unique areas: each side has a matching partner with the exact same dimensions and area. This means that $SA = 2(lh + hw + lw)$.

TIP: A net is a 2-dimensional figure that can be folded to make a 3-dimensional figure.

Triangles and circles can also be used to create nets for an entire host of 3-dimensional objects. You can find the surface area for any prism by breaking it down into its net and calculating the sum of the areas of the individual faces. The following is a 2-dimensional triangular pyramid net and its corresponding 3-dimensional rendering:

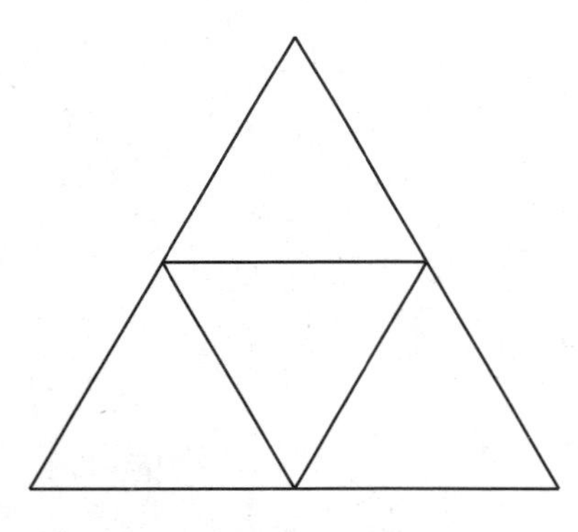

2D Net of Triangular Pyramid

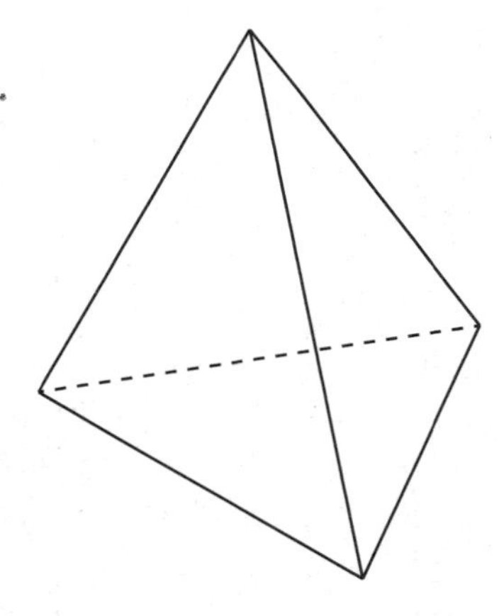

3D Rendering of Triangular Pyramid

Before you learn formulas for the surface area of a few more shapes, see how well your mind can translate between two and three dimensions in the following exercises:

Practice 2

Use the following 2D nets to create 3-dimensional shapes. If you have trouble drawing in 3D, be as specific as possible in describing the shape that the net will create.

1.

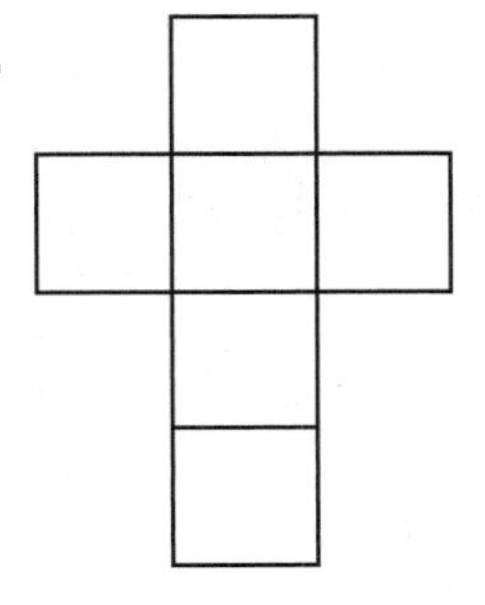

2.

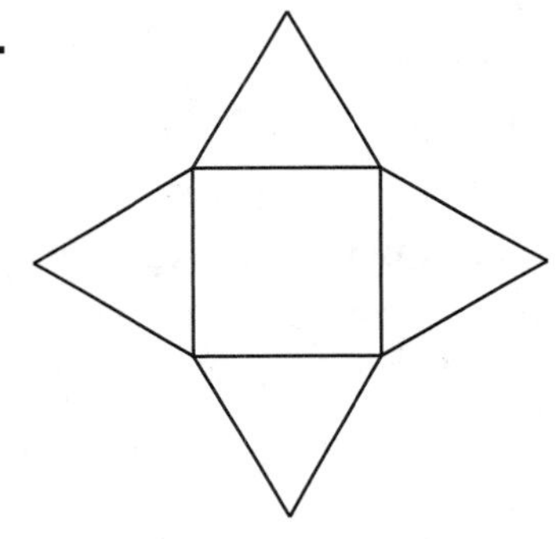

3.

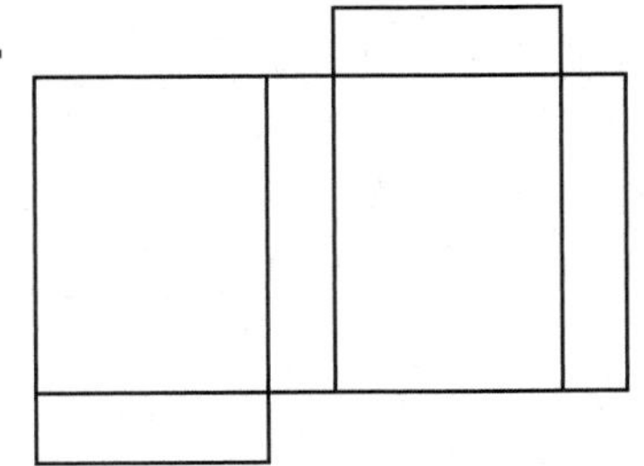

4.

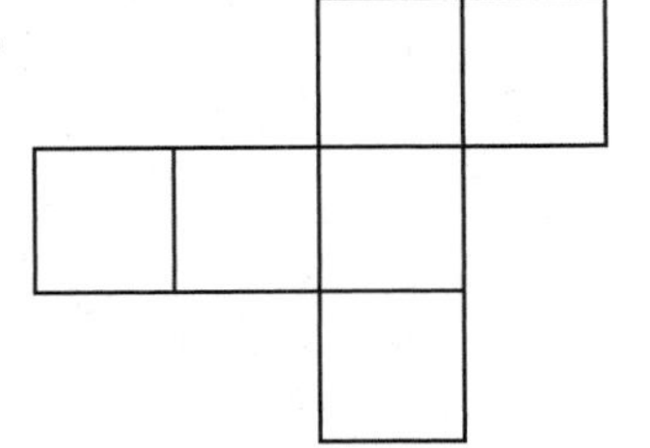

Create 2-dimensional nets from the following 3-dimensional illustrations. (There is more than one way to create a 2D net for the same figure, so if you're feeling ambitious, create as many nets as possible for each illustration.)

5.

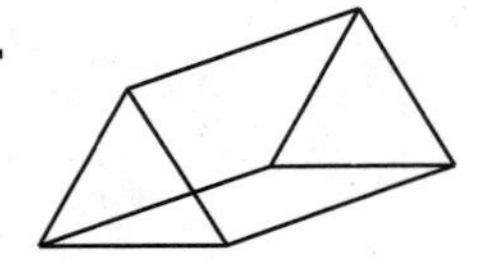

6.

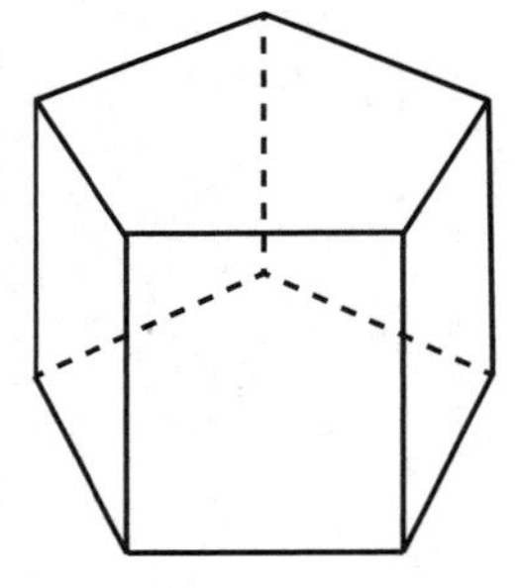

7. This is a tough one! Hint: Think about cutting up the length of the cylinder and then cutting almost all the way around the circumference of the top and bottom circles.

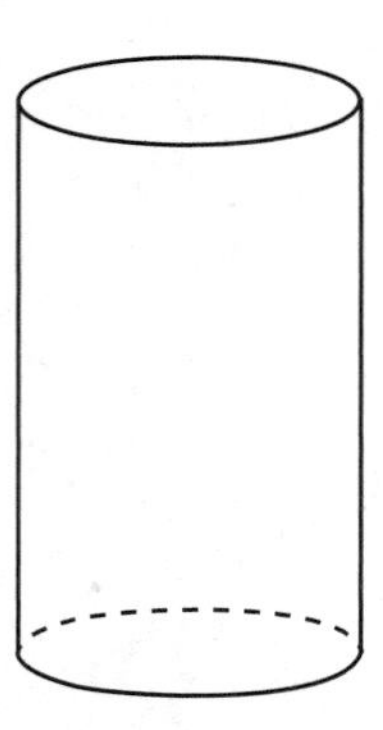

8. Note: Try to create as many unique nets for this cube as possible. Then make a hypothesis for the area formula of this 2-dimensional figure. (If correct, your answer will match the surface area formula for cubes, which is covered in the next section.)

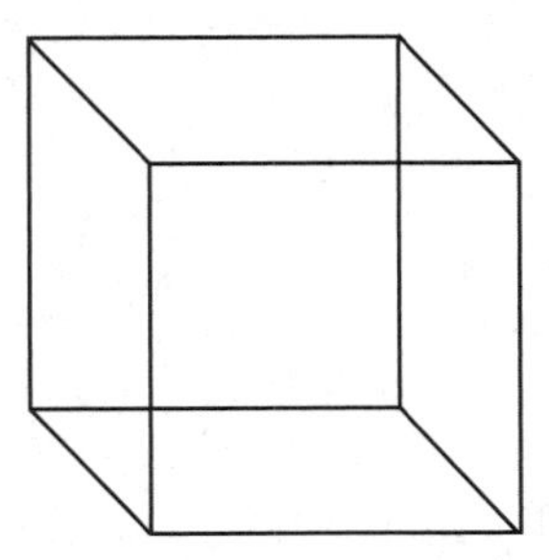

Surface Area of Cubes

The surface area of a cube is an easier formula to remember.

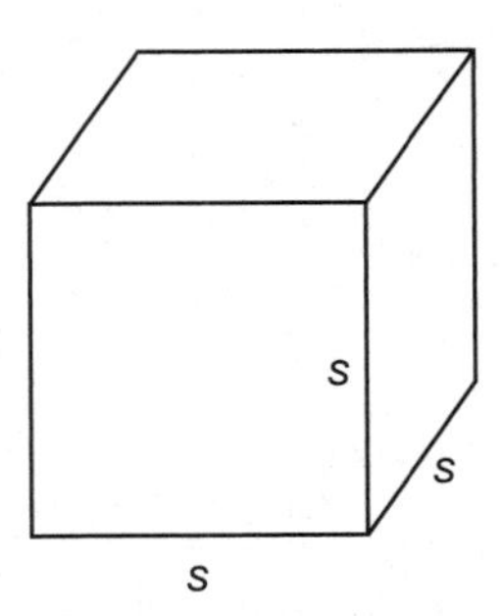

In this figure you can see that all of the edges of the cube are labeled s. In a cube, each of the six faces has an area of s^2, since the length, width, and height are all s units long. Since there are six faces, each with an area of s^2, the formula for the surface area for a cube is $SA = 6s^2$. When you're using this formula, remember to follow the proper order of operations by squaring the side length *first*, before multiplying by 6.

TIP: The surface area of a cube is: $SA = 6s^2$.

STANDARD ALERT!

Now that you have a few surface area formulas under your belt, test your ability to master Standard 7.G.B.6 with these real-world problems.

Practice 3

1. Thy is using solid 18kt gold leaf to cover this box for a dollhouse exhibit at an art museum. If Thy charges $2.80 per square inch of 18kt-gold-leaf coverage, how much will he charge the museum to fully cover this box?

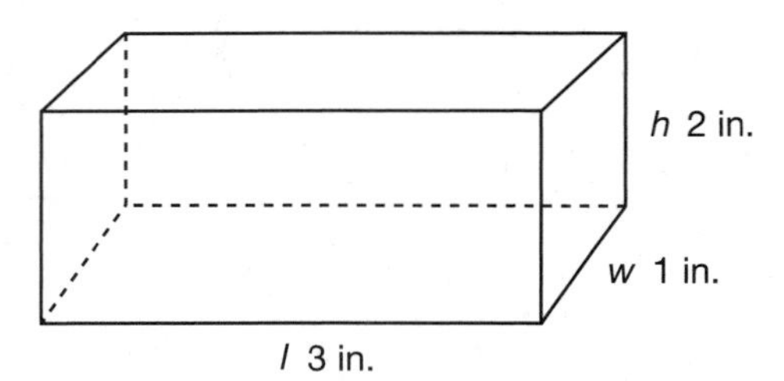

2. This 4-foot cube is going to be a life-sized die made out of foam to be used in a play. The die will be covered in custom-ordered chartreuse fabric, and Karl needs to know how many square feet he must order. Give Karl a hand!

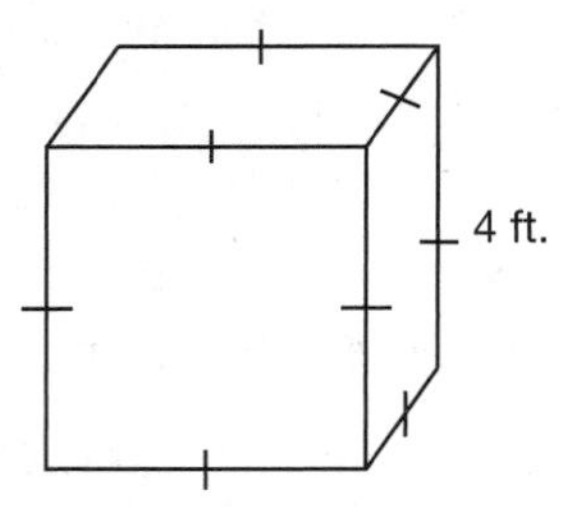

3. The following is a platform for a summer Speaker in the Park series. How many feet of plywood will it take to make this box?

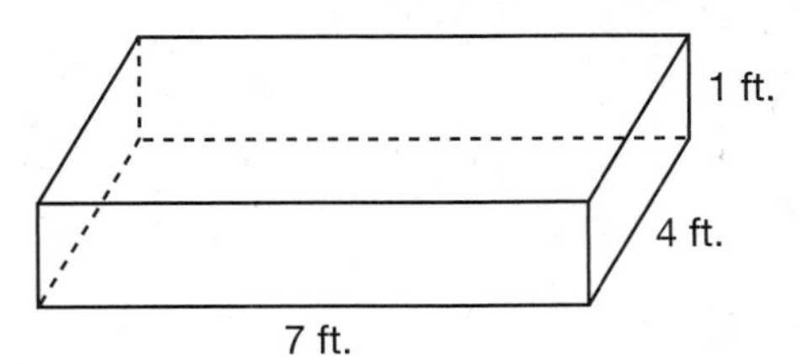

4. Fernando needs to treat all of the inside faces of his wooden storage shed with termite insecticide. If one bottle covers 40 square feet, how many bottles will he need to purchase?

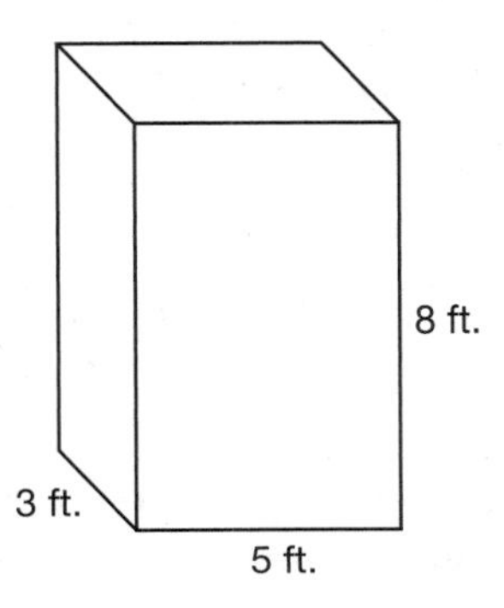

5. Looking at the following trapezoidal prism, you might recognize that you cannot use the rectangular or cube surface area prisms for this problem. Break this trapezoid down into a 2-dimensional net, label each face with dimensions, and calculate the surface area of the prism by adding all of the individual areas. (Hint: Your net should have six faces, but it will not have three congruent pairs of faces.)

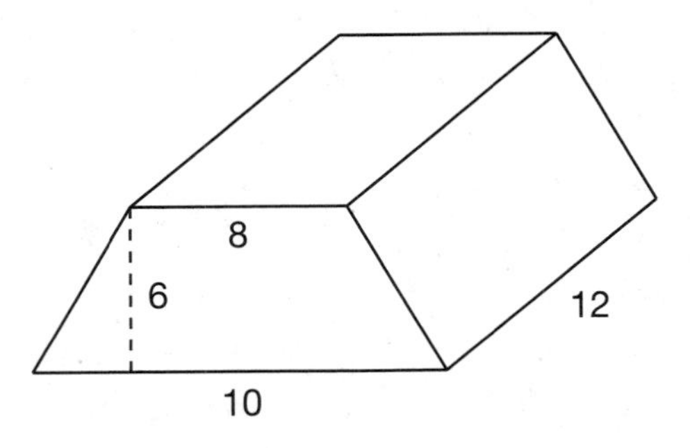

6. The surface area of a rectangular prism is 288 in.^{2}. If the length of the front face is 12 inches and the height of the front face is 6 inches, what is the width of the rectangular prism?

7. The surface area of a cube is 486 ft.^{2}. Find the length of each edge.

Surface Area of Cylinders

In the second practice set, you were given the challenging task of creating a net from a cylinder. Now you will get to see how well you did! We are going to look at a cylinder net in order to understand the formula for the surface area of cylinders. Follow along!

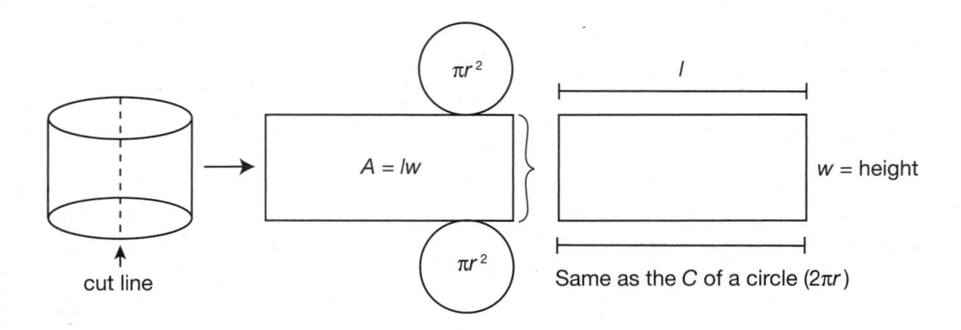

In the leftmost illustration, you can see that there are two parallel circular bases on the top and bottom of the cylinder. In the center illustration, the cardboard cylinder has been cut and unrolled, so that the cylindrical part is now a flattened rectangle with an area of lw. The areas of the circular top and bottom can each be found by using $A = \pi r^2$. In the rightmost illustration, it is shown that the *length* of the flattened rectangle is actually equal to the circumference of the circular bases. The *width* of the rectangle section of the cylinder is actually the height of the cylinder. Therefore, the formula (*circumference*)(*height*) is used to identify the area of the rectangular section of the cylinder.

In summary, the surface area of a cylinder is equal to: (area of two circular bases) + (area of the rectangular portion). The area of the rectangular portion is found by multiplying the *circumference* of the circular base by the *height* of the cylinder.

TIP: The surface area of a cylinder is:

$SA = 2\pi r^2 + (2\pi r)(h)$

Practice 4

Find the surface area of each of the cylinders that are drawn or described next. Use 3.14 for π and round your answers to the nearest tenth.

1.

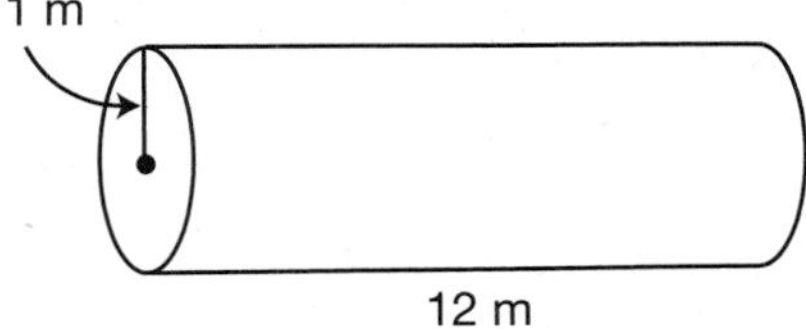

2.

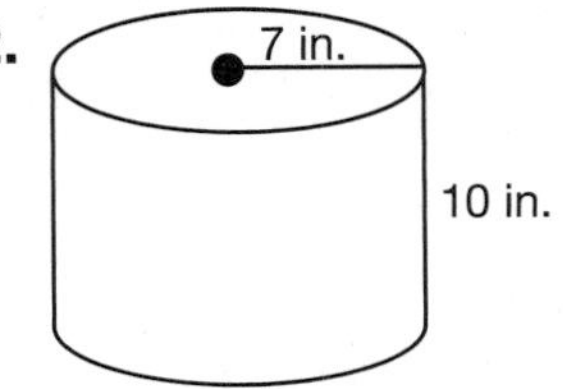

3. Cylinder: $r = 6$ ft., $h = 16$ ft.

4. Cylinder: $d = 18$ cm, $h = 3$ cm

Answers

Practice 1

1. A rectangular prism has 6 faces.
2. An octagonal prism has 10 faces.
3. A triangular prism has 9 edges.
4. A triangular prism has 6 vertices.
5. Rectangular prism
6. Trapezoidal prism
7. Trapezoidal prism (the bases are the front-facing and back-facing trapezoids)
8. Pentagonal prism

Practice 2

1.

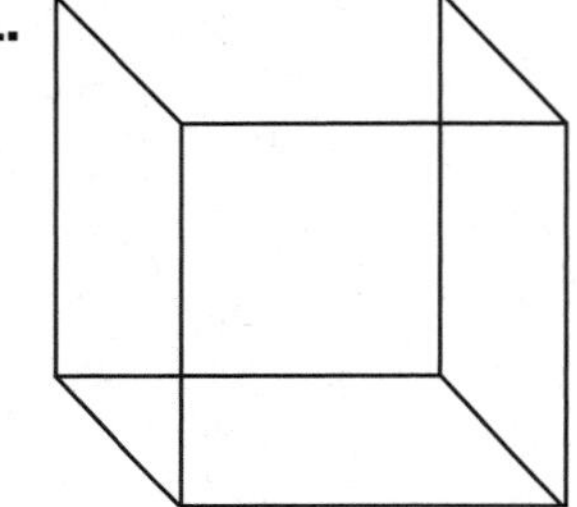

2.

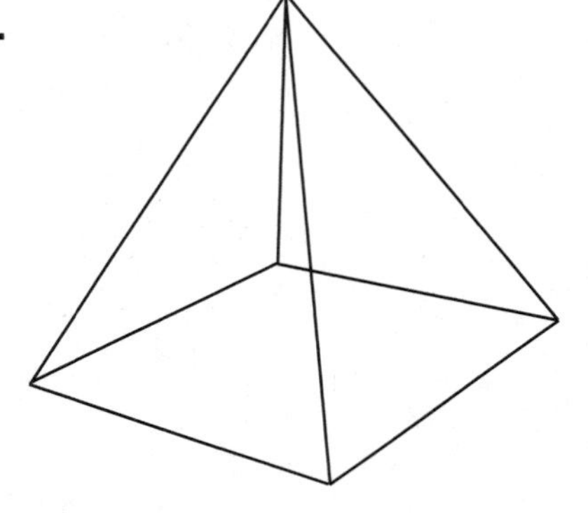

3.

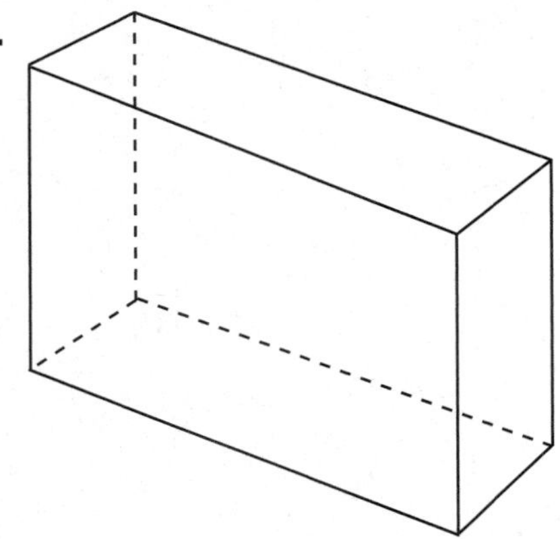

4.

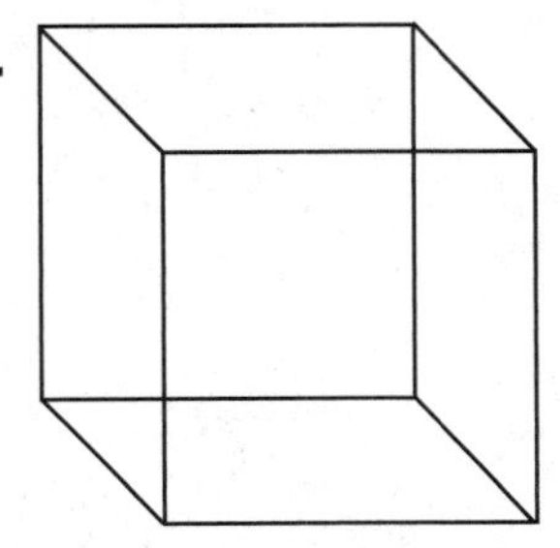

5.

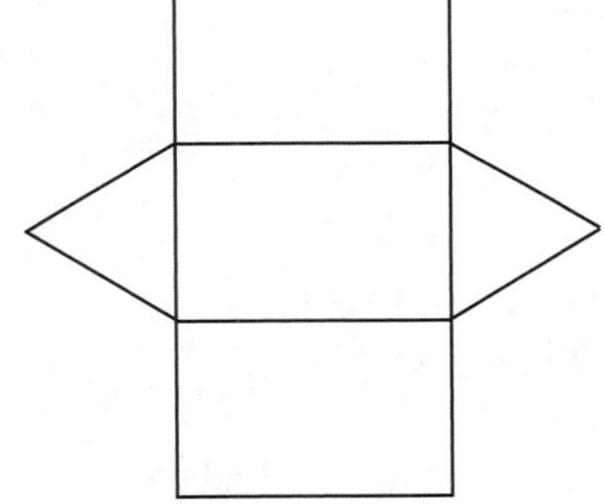

6.

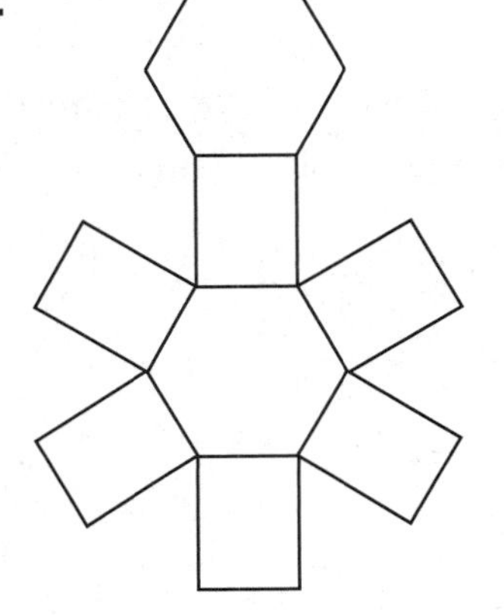

7.

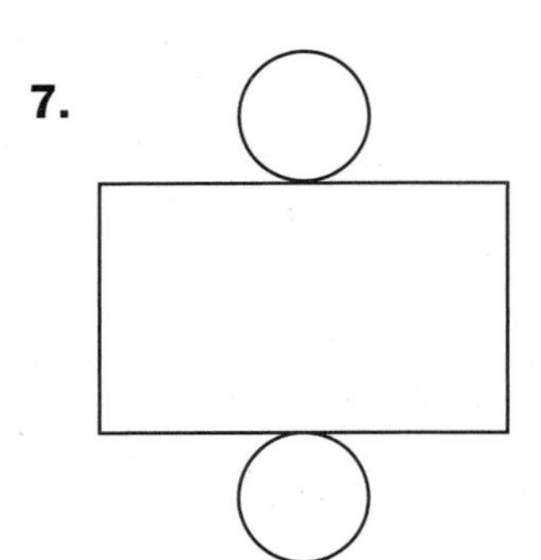

8. Here are six different cube nets.

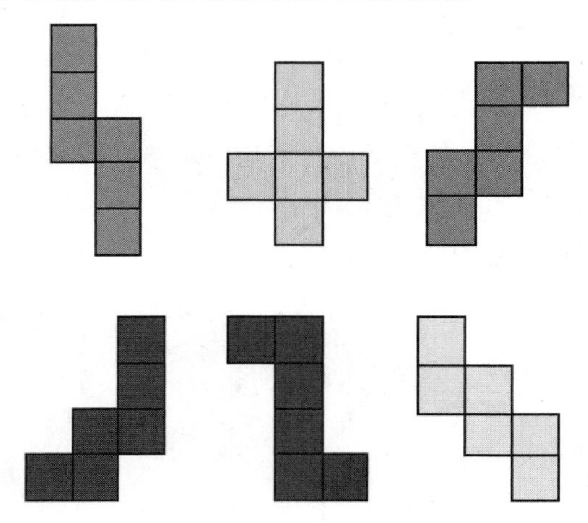

Practice 3

1. $SA = 2(lh + hw + lw) = 2(3 \times 2 + 2 \times 1 + 3 \times 1) = 2(11) = 22$ in.^{2} Multiply 22 square inches by the price of \$2.80 per square inch to get the total cost: \$2.80 × 22 = \$61.60.
2. $SA = 6s^2 = 6(4)(4) = 96$ ft.^{2} Karl needs to buy 96 ft.^{2} of fabric.
3. $SA = 2(lh + hw + lw) = 2(7 \times 1 + 4 \times 1 + 7 \times 4) = 2(39) = 78$ ft.^{2} It will take 78 ft.^{2} of plywood to make the box.
4. $SA = 2(lh + hw + lw) = 2(5 \times 8 + 5 \times 3 + 3 \times 8) = 2(79) = 158$ ft.^{2} Since the surface area of the interior of the shed is 158 ft.^{2}, Fernando will need four bottles of termite insecticide.

5. The following illustration is a net of the given trapezoid.

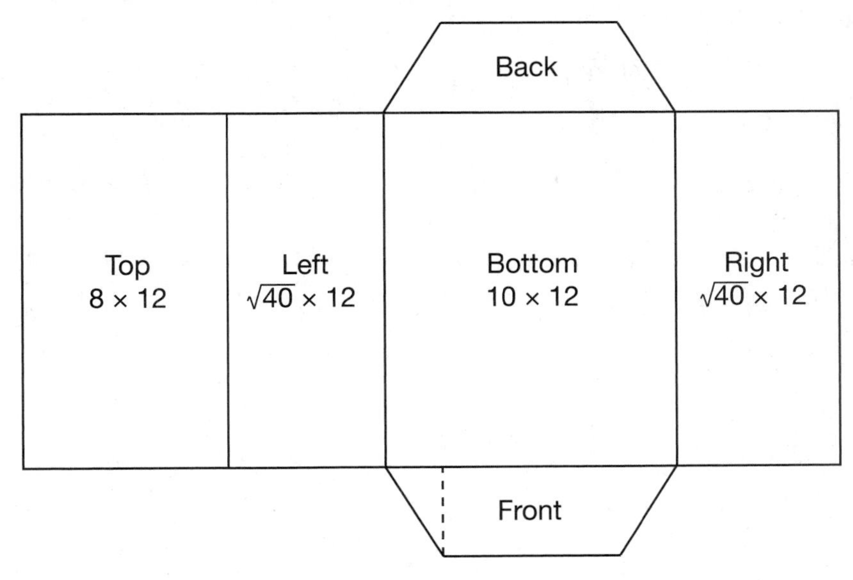

Front and back trapezoids: $6(\frac{8+10}{2}) = 6(9) = 54$, 108 units^2 for both faces.
Top: $8(12) = 96$ units^2
Bottom: $10(12) = 120$ units^2
Right and left sides: $12(\sqrt{40}) \approx 76$, 152 units^2 for both sides
Total surface area = $108 + 96 + 120 + 152 = 476$ units^2

6. $SA = 2(lh + hw + lw)$
$288 = 2(12 \times 6 + 6w + 12w)$
$288 = 2(72 + 18w)$
$288 = 144 + 36w$
$144 = 36w$
$w = 4$ inches

7. $SA = 6s^2$
$486 = 6s^2$
$81 = s^2$
$s = 9$ feet

Practice 4

1. $SA = 2\pi r^2 + (2\pi r)(h) = 2(3.14)1^2 + (2)(3.14)(1)(12) = 6.28 + 6.28(12) = 81.6\text{ m}^2$
2. $SA = 2\pi r^2 + (2\pi r)(h) = 2(3.14)(7^2) + (2)(3.14)(7)(10) = 307.72 + 439.6 = 747.3\text{ in.}^2$
3. $SA = 2\pi r^2 + (2\pi r)(h) = 2(3.14)(6^2) + (2)(3.14)(6)(16) = 226.08 + 602.88 = 829\text{ ft.}^2$
4. $SA = 2\pi r^2 + (2\pi r)(h) = 2(3.14)(9^2) + (2)(3.14)(9)(3) = 508.68 + 169.56 = 678.2\text{ cm}^2$

18

Turn Up the Volume

Creative mathematicians now, as in the past, are inspired by the art of mathematics rather than by any prospect of ultimate usefulness.
—Eric Temple Bell

In this lesson you will learn what volume is and learn the volume formulas for a variety of 3D figures. You will also get some experience applying volume formulas to real-world problems.

STANDARD PREVIEW

Whether you're looking to fill a pool with water, a shipping container with merchandise, or an industrial walk-in freezer with icy air, volume is going to be a concept that will present itself to you from time to time in the workplace and in your home. That's why the CCSS have included volume in three separate standards for grades 6–8. ***From applying the volume formula to fractional edge lengths, to real-world problem-solving skills with right prisms, to applying volume formulas to cones, cylinders, and spheres, standards 6.G.A.2, 7.G.B.6, and 8.G.B.9 are going to keep us busy in this lesson, so let's get started!***

What Is Volume?

"How much volume would you like?" This really should be the question that you are asked when you're ordering a glass of juice in a restaurant. It is not really the size of the glass that you are interested in, but the *amount* of juice that goes into that glass. **Volume** refers to the amount of space occupied inside a three-dimensional object. Similarly to how area is measured in squares, volume is measured in cubes. Think about neatly arranging one-inch cubes, side-by-side, in a small shoebox. The total number of cubes that would fill the box would be the volume of the box in cubic inches.

TIP: *Volume* is the amount of space enclosed by a three-dimensional object.

Practice 1

For questions 1 through 6, state whether the concept of perimeter, area, surface area, or volume would be used to investigate the real-world situation that Dani is in.

1. Dani needs to paint the outside of a box black so that it can be used as a prop on stage for the school play.

2. Dani needs to buy the right-sized air-conditioner to cool down her studio apartment.

3. Dani was late for soccer practice and had to run 5 full laps around the field.

4. Dani wonders how much it will cost to fill the pool with water if each gallon from her hose costs $0.07.

5. Dani is buying fabric to make a tablecloth.

6. Dani wants to put a tile border around the window in her living room.

Finding the Volume of Rectangular Prisms

To understand how volume is calculated, read on while looking at the rectangular prism in the next figure.

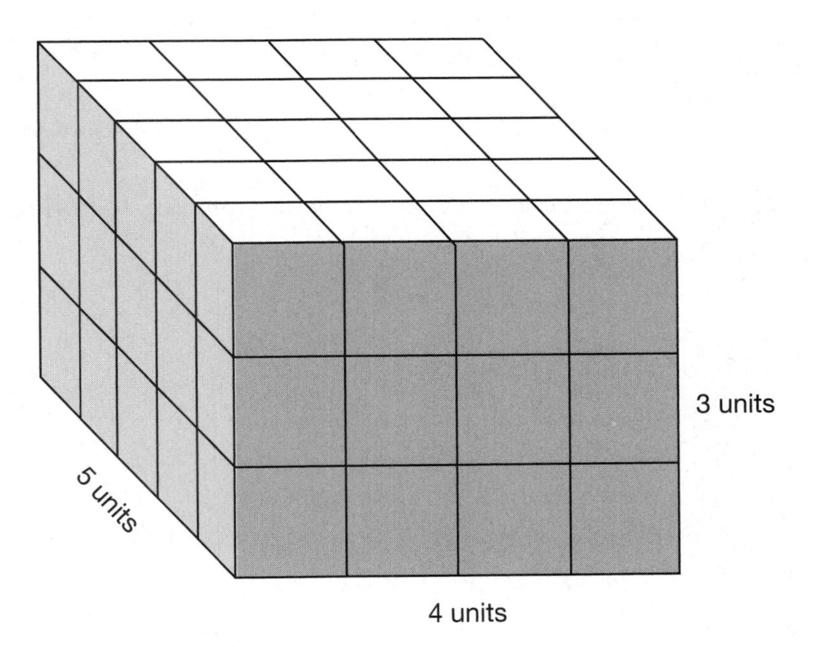

The face of the rectangular prism, which is shaded the darkest, has a length of 4 units and a height of 3 units. Therefore the area of the front face is $4 \times 3 = 12$ square units. The width of this rectangular prism is 5 units. Notice that for each of the 12 cubes in the front face there is a row of 5 cubes extending toward the back of the prism. Knowing this, you can find the volume by multiplying the number of squares in the front face by the number of cubes behind it. Therefore, the number of cubes that make up this prism is $12 \times 5 = 60$, and its volume is 60 cubic units, which is written 60 units^3. We just found the **volume** of this **rectangular prism** by multiplying the (**length**)(**width**)(**height**). (Volume is always expressed with an exponent of 3 after the units, to represent the three different dimensions it measures.)

TIP: Volume of a rectangular prism = (*length*)(*width*)(*height*)

Finding the Volume of Cubes

In cubes, the length, width, and height are all the same length, *s*. Therefore, the formula for finding the **volume of a cube** is (*side*)(*side*)(*side*) or **(*side*)^3**.

TIP: Volume of a cube = s^3.

Refer to the next figure while you look through the following two examples.

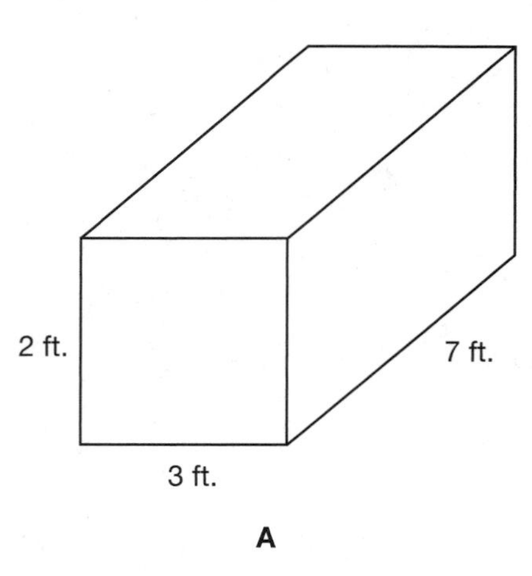

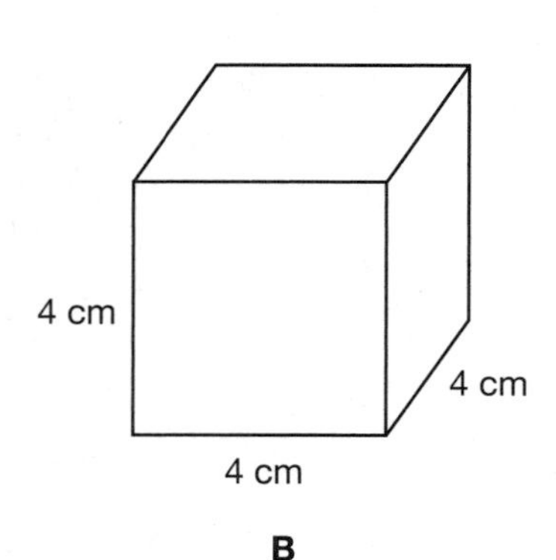

Example 1: Find the volume of the rectangular prism in Figure A.
Solution: $V = (length)(width)(height) = (2)(3)(7) = 42$ ft.^{3}.

Example 2: Find the volume of the cube in Figure B.
Solution: $V = s^3 = 4^3 = 64$ cm^3.

Practice 2

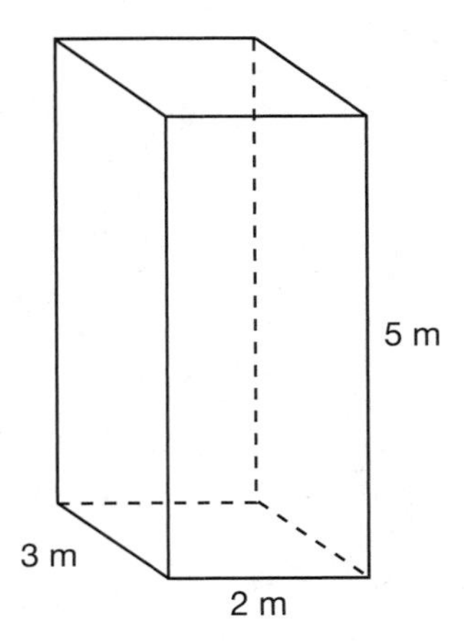

1. The president of the Centerport Aquarium wants to install a custom-made tank in the lobby to showcase some of the more exotic species of fish that the aquarium has. The tank she is considering is a rectangular prism that is 3 meters long by 2 meters wide by 5 meters tall. How many cubic meters of water will this tank hold if filled to the very top?

2. The firm designing the tank for the Centerport Aquarium suggests that if the length and width remain the same but the height is increased by 2 meters, the tank will span two floors and be more dramatic. Make a sketch that represents what the additional 2 meters of height (with 3 meters of length and 2 meters of width) would look like. Parcel it into square meters in order to determine how many more cubic meters of water the taller tank would contain. Check your answer using the formula $V = l \times w \times h$ with the new dimensions, and compare this to your original volume from question 1.

3. Shirley is an influential member of the Centerport Aquarium Board of Trustees, and she doesn't like the idea of such a tall tank. She'd like to consider a tank that is a cube with 3-meter side lengths. She argues that her tank will be much more stable and have almost the same volume as the proposed tank. How many cubic meters of water would Shirley's tank hold, and how does it compare to the original tank proposal?

4. You are now going to walk through an investigation of how volume changes as side length changes. If the side lengths of a cube are all doubled, how do you think the volume of that new cube will compare to the volume of the original cube?

5. Using the cube described in question 3, double the side length and find the new volume.

6. Check your answer to question 4. Were you correct? If not, correct your answer and explain why the volume of a cube does not simply double when its side length doubles.

Finding the Volume of Cylinders and Triangular Prisms

STANDARD ALERT!

A goal of the CCSS is that students improve their critical thinking and problem-solving skills. This means students should be able to use previous knowledge to create new models and formulas to apply to new situations. After all, life is just one long series of new situations arising! Standard 7.G.B.6 does not mention formulas when it stresses that students will solve real-world problems involving volume for 3-dimensional objects composed of triangles, quadrilaterals, polygons, cubes, and right prisms. Try to grasp not just the formulas below but the underlying logic too. Do so until you feel comfortable applying it to new composite figures that you have never encountered before.

Given *any* right prism, if you can calculate the area of one of its congruent bases, then multiplying that area by the height of the prism will give you the volume. To understand this, look at the following figure.

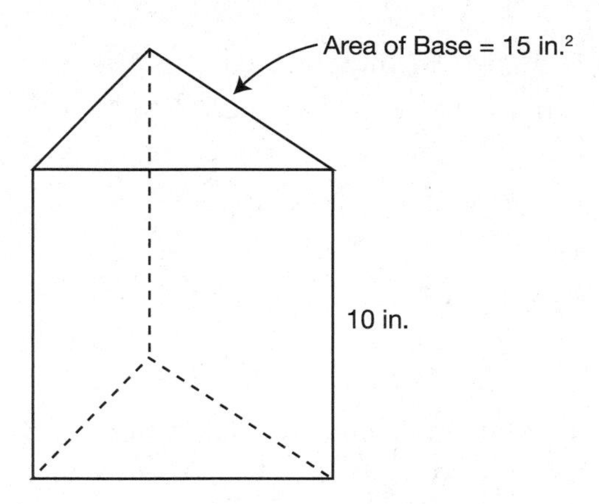

If you know that the area of the triangular base is 15 in.^{2}, then you would multiply that by 10 inches to get the total volume of 150 in.^{3}.

TIP: **Volume of any prism = (*area of base*)(*height*)**

This formula is useful because it can be applied to cylinders and other various prisms. You can even use it for rectangular prisms or cubes if you forget those formulas.

Refer to the next figure while you look through the following two examples.

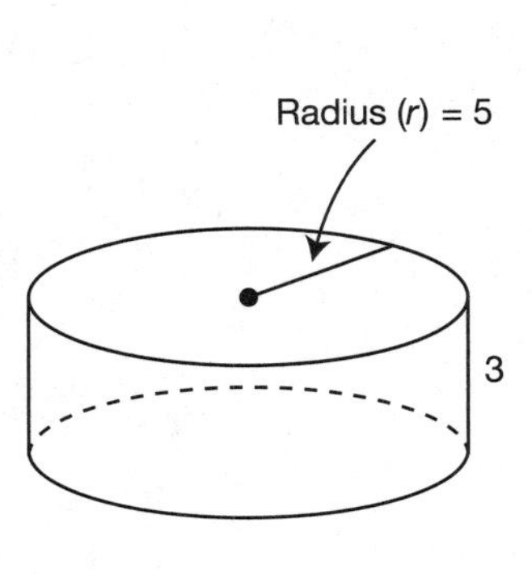

A

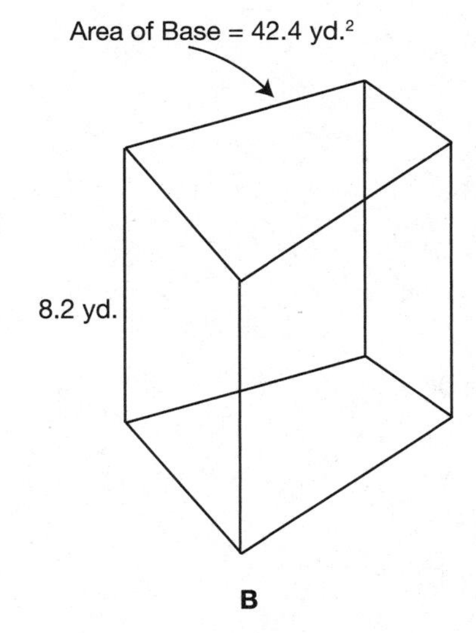

B

Example 1: Find the volume of the cylinder in Figure A.
Solution: $V = (\textit{area of base})(\textit{height})$. In this case, the area of the base is πr^2 since the base is a circle. $V = (\pi r^2)(h) = (\pi 5^2)(3) = 235.5$ m^3.

Example 2: Figure B contains a base that is an irregularly shaped prism with an area of 42.4 yd.^{2} and a height of 8.2 yards. Find its volume.
Solution: $V = (\textit{area of base})(\textit{height}) = (42.4)(8.2) = 347.68$ yd.^{3}.

Practice 3

1. Alexis uses the following cylindrical containers to grow bean sprouts for local restaurants. She just bought some new seeds, whose packaging suggests using one quarter of a cup of seeds for every 200 cubic inches of growing space. Find the volume of this cylindrical container, and determine how many cups of these new seeds Alexis should use in each container.

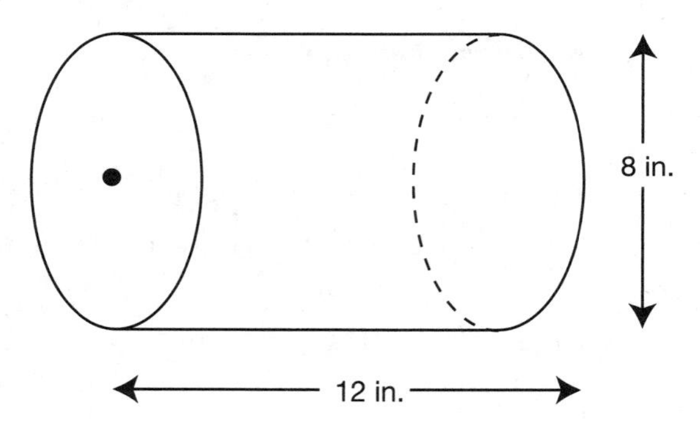

2. The following illustration represents a proposed ramp to be added to a skate park in Venice Beach. Determine how many cubic feet of concrete would be needed to create a ramp of this size. (*Hint:* Use the trapezoid as your base. If you forgot the formula for a trapezoid's area, this is a great opportunity to freshen up on that.)

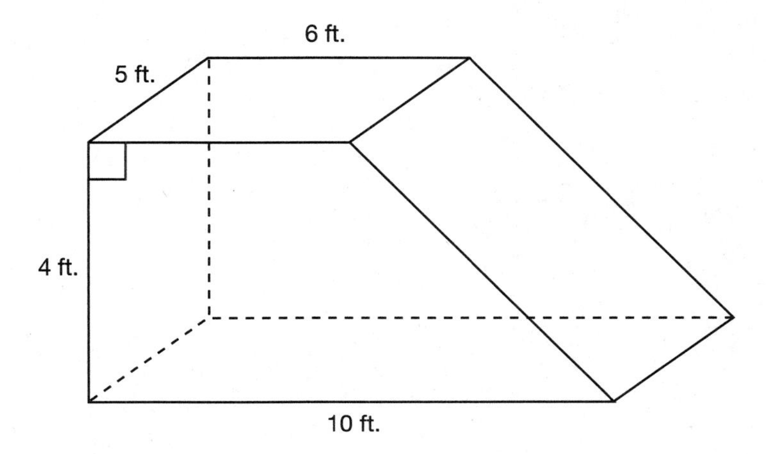

3. Lisa is an occupational therapist who regularly uses wedges like this to help stroke patients relearn everyday tasks. The wedge is made of a specialized memory foam that is just the right density. How many cubic centimeters of specialized memory foam are needed to produce this particular wedge? (*Hint:* Think of this shape as having a base that is a right triangle.)

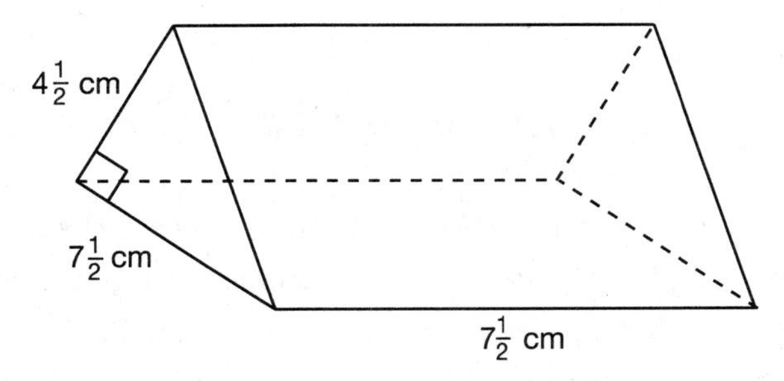

4.

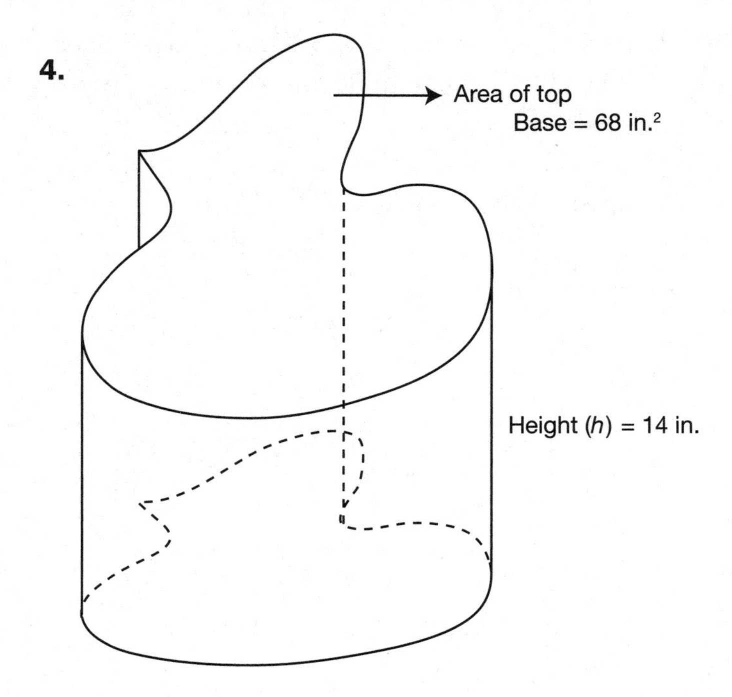

Can the formula given be used for this organically shaped vase? It has a curvy top surface, but the sides are perfectly straight. If the volume can be identified, find it. If it can't be calculated, explain why.

Finding the Volume of Pyramids, Cones, and Spheres

STANDARD ALERT!

You've probably noticed that the past few lessons have been packed with real-world applications of area, surface area, and volume to right prisms. After all, this is what Standard 7.G.B.6 called for. For Standard 8.G.B.9, we are going to give you an opportunity to become familiar with applying the volume formulas to cones, cylinders, and spheres.

Last, we will look at how to calculate the volume of pyramids, cones, and spheres. A **pyramid** is a three-dimensional shape with a regular polygon base and congruent triangular sides that meet at a common point. Pyramids are named by their base: Illustration A is a triangular pyramid, illustration B is a rectangular pyramid, and illustration C is a hexagonal pyramid.

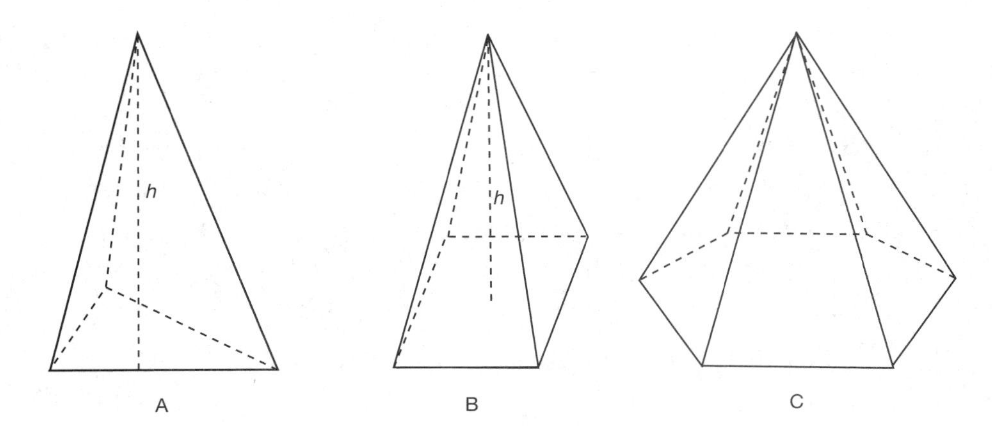

As long as you are given, or can calculate, the area of the base of a pyramid, and you know the perpendicular height from the base to the vertex, you can determine the **volume of a pyramid** by using $V = \frac{1}{3}$(*area of base*)(*height*).

TIP: Volume of Pyramid = ($\frac{1}{3}$)(*area of base*)(*height*).

The **volume of a cone** is also $\frac{1}{3}$ the product of its base area and height. However, in the case of cones, the area of the base is the formula for the area of a circle.

TIP: Volume of cone = ($\frac{1}{3}$)(πr^2)(*height*).

Spheres have only one measurement to use in formulas—their radii. In order to calculate the **volume of a sphere**, use the following formula.

TIP: Volume of sphere = $(\frac{4}{3})\pi r^3$.

Practice 4

Find the volume of the following figures.

1. Find the number of cubic centimeters of jade that is in the following stone paperweight.

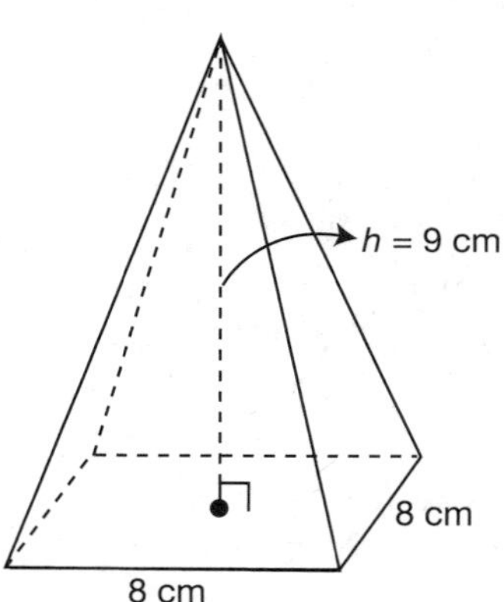

2. The inverted cone pictured here represents a pile of chicken feed that was made by a huge conveyor belt. How many cubic meters of chicken feed are there?

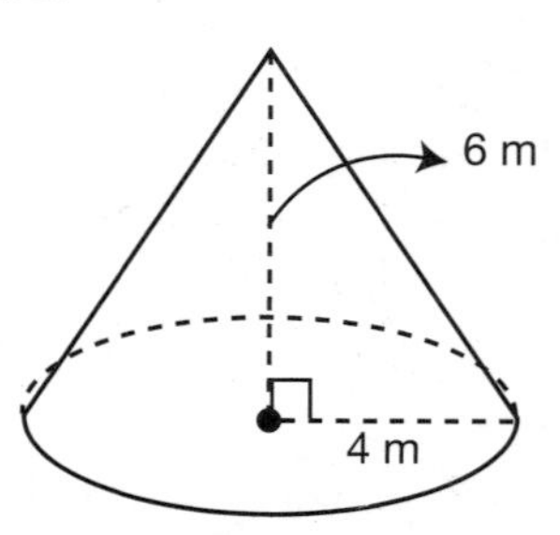

3. Note, the area of the base is given as b in the following pyramid:

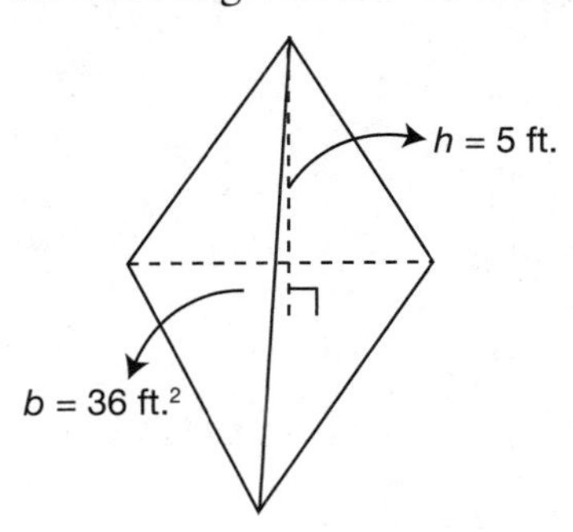

4. How many cubic inches of helium will it take to fill up the following ball? Will it require more or less than one cubic foot of helium? (Hint: Think about the edge length of a cubic food in terms of inches.)

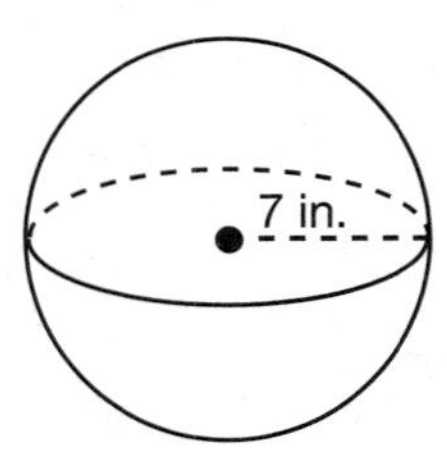

Answers

Practice 1

1. Surface area
2. Volume—how many cubic feet of air does the air-conditioner need to cool?
3. Perimeter
4. Volume—how many cubic gallons of water does the pool hold?
5. Area
6. Perimeter

Practice 2

1. $V = (length)(width)(height) = (2)(3)(5) = 30 \text{ m}^3$.
2. The additional 2 meters on the top of the tank will look like 2 rows of 6 cubic meters each, so it will have 12 m^3 more.

   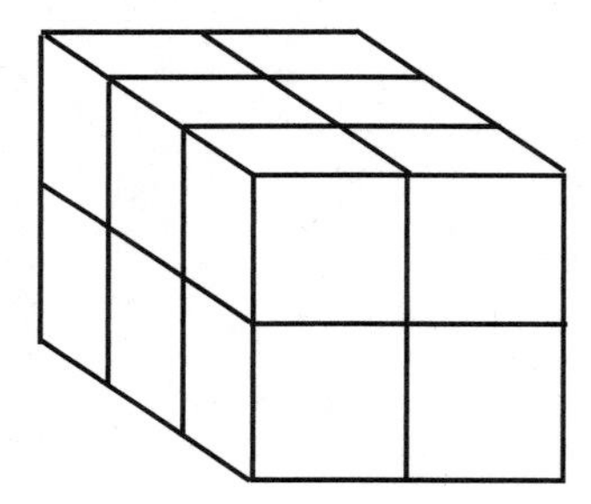

   $V = (length)(width)(height) = (2)(3)(7) = 42 \text{ m}^3$. The new prism will have $42 - 30 = 12$ more cubic meters than the original prism.
3. $V = s^3 = 3^3 = 27 \text{ yd.}^3$. It is only 3m^3 smaller than the original tank proposal.
4. Answers will vary. Many students will think that when the edge lengths of a cube double, the volume will also double, but this is false. $V = s^3$. Since the side length is doubled, replace s with $2s$ and use the formula for volume again: $V = (2s)^3 = 2^3s^3 = 8s^3$. Therefore, the new volume will be 8 times the original volume.
5. $V = s^3 = 6^3 = 216 \text{ yd.}^3$.
6. It is a common mistake to think that when the side length doubles, the volume will also double. However, since the side length is multiplied by itself three times, the volume of the new prism will be 2^3 or 8 times bigger.

Practice 3

1. $V = (\textit{area of circular base})(\textit{prism height})$
   $V = (\pi r^2)(\textit{prism height})$
   $V = (\pi 4^2)(12) = 192\pi$ in.^{3} or 602.88 in.^{3}. Since the seeds call for one quarter of a cup of seeds for every 200 cubic inches of growing space, Alexis should use $\frac{3}{4}$ of a cup of seeds.
2. $V = (\textit{area of trapezoidal base})(\textit{prism height})$
   $V = [(\frac{1}{2})(h)(b_1 + b_2)](\textit{prism height})$
   $V = (\frac{1}{2})(4)(10 + 6)(5) = 160$ ft.^{3} of cement would be needed
3. $V = (\textit{area of triangular base})(\textit{prism height})$
   $V = [(\frac{1}{2})(bh)](\textit{prism height})$
   $V = [(4\frac{1}{2})(7\frac{1}{2})(7\frac{1}{2})] \approx 127$ cm^3 of specialized memory foam are needed.
4. Since the area of the irregular top of the vase is given, and since its sides are straight, we can use the formula, (*area of base*)(*height*), to calculate the volume of the vase: $68 \times 14 = 952$ in.^{3}.

Practice 4

1. $\textit{Volume of pyramid} = (\frac{1}{3})(\textit{area of base})(\textit{height})$
   $V = (\frac{1}{3})(8 \times 8)(9) = 192$ cm^3 of jade
2. $\textit{Volume of cone} = (\frac{1}{3})(\pi r^2)(\textit{height})$
   $V = (\frac{1}{3})(\pi 4^2)(6) = 32\pi$ m^3 or 100.48 m^3 of chicken feed
3. $\textit{Volume of pyramid} = (\frac{1}{3})(\textit{area of base})(\textit{height})$
   $V = (\frac{1}{3})(36)(5) = 60$ ft.^{3}
4. $\textit{Volume of sphere} = (\frac{4}{3})\pi r^3 = (\frac{4}{3})(\pi)(7^3) = 457.33\pi$ in.^{3} or 1,436 in.^{3} 1 cubic foot contains $12 \times 12 \times 12 = 1{,}728$ cubic inches, so it will take less than a cubic foot of helium to inflate this ball.

19

Coordinate Geometry

The shortest distance between two points is under construction.
—Bill Sanderson

In this lesson you will review the coordinate plane, where you will then draw polygons with coordinate pairs for vertices. You will learn how to use coordinate pairs to find side lengths for vertical and horizontal sides of polygons and also learn how to apply the Pythagorean theorem to find the lengths between any two points. This lesson includes a bonus section on finding the midpoint of a line segment in the coordinate plane.

STANDARD PREVIEW

Once people spend a little time in New York City, they come to understand that the streets fall on a grid system. A grid of streets makes it fairly straightforward to navigate around this huge metropolis. The larger streets that run north and south in Manhattan are named Avenues, and the smaller streets that run east and west are named Streets. This means that if Jessie says she is at the corner of 2nd Avenue and 48th Street, she just gave her location as an (x,y) coordinate, which could be plotted on a coordinate plane. If Niko is at the corner of 2nd Avenue and 68th Street at the same time, you would probably be able to figure out that Jessie and Niko are 20 blocks from each other. Plotting the streets of Manhattan on a coordinate plane is just one real-world application of coordinate planes, but they can be used in many design, art, and construction careers. ***In this lesson we are going to focus on Standards 6.G.A.3 and 8.G.B.8 by plotting polygons on coordinate grids and finding the distance between coordinate pairs.***

What Is the Coordinate Plane?

A **coordinate plane** is similar to a city map, except it is a special type of map used in mathematics to plot points, lines, and shapes. The coordinate plane is made up of two main axes that intersect at the **origin**, to form a right angle. The ***x*-axis** is the main horizontal axis, and the ***y*-axis** is the main vertical axis. These perpendicular axes divide the coordinate plane into four sections, or **quadrants**. The quadrants are numbered in counterclockwise order, beginning with the top-right quadrant. See the following figure to understand all the terms that have just been presented.

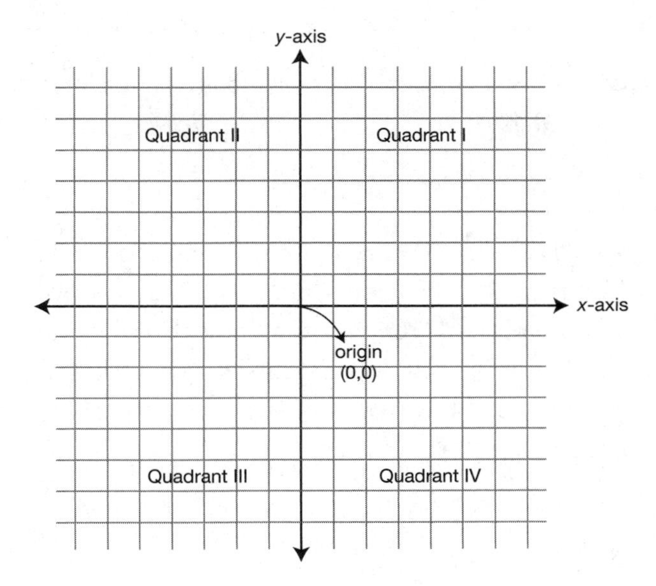

In order to give a friend directions, you need to know from where she is starting. In a coordinate plane, the starting point is the **origin**, or center point. (This makes sense, since "origin" means the place where something begins.) The origin is located at (0,0). Both the x- and y-axes are numbered like number lines. Starting from the origin, positive movements are done by moving right or up and negative movements are indicated by moving left or down.

TIP: The *origin* is the center point of a coordinate plane. It is located at (0,0), where the *x*-axis and *y*-axis intersect.

Plotting Points in the Coordinate Plane

Points in the coordinate plane have an x-coordinate and a y-coordinate that show where the point is in relation to the origin. Points are presented in the form (x,y). The first number represents the point's *horizontal* distance and direction from the origin. The second number represents the point's *vertical* distance and direction from the origin.

TIP: A *coordinate pair* is a point on a coordinate graph. It is an ordered pair of numbers written as (*x*,*y*).

For example, point A in the next figure is located at (–3,5). Notice that this point is 3 units to the *left* of the origin and 5 units *above* it. Similarly, point B in the figure at (2,–4) represents a point that is 2 units to the *right* of the origin and 4 units *below* it. When plotting points that have 0 in them, the point will sit on one of the axes since either its vertical or its horizontal movement will be zero units.

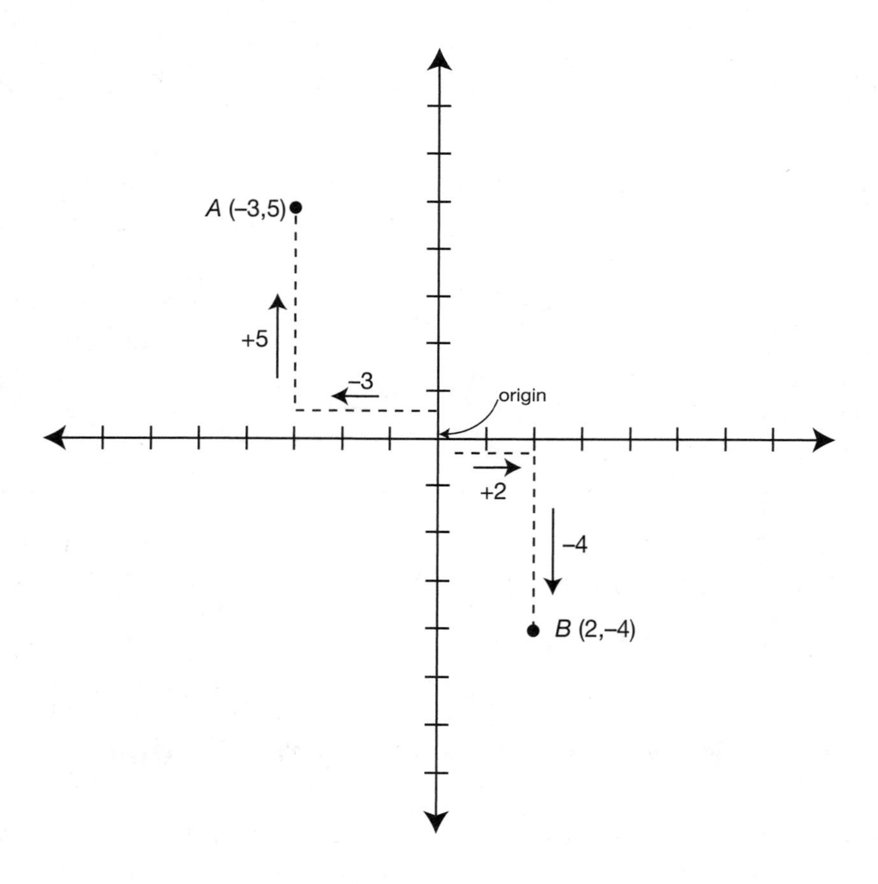

Since the x-coordinate always comes first in coordinate pairs, it is necessary that your first movement from the origin be horizontal. To plot a point, start at the origin, and move x units to the left if the x-coordinate is negative, or x units to the right if it is positive. Next, look at the y-coordinate. Move y units up if it is positive, or y units down if it is negative.

Practice 1

For questions 1 through 6, graph and label the following points on the given coordinate plane.

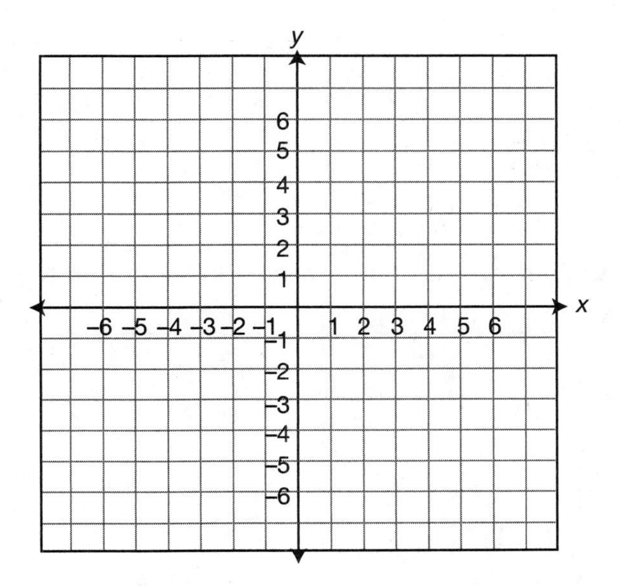

1. A (–3,5)

2. B (5,–3)

3. C (0,4)

4. D (2.5,6)

5. E (–4.5,–6)

6. F (5,0)

For questions 7 through 12, identify what quadrant the point lies in. If the point is not in a quadrant, determine what axis it sits on.

7. (–2,5)

8. (1,–8)

9. (0,3.75)

10. (12,16)

11. (–7,–7)

12. (9,0)

For questions 13 through 20, write the coordinate for each point.

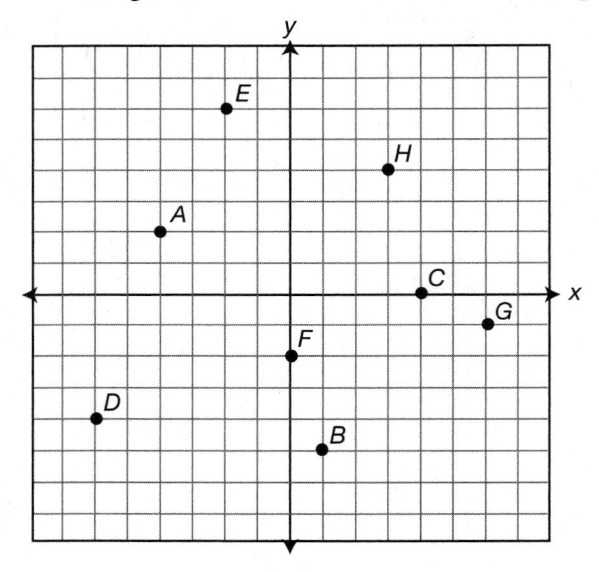

13. *A*

14. *B*

15. *C*

16. *D*

17. E

18. F

19. G

20. H

Drawing Polygons in the Coordinate Plane

STANDARD ALERT!

The previous section was a review of the coordinate plane so that your graphing skills would be sharp enough to tackle Standard 6.G.A.3. This standard has you drawing polygons in the coordinate plane by using coordinate pairs for the vertices. It also requires students to determine the lengths of horizontal and vertical sides in polygons on the coordinate plane. Grab a ruler and read on!

Why Use a Coordinate Plane?

An urban planner might draw plans for a new housing development or industrial complex on a coordinate grid. This would help him or her to organize the layout carefully and be able to easily see how various spaces relate to one another and fit into a predetermined amount of space. Scale would be used to let each box on a coordinate grid represent a fixed dimension like 10 feet, so that all the different parts of the whole could be appropriately drawn to scale and visualized on a single sheet of paper. Interior designers, directors, and many other professionals also use this technique so that they can easily change the layout of objects in a space on paper, rather than lugging around couches and stages in order to find the best use of space. You are going to get a taste of this skill by learning how to plot shapes on a coordinate plane and how to find the lengths of their vertical and horizontal sides.

Plotting a Polygon on a Coordinate Plane

Let's begin by plotting the rectangle *TALK*, which has vertices *T* (1,4), *A* (5,4), *L* (5,–2), and *K* (1,–2). The first step is to plot the coordinates and label each one with the appropriate letter. These coordinates are the vertices of our polygon, so the next step is to connect the coordinates *in the order that they are listed*. That will give you three sides of the rectangle, and by connecting the first coordinate, *T*, with the last coordinate, *K*, you will draw the last side. *TALK* is shown below, and the coordinate pairs have been included in this illustration to help with our investigation of side length:

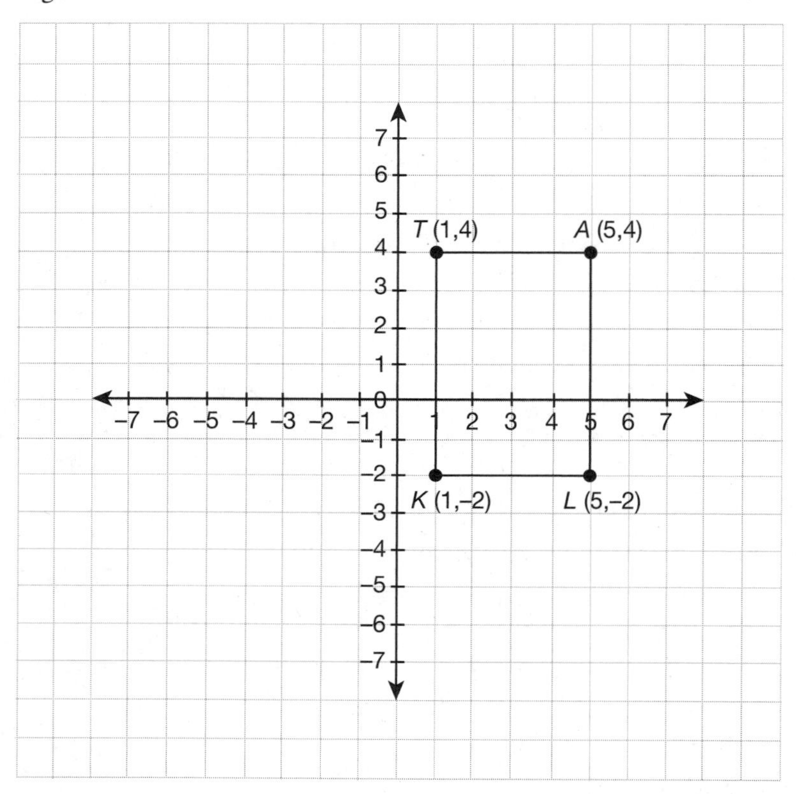

Finding Vertical and Horizontal Side Lengths

Let's now take a closer look at vertices *T* (1,4), *A* (5,4), L (5,–2), and *K* (1,–2). Notice that *T* (1,4) and *A* (5,4) have the same *y*-coordinate of 4. This means that they are going to be the same height off the *y*-axis and will therefore create a horizontal line segment. Next, look at the *x*-coordinates of 1 and 5. In order to see how long $\overline{TA}$ is, subtract the smaller *x*-coordinate

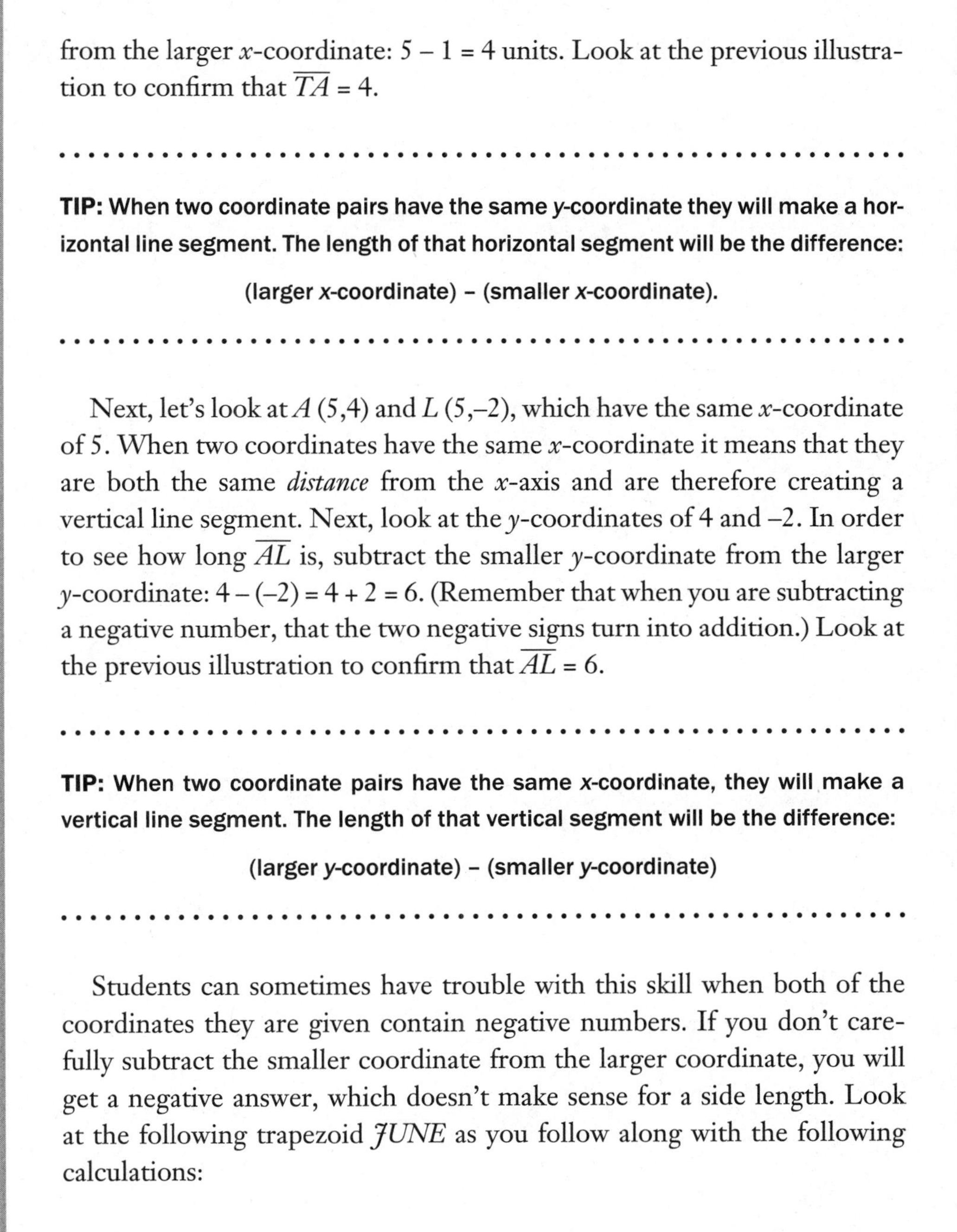

from the larger x-coordinate: 5 – 1 = 4 units. Look at the previous illustration to confirm that $\overline{TA}$ = 4.

TIP: When two coordinate pairs have the same *y*-coordinate they will make a horizontal line segment. The length of that horizontal segment will be the difference:

(larger *x*-coordinate) – (smaller *x*-coordinate).

Next, let's look at A (5,4) and L (5,–2), which have the same x-coordinate of 5. When two coordinates have the same x-coordinate it means that they are both the same *distance* from the x-axis and are therefore creating a vertical line segment. Next, look at the y-coordinates of 4 and –2. In order to see how long $\overline{AL}$ is, subtract the smaller y-coordinate from the larger y-coordinate: 4 – (–2) = 4 + 2 = 6. (Remember that when you are subtracting a negative number, that the two negative signs turn into addition.) Look at the previous illustration to confirm that $\overline{AL}$ = 6.

TIP: When two coordinate pairs have the same *x*-coordinate, they will make a vertical line segment. The length of that vertical segment will be the difference:

(larger *y*-coordinate) – (smaller *y*-coordinate)

Students can sometimes have trouble with this skill when both of the coordinates they are given contain negative numbers. If you don't carefully subtract the smaller coordinate from the larger coordinate, you will get a negative answer, which doesn't make sense for a side length. Look at the following trapezoid *JUNE* as you follow along with the following calculations:

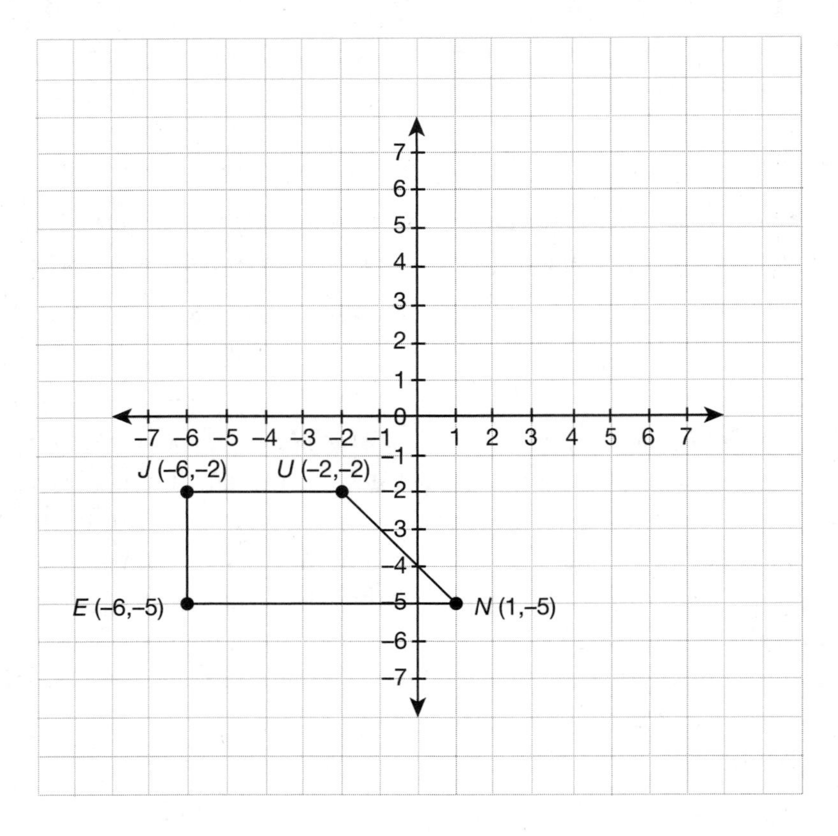

Side length of $\overline{JU}$ using J (–6,–2) and U (–2,–2):

Since J and U both have the same y-coordinate, subtract the smaller x-coordinate (–6) from the larger x-coordinate (–2) to find the side length:

$\overline{JU} = -2 - (-6) = 4$

Side length of $\overline{JE}$ using J (–6,–2) and E (–6,–5):

Since J and E both have the same x-coordinate, subtract the smaller y-coordinate (–5) from the larger y-coordinate (–2) to find the side length:

$\overline{JE} = -2 - (-5) = 3$

After looking at the calculations, refer to trapezoid $JUNE$ to confirm that $\overline{JU} = 4$ and $\overline{JE} = 3$. Do you think you can use the methods given above to determine the length of side $\overline{UN}$? Since points U and N do not have the same x-coordinates or y-coordinates, the previous methods do not apply when calculating the length of this line segment. Before we learn how to find the length of diagonal line segments, practice your proficiency at Standard 6.G.A.3 with the following practice problems:

Practice 2

Plot the following vertices and connect them to discover what kind of polygon they make. When explaining the polygon, make sure you name it as specifically as possible. (Hint: Look for parallel sides and/or congruent sides.)

1. $A\,(-4,1)$, $P\,(0,4)$, $R\,(4,1)$, $I\,(2,-4)$, $L\,(-2,-4)$

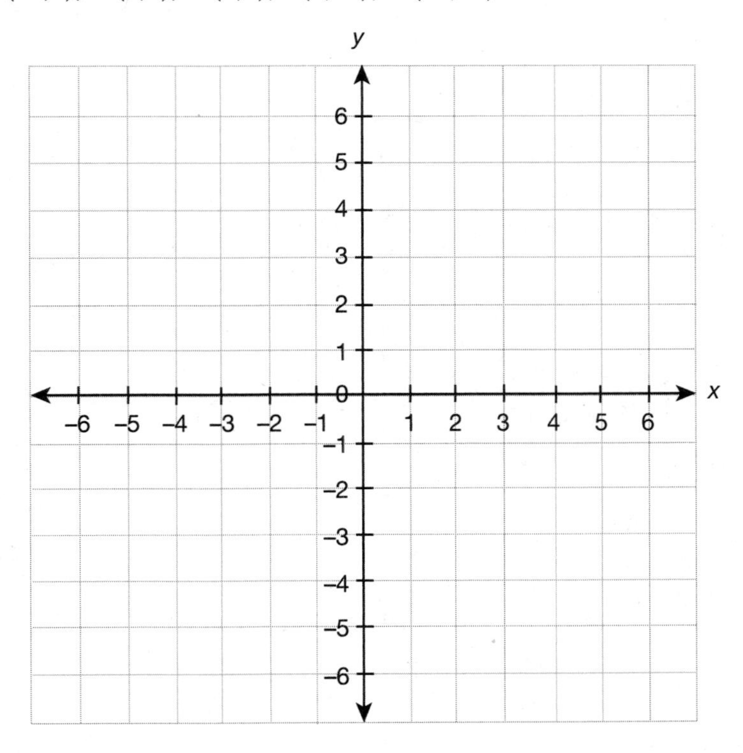

2. $M(-5,3)$, $A(1,3)$, $E(-2,-5)$

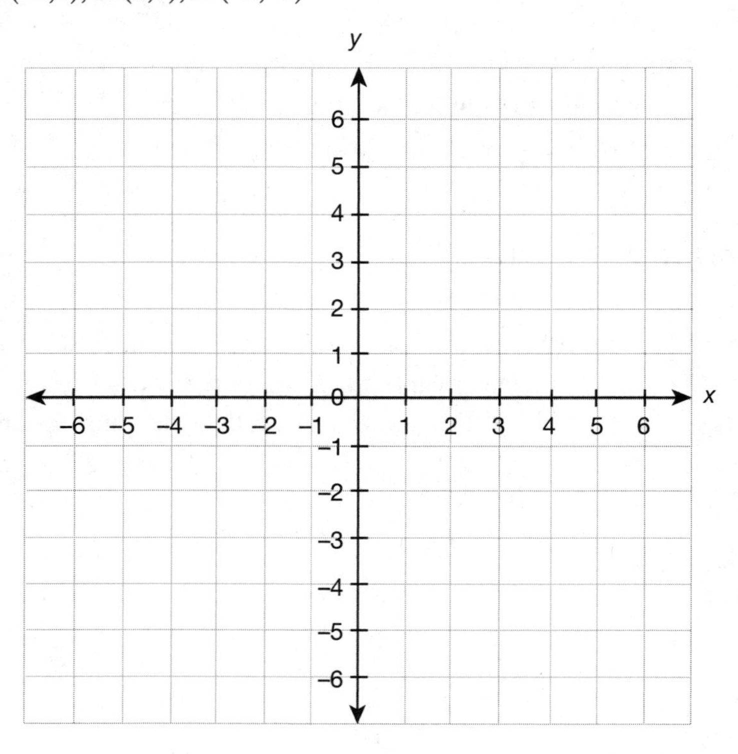

3. $L(-3,3)$, $O(3,3)$, $V(1,-5)$, $E(-5,-5)$

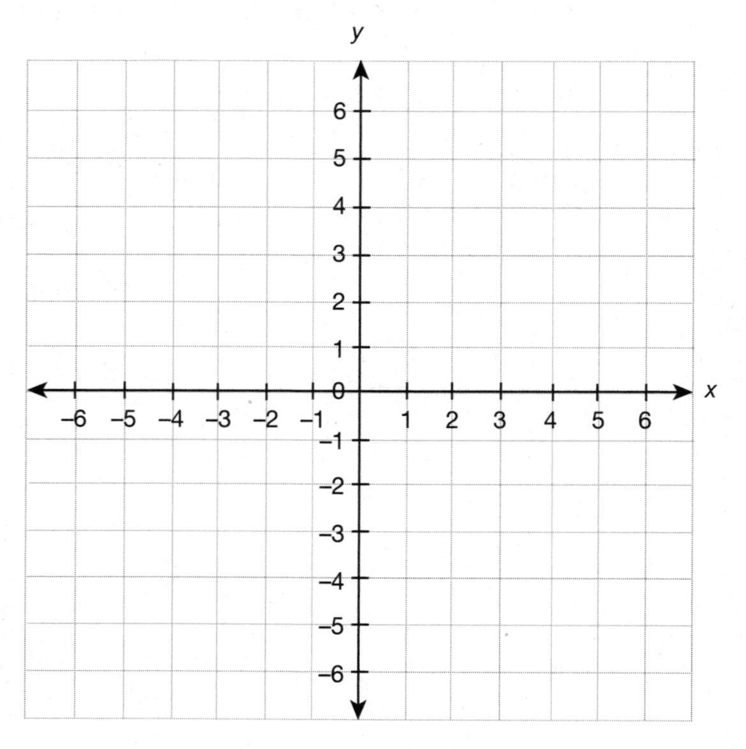

4. $Y(-3,2)$, $O(4,2)$, $U(4,-5)$

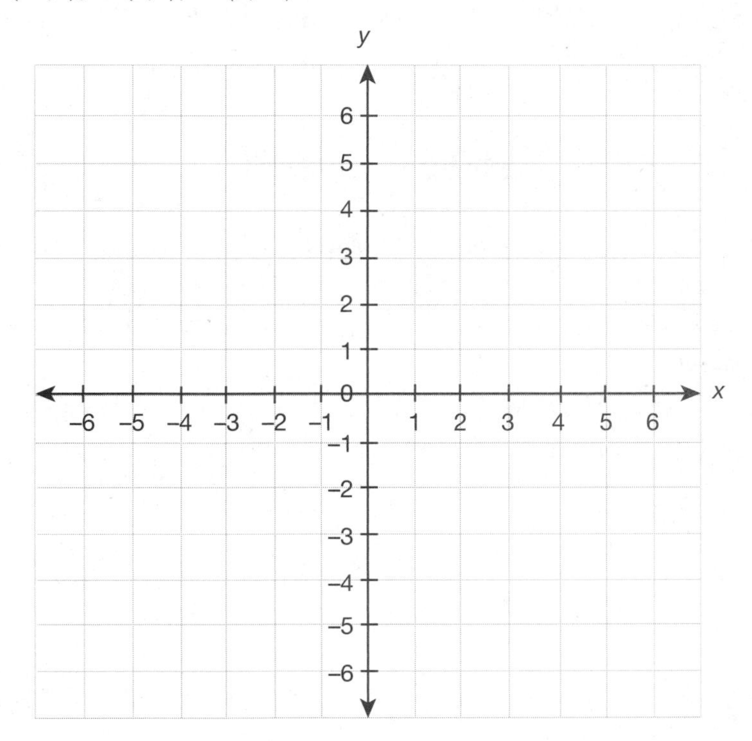

Answer the following questions using the methods presented in this section.

5. Given vertices for pentagon $S(-4,1)$, $A(0,4)$, $R(4,1)$, $D(2,-4)$, $I(-2,-4)$, which are the only two points that will make a horizontal or vertical side? How do you know this? What is the length of that side?

6. What kind of shape did question 3 make? Which side lengths can you calculate? Find those side lengths.

7. Find the length of the two legs of triangle *YOU* from question 4. What does this tell you about what kind of triangle it is?

Finding the Distance Between Any Two Points

The distance between two points in a coordinate plane is the shortest route between two points. In the previous section, you learned how to find the distance between two points that have the same x- or y-coordinates, and in this section you will build upon that and learn how to find the distance between any two points.

STANDARD ALERT!

Now that you're proficient with Standard 6.G.A.3, it's time for us to move on to Standard 8.G.B.8, where you will apply the Pythagorean theorem to find the distance between any two points in a coordinate system.

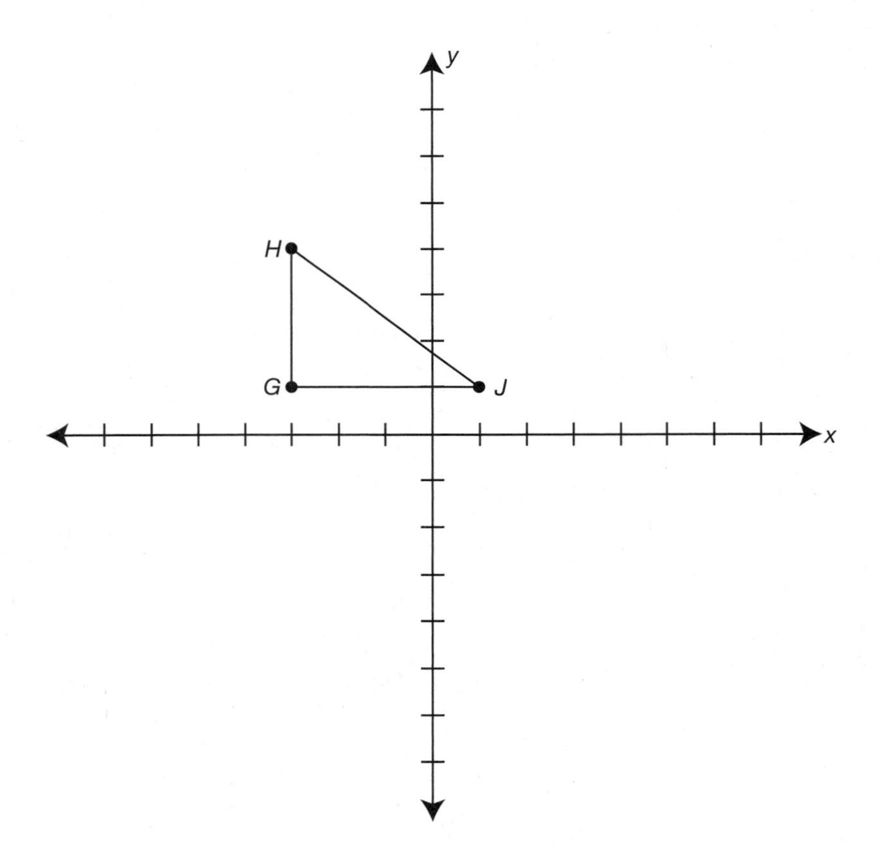

Hopefully, you could easily apply what you learned in the previous section to find the length of $\overline{GJ}$ and $\overline{GH}$ above. Unfortunately, it is not as easy to calculate the diagonal distance between points H and J. However, since connecting these three points makes a right triangle, we can use the Pythagorean theorem to calculate the distance between *H* and *J*.

$(\overline{HG})^2 + (\overline{GJ})^2 = (\overline{HJ})^2$
$(3)^2 + (4)^2 = (\overline{HJ})^2$
$9 + 16 = (\overline{HJ})^2$
$25 = (\overline{HJ})^2$
$\overline{HJ} = 5$

It would be a lot of work to have to plot your points, connect them, and use the Pythagorean theorem every time you needed to find the distance between two points! Let us figure out a formula to use by taking a look at the next figure.

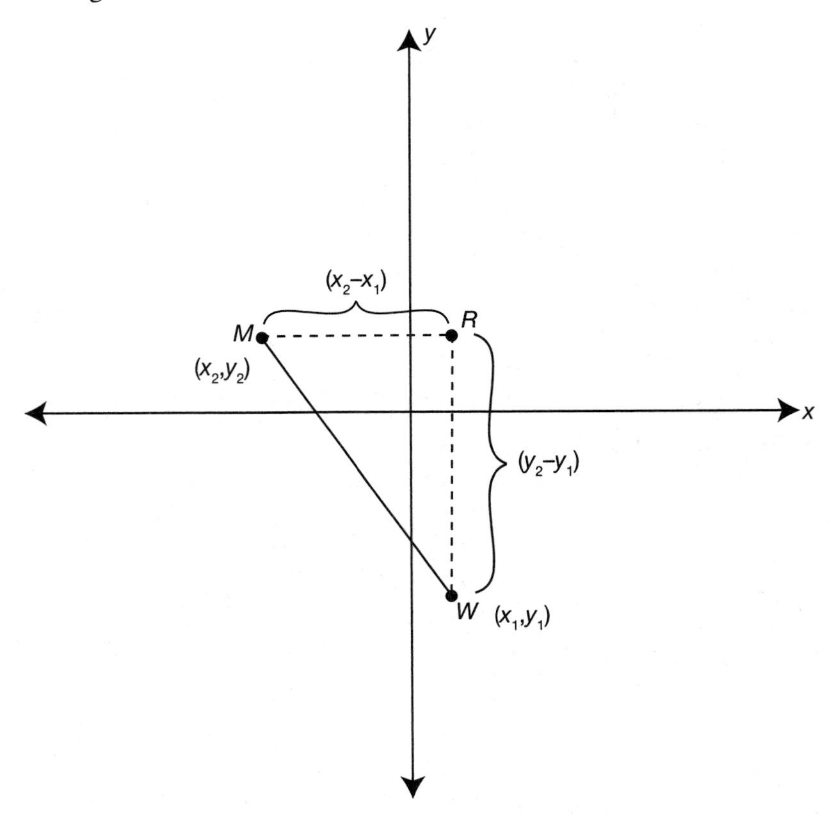

In this figure, point *W* is labeled (x_1, y_1) and point *M* is labeled (x_2, y_2). The distance we are trying to solve for is the solid line between points *M* and *W*. The *dotted vertical* line $\overline{WR}$ represents the vertical distance between the points, which can be calculated with $(y_2 - y_1)$. The dotted horizontal line $\overline{MR}$ represents the horizontal distance between the points, which can be calculated with $(x_2 - x_1)$. Since $(x_2 - x_1)$ and $(y_2 - y_1)$ represent the length of two legs of a right triangle, we can use these algebraic expressions in the Pythagorean theorem in order to come up with a formula for distance.

$$c^2 = a^2 + b^2$$
$$(\overline{MW})^2 = (x_2 - x_1)^2 + (y_2 - y_1)^2$$

In order to solve for the length of $\overline{MW}$ we need to take the square root of both sides of this equation:

$$\sqrt{(\overline{MW})^2} = \sqrt{((x_2 - x_1)^2 + (y_2 - y_1)^2)}$$
$$\overline{MW} = \sqrt{(x_2 - x_1)^2 + (y_2 - y_1)^2}$$

You can generalize this discovery to find the distance between any two points in the coordinate plane.

TIP: The distance, d, between any two points $A(x_1,y_1)$ and $B(x_2,y_2)$ in the coordinate plane is $d = \sqrt{(x_2 - x_1)^2 + (y_2 - y_1)^2}$

When you're looking for the distance between two points, it is not important which point you chose to be your (x_1,y_1) and which point to be your (x_2,y_2). In fact, the order in which you use x_1 and x_2 or y_1 and y_2 is not important since when the difference of these two numbers is squared, the result will always be positive. When you're working with the distance formula, it is important to remember that the square root of a sum of two terms is *not* equal to the sum of the individual square roots of those terms: $\sqrt{a^2 + b^2} \neq \sqrt{a^2} + \sqrt{b^2}$. Therefore, it is necessary that you always first find the sum of $(x_2 - x_1)^2$ and $(y_2 - y_1)^2$ *before* taking the square root.

Example: Find the distance between the points (3,8) and (–5,12).
Solution: Let (3,8) be (x_1,y_1); and let (–5,12) be (x_2,y_2).

$$d=\sqrt{(x_2-x_1)^2+(y_2-y_1)^2}$$
$$d=\sqrt{(-5-3)^2+(12-8)^2}$$
$$d=\sqrt{(-8)^2+(4)^2}$$
$$d=\sqrt{64+16}$$
$$d=\sqrt{80}$$
$$d=8.9$$

Practice 3

1. The distance formula $d=\sqrt{(x_2-x_1)^2+(y_2-y_1)^2}$, comes from which theorem?

2. True or false: $d=\sqrt{(x_2-x_1)^2+(y_2-y_1)^2} = \sqrt{(x_2-x_1)^2} + \sqrt{(y_2-y_1)^2}$

3. True or false: The distance between any two points can always be counted accurately as long as the points are plotted on graph paper.

For questions 4 through 7, calculate the distance between the given coordinates.

4. (3,5) and (–1,5)

5. (3,–2) and (–5,4)

6. (–3,–8) and (–6,–4)

7. (–1,–6) and (4,6)

STANDARD ALERT!

That wraps up all of our work with the CCSS middle school geometry standards for this lesson. The following section on finding the midpoint between two coordinates is an extension of these standards. It is not required for you to know the following material until high school, but you are more than welcome to stick around and learn this concept ahead of schedule. Psst . . . The Midpoint Formula is actually a lot easier to work with than the Distance Formula, so why not read on?!

Finding the Midpoint Between Two Points

The midpoint of two coordinate pairs is the point that is equal in distance from both of the coordinate pairs. Since taking the average of two numbers finds the middle number, it makes sense that the midpoint formula involves finding the average. Consider points A, at (x_1,y_1), and B, at (x_2,y_2). The x-coordinate that will be a midpoint, equidistant from x_1 and x_2, will be the average of these coordinates: $\frac{x_1 + x_2}{2}$. The same goes for finding the y-coordinate, which will be equidistant from both y_1 and y_2: $\frac{y_1 + y_2}{2}$. These two expressions are combined to create the midpoint formula.

TIP: The *midpoint* between any two points $A(x_1,y_1)$ and $B(x_2,y_2)$ is $(\frac{x_1 + x_2}{2}, \frac{y_1 + y_2}{2})$

Example: Find the midpoint between the points (3,8) and (–5,12).
Solution: Let (3,8) be (x_1,y_1), and let (–5,12) be (x_2,y_2).

$$\text{midpoint} = (\frac{x_1 + x_2}{2}, \frac{y_1 + y_2}{2})$$

$$\text{midpoint} = (\frac{3 + -5}{2}, \frac{8 + 12}{2})$$

$$\text{midpoint} = (\frac{-2}{2}, \frac{20}{2})$$

$$\text{midpoint} = (-1,10)$$

Sometimes, the midpoint and one of the endpoints will be given, and you will need to find the coordinates of the other endpoint.

Example: Find the coordinates of the endpoint of a line segment that has one endpoint at (11,6) and a midpoint at (7,–4).
Solution:

$$\text{midpoint} = (\frac{x_1 + x_2}{2}, \frac{y_1 + y_2}{2})$$

$$(7,-4) = (\frac{11 + x_2}{2}, \frac{6 + y_2}{2})$$

$$(7,-4) \text{ must equal } (\frac{14}{2}, \frac{-8}{2})$$

So working backwards you can reason:

$$(7,-4) = (\frac{11 + 3}{2}, \frac{6 + -14}{2})$$

$$\text{endpoint} = (3,-14)$$

Practice 4

Find the midpoint between the coordinate pairs for questions 1 through 5.

1. (2,5) and (–10,11)
2. (–3,6) and (1,8)
3. (–9,2) and (–1,–6)
4. (8,–5) and (0,–5)
5. (1,–3) and (–8,3)
6. Find the second endpoint of a line segment that has one endpoint at (–5,8) and a midpoint at (3,–2).

Answers

Practice 1

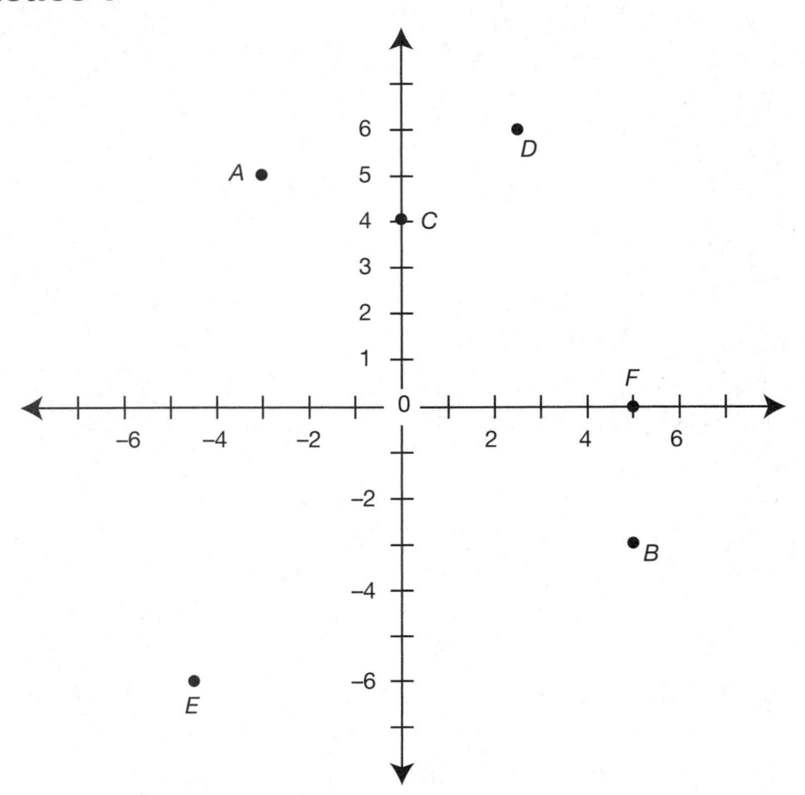

7. Quadrant II
8. Quadrant IV
9. y-axis
10. Quadrant I
11. Quadrant III
12. x-axis
13. (–4,2)
14. (1,–5)
15. (4,0)
16. (–6,–4)
17. (–2,6)
18. (0,–2)
19. (6,–1)
20. (3,4)

Practice 2

1.

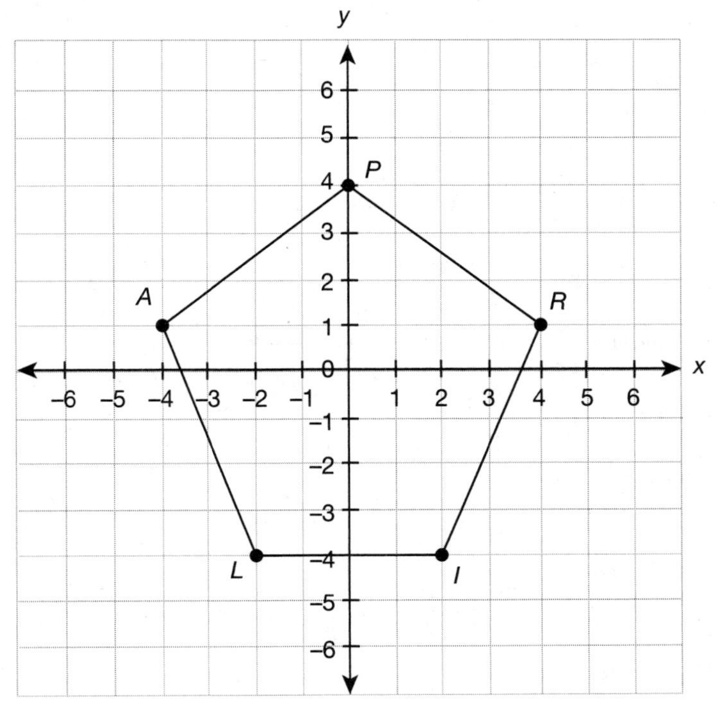

Pentagon

2.

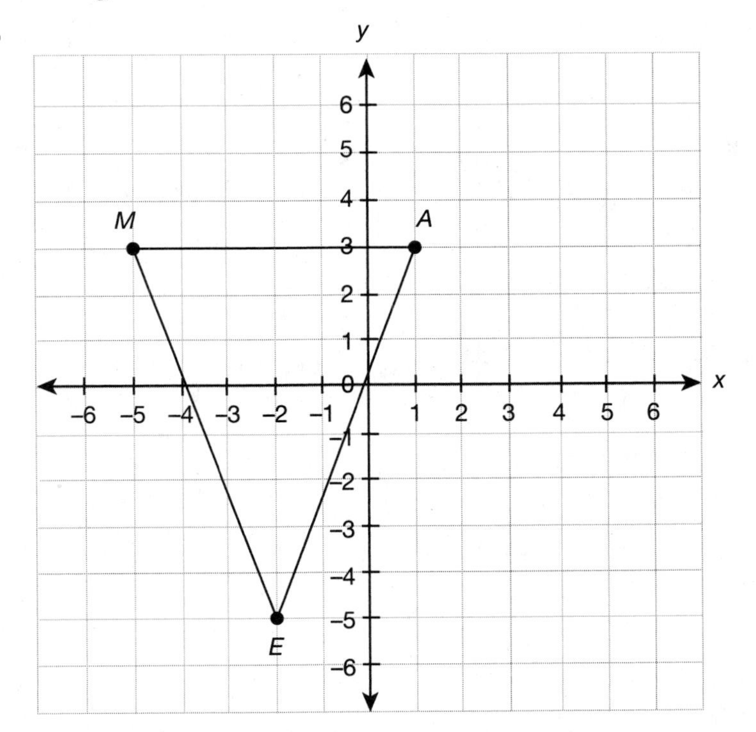

Acute isosceles triangle

3.

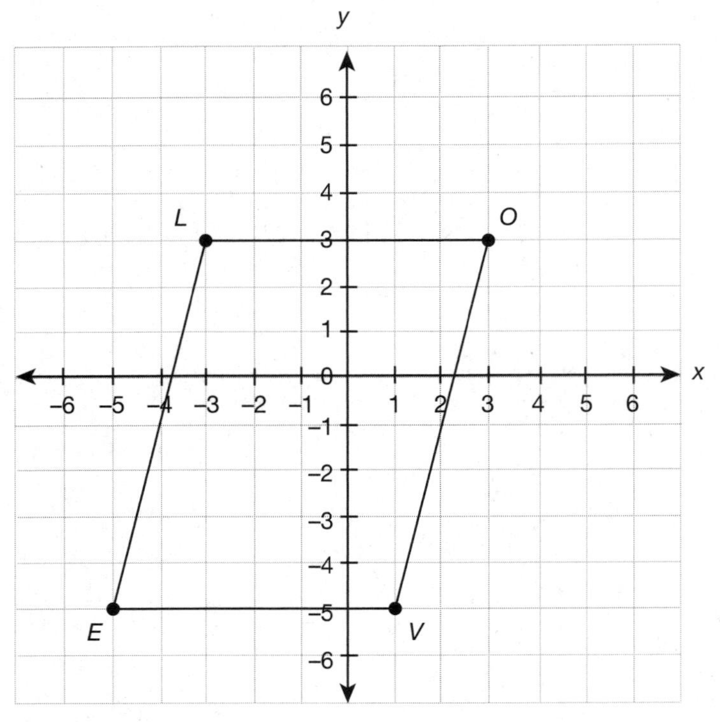

Parallelogram

4.

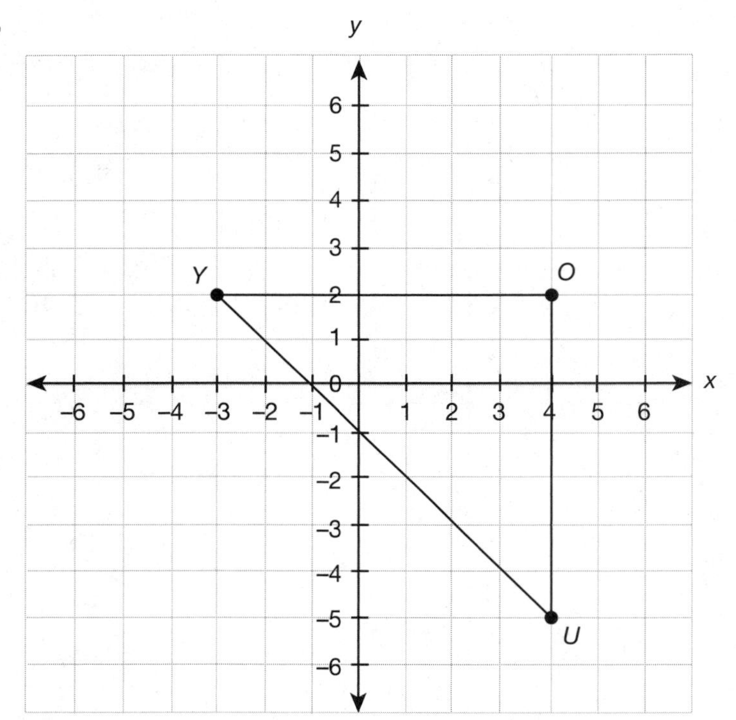

Right isosceles triangle

5. D (2,–4) and I (–2,–4) make the only horizontal side, since they are the only two coordinates that have the same x- or y-coordinates. $\overline{DI} = 4$.
6. The vertices in question 3 make a parallelogram. $\overline{LO} = 6$ and $\overline{EV} = 6$.
7. Question 4 forms a right isosceles triangle because $\overline{YO} = 7$ and $\overline{OU} = 7$.

Practice 3

1. The distance formula comes from the Pythagorean theorem.
2. False: You must find the sum inside the square root symbol before taking the square root.
3. False: Unless the coordinate pairs have the same x- or y-coordinate, the distance between them will be a diagonal line, which cannot be counted by hand.
4. $d = 4$
5. $d = 10$
6. $d = 5$
7. $d = 13$

Practice 4

1. (–4,8)
2. (–1,7)
3. (–5,–2)
4. (4,–5)
5. (–3.5,0)
6. midpoint $= (\frac{x_1 + x_2}{2}, \frac{y_1 + y_2}{2})$

   $(3,-2) = (\frac{-5 + x_2}{2}, \frac{8 + y_2}{2})$

   $(3,-2)$ must equal $(\frac{6}{2}, \frac{-4}{2})$

   So working backwards you can reason:

   $(3,-2) = (\frac{-5 + 11}{2}, \frac{8 + -12}{2})$

   endpoint = (11,–12)

20

Transformations: Reflections, Translations, Rotations, and Dilations

I never got a pass mark in math. . . . Just imagine—
mathematicians now use my prints to illustrate their books.
—M. C. ESCHER

In this lesson you will learn about the four ways that polygons can be transformed on the coordinate plane. You will also investigate how each type of transformation preserves or changes the attributes of polygons, such as general shape, angle measurements, side lengths, and orientation.

STANDARD PREVIEW

"Mirror, mirror on the wall, what's the finest math of all?" For now we're going to answer that question with "transformations!" Mirrors are one way to get a reflection, but *reflections* happen on the coordinate plane as well. So do three other types of transformations: translations, *rotations*, and *dilations*. You probably already know more than you realize about transformations, since you come across them every day. Architects use reflections to create spaces that are pleasing to be in, while designers and artists use rotations, translations, and dilations to create beautiful décor or thought-provoking artwork. In this lesson we start with ***Standard 8.G.A.1 by studying the properties of rotations, reflections, and translations.*** We are also going to cover ***Standard 8.G.A.3 by using coordinates to describe how reflections, rotations, reflections, and translations alter the appearance of 2-dimensional figures.*** We will also be laying a solid foundation to cover ***Standards 8.G.A.2 and 8.G.A.4, which deal with congruence and similarity***, in later lessons.

What Are Geometric Transformations?

When something is "transformed" it is changed in one way or another. In geometry, points, lines, and even entire polygons can be transformed on a coordinate grid. Sometimes, a transformation will result in movement in one or two directions, but the size and appearance of the original figure will remain the same. Some transformations will create a mirror image of an object, and other transformations will change a figure's size.

How Are Transformed Images Named?

Regardless of which type of transformation is being used, all transformations use the same method for naming the original image and the transformed image. The original image is called the **preimage** and the new figure is called the **image**. Points in transformations have a specific way of being named. When a point undergoes a transformation, it keeps the same letter, but it is followed by a prime mark ('). For example, after point *Q* is transformed, it becomes *Q*', which is said, "*Q* prime." Look at the transformation of polygon *ABCD* in the next figure. Notice that the image it creates is named *A'B'C'D'*.

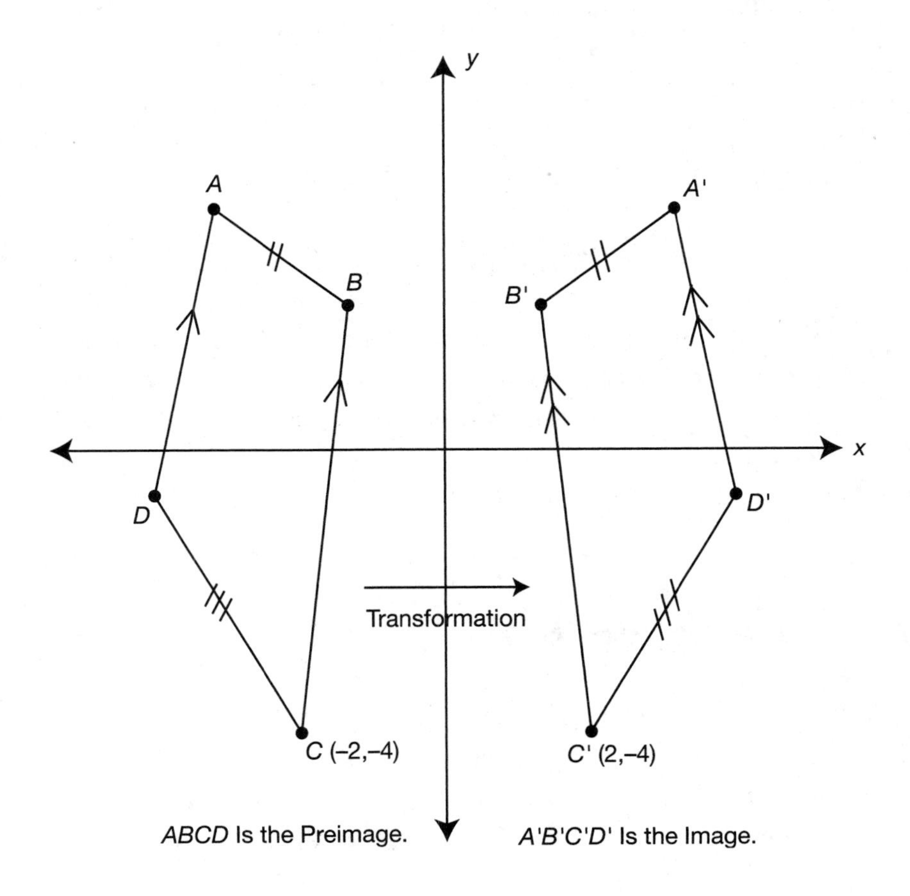

TIP: A transformation is a shift of a *preimage* into a new figure called the *image*. Transformed points are named with the same variable as the original point, but are followed with a prime mark (').

Before we move on to the specific types of transformations, let's first look at which traits reflections, translations, and rotations have. The following table pairs a specific example of what is happening with *ABCD* with the general rule of what holds true for *all* reflections, translation, and rotations:

What happens to:	In the ABCD to A'B'C'D' transformation:	In all reflections, translations, and rotations:
Line segments	The line segment $\overline{AB}$ transformed to the congruent line segment $A'B'$	Line segments always reflect/translate/rotate to congruent line segments
Angles	∠A transformed to the congruent ∠A'	Angles always reflect/translate/rotate to congruent angles
Parallel Lines	The parallel line segments $\overline{AD}$ and $\overline{BC}$ transformed to the parallel line segments $\overline{A'D'}$ and $\overline{B'C'}$	Parallel line segments always reflect/translate/rotate to parallel line segments

STANDARD ALERT!

The previous table contains all that Standard 8.G.A.1 wants students to know. Make sure you understand that for all rotations, translations, and rotations, line segments, angles, and parallel lines remain intact.

What Is a Reflection?

A **reflection** is a transformation that flips a point or object over a given line. While the shape of the object will be the same, it will be a mirror image of itself. It is common that points, lines, or polygons will be reflected over the x-axis or y-axis, but they can be reflected over other lines as well. Polygon $ABCD$ in the preceding figure underwent a reflection over the y-axis to become $A'B'C'D'$. In order to reflect a point over the y-axis, its y-coordinate stays the same, and its x-coordinate changes signs. Point $A(x,y)$ will become $A'(-x,y)$ after being reflected over the y-axis.

TIP: Point $A(x,y)$ reflected over the y-axis becomes $A'(-x,y)$.

Similarly, in order to reflect a point over the x-axis, its x-coordinate will stay the same, and its y-coordinate will change signs. Point $A(x,y)$ will become $A'(\text{x},-y)$ after being reflected over the x-axis.

TIP: Point $A(x,y)$ reflected over the x-axis will become $A'(x,-y)$.

This figure demonstrates the reflection of point P over the x-axis.

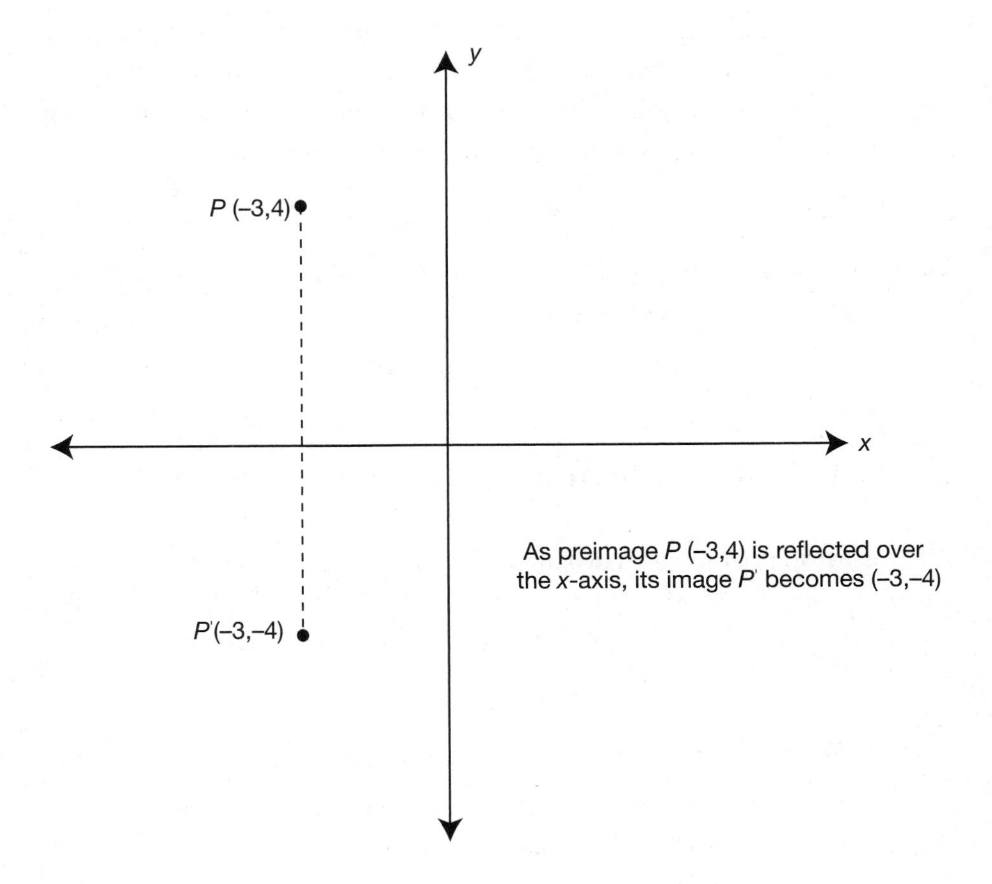

How Do Reflections Change an Image?

How did polygon $ABCD$ change when it was *reflected* in the previous illustration?

Shape: Does it still have the same general shape? Yes, it is still a long and fairly flat trapezoid with one side that is nearly perpendicular to the other two sides, and one long base and leg pair that makes a pointy angle.

Size: Is $A'B'C'D'$ the same size as $ABCD$? Yes $\overline{AB}$ is the same length as $\overline{A'B'}$ and the remaining sides are all the same sizes as well.

Orientation: Does $A'B'C'D'$ have the same orientation as $ABCD$? Two items have the same orientation if their defining characteristics are all in the same place. For example, cars in the United States have the steering wheel on the left and the glove compartment on the right, so they all have the same orientation in that regard. Polygons $ABCD$ and $A'B'C'D'$ do *not* have the same orientation after being reflected. Notice that the sharp pointy angle is on the right in $ABCD$ and is on the left in $A'B'C'D'$. Even if the two polygons were swiveled so that their longest sides were on the bottom, you can notice that their pointy angles would be facing opposite directions.

TIP: Reflected polygon images have the same shapes and size of their preimage but do not have the same orientation.

What Is a Translation?

A **translation** is a transformation that slides a point or object horizontally, vertically, or in both directions, which creates a diagonal shift. When a figure is translated, it will look exactly the same, but will just have a different location.

TIP: When a point is *translated*, it is shifted horizontally, vertically, or diagonally. Translated figures keep the same appearance and size.

The notation $T_{h,k}$ is used to give translation instructions. The subscript h shows the movement that will be applied to the x-coordinate and the subscript k shows the movement that will be applied to the y-coordinate. These numbers are each added to the x-coordinate and y-coordinate in the preimage.

TIP: Point $A(x,y)$ translated using $T_{h,k}$ becomes $A'((x + h), (y + k))$.

Example: Translate ΔVIC with vertices $V(-5,4)$, $I(1,6)$, and $C(-1,2)$ using translation $T_{5,-7}$.
Solution: In order to perform $T_{5,-7}$, add 5 to all the x-coordinates and subtract 7 from all the y-coordinates:

$V(-5,4) \rightarrow V'(0,-3)$
$I(1,6) \rightarrow I'(6,-1)$
$C(-1,2) \rightarrow C'(4,-5)$

This translation is illustrated in the next figure.

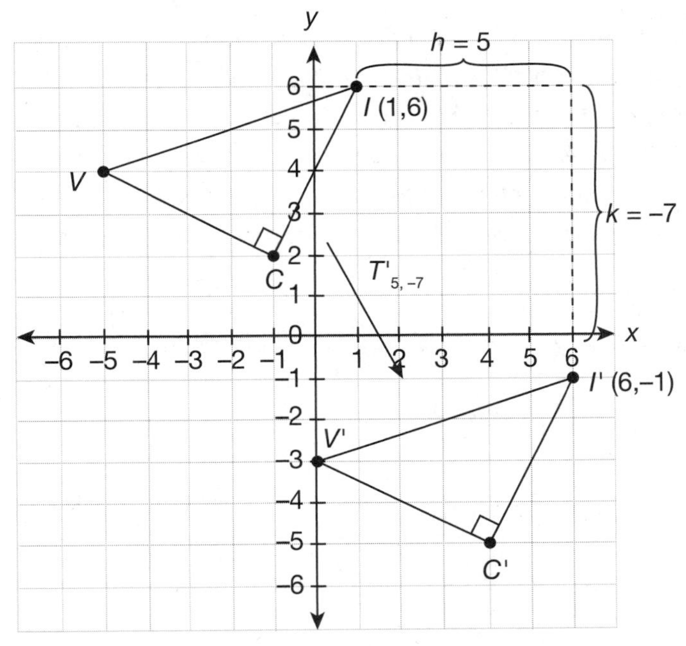

ΔVIC is translated using $T'_{5,-7}$ and becomes $\Delta V'I'C'$

How Do Translations Change an Image?

How did triangle *VIC* change when it was *translated* in the previous illustration?

Shape: Does it still have the same general shape? Yes, it is still a right triangle.
Size: Is $V'I'C'$ the same size as VIC? Yes, the distance formula could be used to verify that all the side lengths are the same.

Orientation: Does $V'I'C'$ have the same orientation as VIC? Yes, the right angle, legs, and hypotenuse are all still in the same relative places.

TIP: Translated polygon images have the same shapes, size, and orientation of their preimage.

STANDARD ALERT!

Before moving on to the third transformation, dilation, it should be pointed out that all of the above rules for transforming (x,y) coordinate pairs and all of the patterns that are being investigated are part of Standard 8.G.A.3. This standard requires students to be able to describe the effects of reflections, translations, dilations, and rotations of 2-dimensional figures using coordinates. We have just one more transformation to go before moving on to our first practice set.

What Is a Dilation?

One type of dilation occurs when the eye doctor puts drops in someone's eyes to dilate her pupils (enlarge them), so that they can be examined. We are going to look at a different type of dilation on the coordinate plane. **Dilation** is a transformation, written D_k, that will either expand or shrink a shape, depending on the scale factor being used. You might remember from an earlier lesson that scale factors are used to create objects that are the same shape but different sizes. Therefore, the reproduced image in a dilation will have the same shape as the original image but a different size. *Note: In this lesson, we are going to investigate dilations where the center of the dilation is at the origin*, (0,0).

When a figure is dilated, each of the coordinates is multiplied by the scale factor k. To perform the dilation D_2, multiply each of the coordinates by 2. The vertices of triangle WIN are dilated below using D_2 to become

$W'I'N'$. Look at how each of the coordinate pairs change, and then observe how the preimage and image are related in the coordinate plane below:

$$W(-2,3) \rightarrow W'(-4,6)$$
$$I(3,2) \rightarrow I'(6,4)$$
$$N(-1,-2) \rightarrow N'(-2,-4)$$

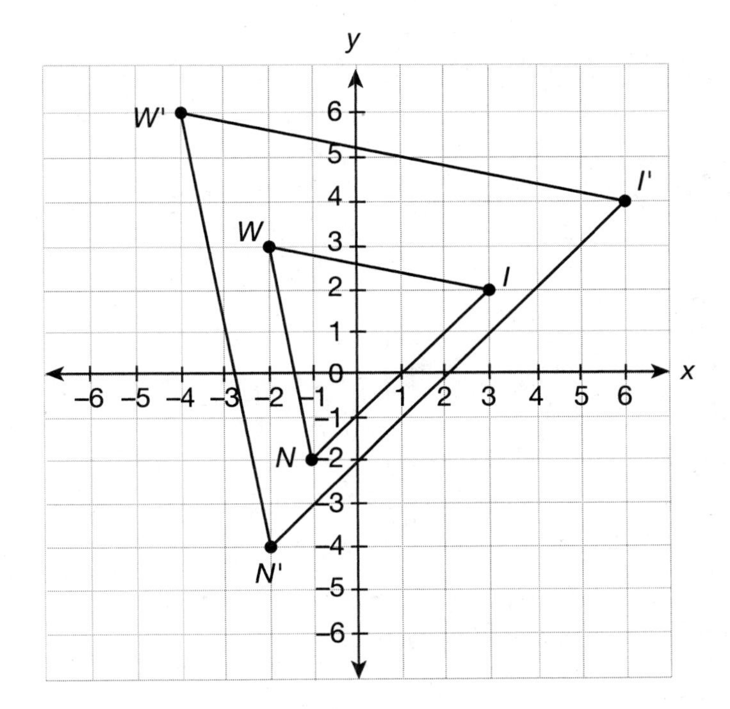

TIP: Point *A* (*x*,*y*) dilated using D_k becomes *A*' (*xk*,*yk*).

As discussed and shown in the previous illustration, a dilated image does not have the same size as its original image. But will the image always be larger than the preimage? Not necessarily. If the scale factor *k* is less than 1, the image will be smaller than the preimage. If the scale factor *k* is greater than 1, the image will be larger than the preimage.

TIP: Figures dilated on the coordinate plane using D_k have the following properties:

- The image will be smaller than the preimage when $k < 0$.
- The image will be larger than the preimage when $k > 0$.
- Angle measures will stay the same.
- Parallel lines will stay parallel.

The following image shows $W''I''N''$ which has been dilated using $k = \frac{1}{2}$.

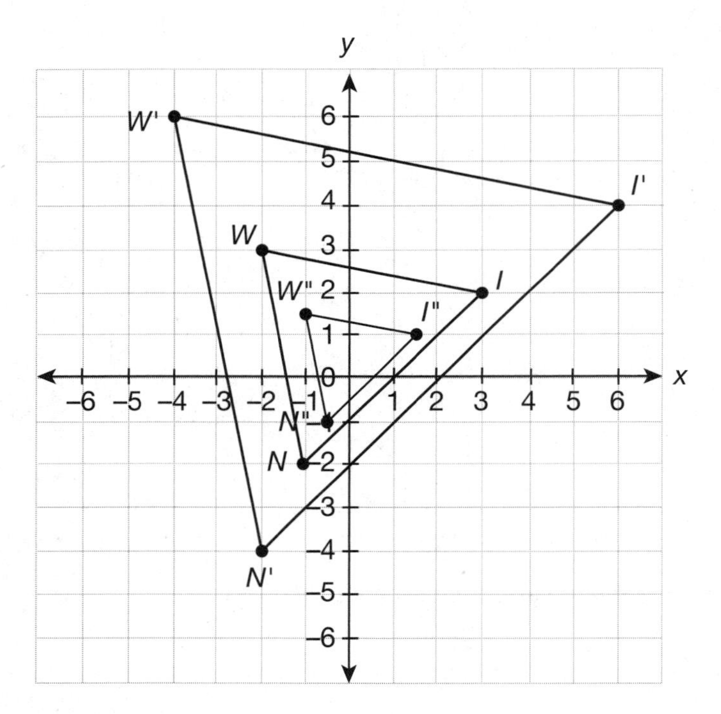

How Do Dilations Change an Image?

How did triangle *WIN* change when it was *dilated* in the previous illustration?

> **Shape:** Does it still have the same general shape? Yes, it is still an isosceles triangle.
> **Size:** Is *W'I'N'* the same size as *WIN*? No, *W'I'N'* is larger than *WIN*, and *W"I"N"* is smaller than *WIN*.

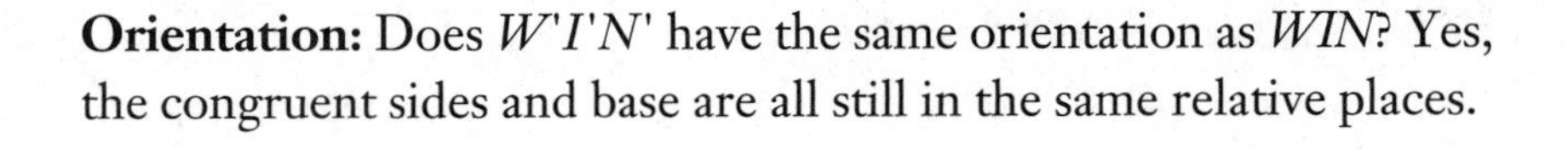

Orientation: Does $W'I'N'$ have the same orientation as WIN? Yes, the congruent sides and base are all still in the same relative places.

TIP: Dilated polygon images have the same shapes and orientation of their preimage, but not the same size.

Practice 1

1. After transforming point W, the new point will be named __________.

2. What transformations do not change the size of an image and which transformations make an image larger or smaller?

3. True or false: A translation will result in an image that is a mirror of the preimage.

4. How do you translate a point $R(x,y)$ using $T_{h,k}$?

5. After reflecting $N(-3,7)$ over the x-axis, what will the coordinates of N' be?

6. After reflecting $M(6,-1)$ over the y-axis, what will the coordinates of M' be?

7. After translating $L(2,-5)$ using $T_{-3,-4}$, what will the coordinates of L' be?

8. Reflect point $J(5,-2)$ over the x-axis to get point J'. Then reflect J' over the y-axis to get J''. What will the coordinate be of J''?

9. What value for k will make an image dilated using D_k smaller? What value for k will make an image dilated using D_k larger?

10. Find the image points for ΔARM using $D_{\frac{1}{2}}$ and the preimage points A (7,8), R (–2,4), and M (3,–6).

What Is a Rotation?

A rotation is a type of translation that swivels a point or polygon around a fixed point, such as the origin. Rotations are done either clockwise or counterclockwise and are done in degrees. Most commonly, figures are rotated 90° or 180°. In the next figure, you can observe how triangle *KAY* in Quadrant I is rotated counterclockwise 90° to form triangle *K'A'Y'*.

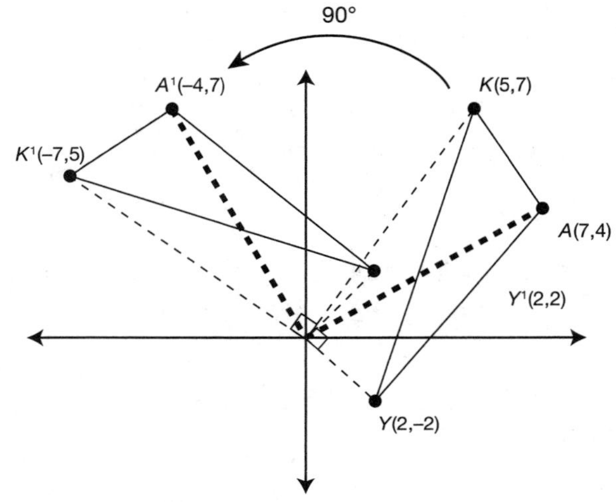

Can you notice what is happening each time a point is rotated 90° counterclockwise?

90° counterclockwise

$K(5,7) \longrightarrow K^1(-7,5)$

$A(7,4) \longrightarrow A^1(-4,7)$

$Y(2,-2) \longrightarrow Y^1(2,2)$

By looking at what happens to points *K*, *A*, and *Y* in this figure as they are rotated counterclockwise 90°, can you guess what the rule is for this type of translation? Did you notice that the preimage *x*-coordinate becomes the image *y*-coordinate? Also, the preimage *y*-coordinate switches signs and becomes the *x*-coordinate in the image point. This relationship is summed up as follows:

TIP: When preimage point *A*(*x*,*y*) is rotated 90° counterclockwise around the origin, the image point will be *A*'(–*y*,*x*).

Look at the next figure to see what happens to points that are rotated 90° clockwise.

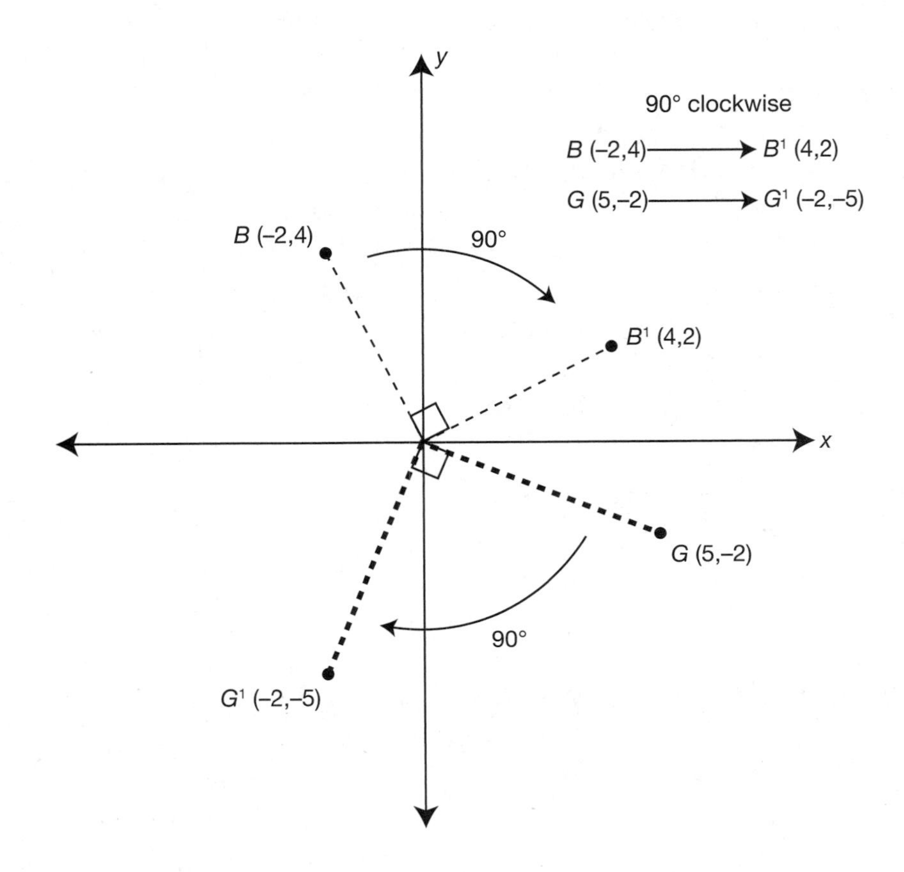

In this case, the preimage *x*-coordinate switches signs and becomes the image *y*-coordinate. The preimage *y*-coordinate becomes the *x*-coordinate in the image point.

TIP: **When preimage point $A(x,y)$ is rotated 90° clockwise around the origin, the image point will be $A'(y,-x)$.**

How Do Rotations Change an Image?

How did triangle *KAY* change when it was *rotated* in the previous illustration?

> **Shape:** Does it still have the same general shape? Yes, it is still a long obtuse triangle.
> **Size:** Is *KAY* the same size as *K'A'Y'*? Yes, the distance formula could be used to verify that all the side lengths are the same.
> **Orientation:** Does *KAY* have the same orientation as *K'A'Y'*? No, these do *not* have the same orientation after being rotated. Notice that the smallest angle points down to the left in *KAY*, while in *K'A'Y'* it points to the right.

TIP: Rotated polygon images have the same shapes and size of their preimage but do not have the same orientation.

Transforming Polygons

You have seen a few examples of polygons undergoing each type of transformation. In this section, we are going to review the characteristics that remain intact with the various types of transformations. Before getting some of your own practice with transforming polygons, let's review what each type of translation does:

- Reflections, translations, and rotations maintain a figure's shape and size.
- Dilation maintains a figure's general shape but changes its size.
- Reflections and rotations change a figure's orientation, while translations and dilations preserve the orientation of an image.
- All four transformations preserve angle sizes and parallel lines.

STANDARD ALERT!

The next practice set gives you another opportunity to practice Standard 8.G.A.3. You will be asked to describe the effects of reflections, translations, dilations, and rotations of 2-dimensional figures using coordinates.

Practice 2

1. True or false: To rotate a point 90° counterclockwise around the origin, switch the order of x and y and change both of their signs.

2. True or false: When $U(8,-3)$ is rotated 90° clockwise, the result will be $U'(-3,-8)$.

For questions 3 through 8, identify the image point based on the transformation given:

3. ΔJIM is going to be reflected over the y-axis. It has preimage vertices $J(0,-6)$, $I(5,-2)$, $M(4,-5)$. Find its image vertices.

4. ΔKAI has preimage vertices $K(4,-3)$, $A(-2,-4)$, $I(-1,0)$. What type of transformation has it undergone if its image vertices are $K'(3,4)$, $A'(4,-2)$, $I'(0,-1)$?

5. ΔMAT has preimage vertices $M(9,-8)$, $A(-4,0)$, $T(0,5)$. What type of transformation has it undergone if its image vertices are $M'(7,-5)$, $A'(-6,3)$, $T'(-2,8)$?

6. Plot preimage ΔJEN and its image $\Delta J'E'N'$ after the transformation $D_{\frac{1}{3}}$: $J(12,-9)$, $E(9,6)$, $N(-9,3)$.

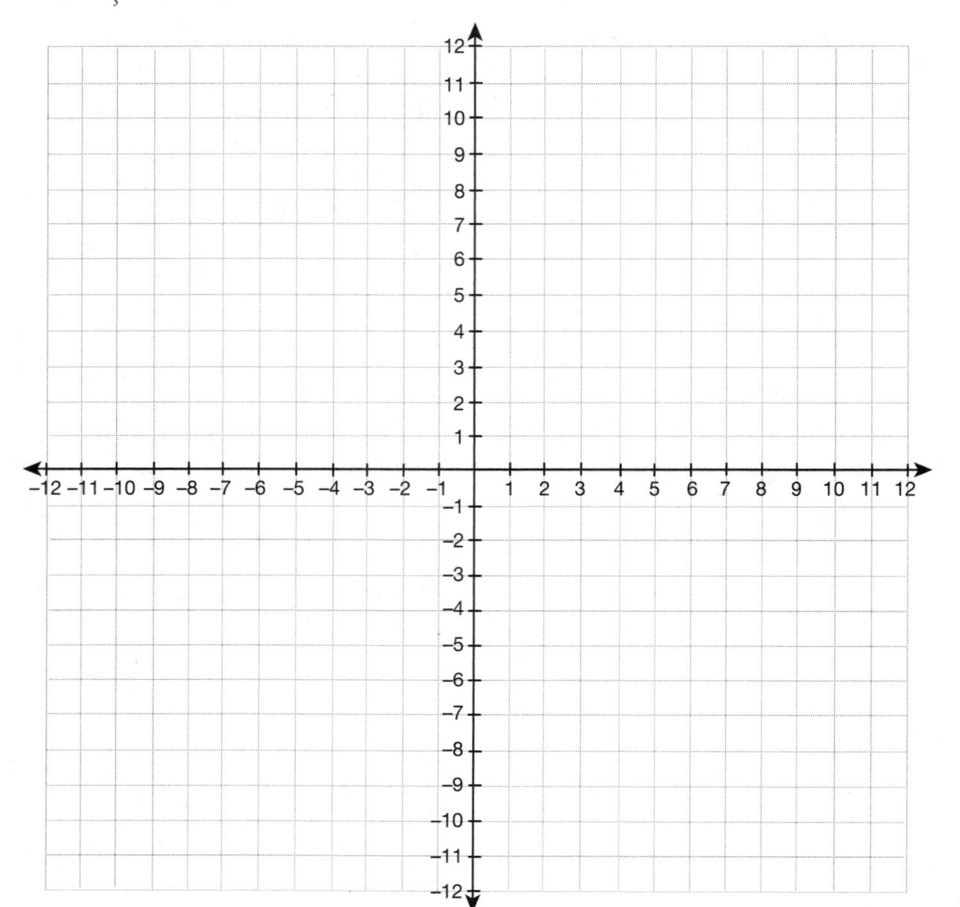

7. **a.** Plot trapezoid *BOYD* with the given vertices: $B(-6,-2)$, $O(-3,-2)$, $Y(-1,-5)$, and $D(-6,-5)$.

b. Now plot $B'O'Y'D'$ by reflecting *BOYD* over the x-axis.

c. Next plot $B''O''Y''D''$ by reflecting $B'O'Y'D'$ over the y-axis. Record the coordinates of $B''O''Y''D''$ here: B''(__,__), O''(__,__), Y''(__,__), D''(__,__).

d. Describe what you notice about the appearance of $B''O''Y''D''$ compared to *BOYD*.

e. Finally, make a hypothesis about the rule for a coordinate pair (x,y) that is reflected over the x-axis and then reflected over the y-axis. Write your rule here: $(x,y) \rightarrow$ (__,__).

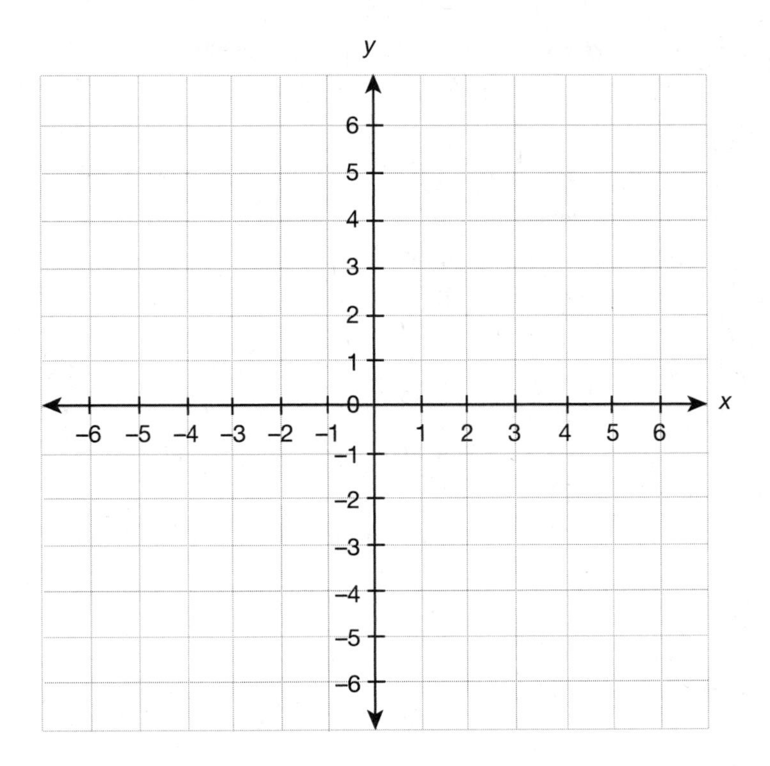

8. **a.** Preimage point $V(9,–3)$ is reflected over the y-axis to find V'(__,__).

b. V' is then rotated 90° counterclockwise to find V''(__,__).

c. Now make a hypothesis about the rule for a coordinate pair (x,y) that is reflected over the y-axis and then rotated 90° counterclockwise. Write your rule here: $(x,y) \rightarrow$ (__,__).

9. Fill in the chart below with "*stays the same*" or "*changes*" to show how each type of translation affects the general shape, angle measurements, side lengths, and orientation of a polygon:

	Reflections	Translations	Dilations	Rotations
Shape				
Angle Measure				
Side lengths				
Orientation				

Answers

Practice 1

1. After transforming point W, the new point will be named W'.
2. Reflections, translations, and rotations do not change the size of an image. Dilations make an image larger or smaller.
3. False: Reflections result in mirror images, but translations do not.
4. $R'((x + h),(y + k))$
5. $N'(-3,-7)$
6. $M'(-6,-1)$
7. $L'((2 + -3), (-5 + -4)) = L'(-1,-9)$
8. $J'(5,2)$ and $J''(-5,2)$
9. When $k < 0$, the dilated image will be smaller. When $k > 0$, the dilated image will be larger.
10. $A'(3.5,4)$, $R'(-1,2)$, and $M'(1.5,-3)$

Practice 2

1. False: The x- and y-coordinates will switch places, but only one of them will change signs. Which one depends on whether the rotation is clockwise or counterclockwise.
2. True.
3. $J'(0,-6)$, $I'(-5,-2)$, $M'(-4,-5)$ (Note that 0 cannot become –0, so coordinates that are on the axes will not change signs in this case.)
4. ΔKAI has undergone a 90° counterclockwise rotation.
5. Each of the vertices in $M'A'T'$ underwent the transformation $T_{-2,3}$.

6. $J'(4,-3), E'(3,2), N'(-3,1)$

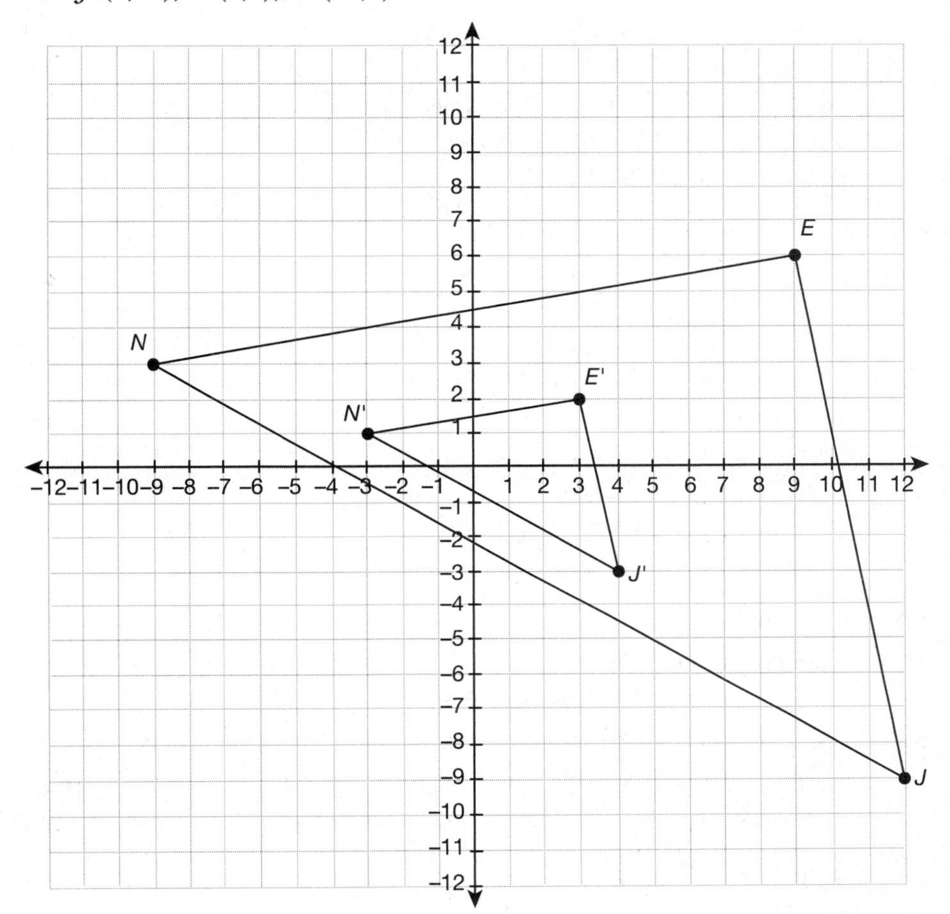
E
N
E'
N'
J'
J
12
11
10
9
8
7
6
5
4
3
2
1
−1
−2
−3
−4
−5
−6
−7
−8
−9
−10
−11
−12
−12 −11 −10 −9 −8 −7 −6 −5 −4 −3 −2 −1
1 2 3 4 5 6 7 8 9 10 11 12

7.

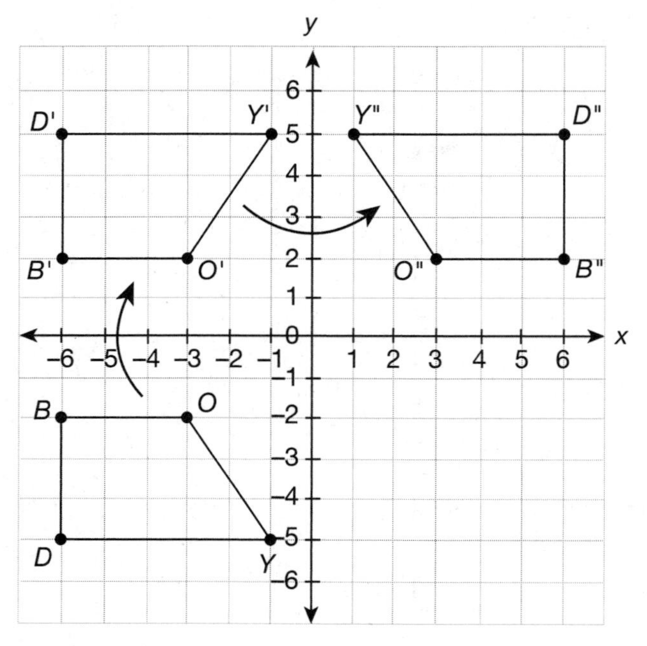

a. See coordinate plane

b. See coordinate plane

c. $B''(6,2)$, $O''(3,2)$, $Y''(1,5)$, $D''(6,5)$.

d. $B''O''Y''D''$ is still a trapezoid with one right angle and one pair of parallel sides. It has been flipped and flipped, so now it is actually a rotation of 180° (clockwise or counterclockwise) of the original preimage *BOYD*.

e. The x- and y-coordinates are in the same positions, but they both changed signs: $(x,y) \rightarrow (-x,-y)$.

8. a. $V(9,-3) \rightarrow V'(-9,-3)$

b. $V'(-9,-3) \rightarrow V''(3,-9)$

c. The x- and y-coordinates have switched positions, and they both changed signs: $(x,y) \rightarrow (-y,-x)$.

9.

	Reflections	Translations	Dilations	Rotations
General shape	stays the same	stays the same	stays the same	stays the same
Angle Measurements	stays the same	stays the same	stays the same	stays the same
Side lengths	stays the same	stays the same	changes	stays the same
Orientation	changes	stays the same	stays the same	changes

21

Applying Transformations to Congruency and Similarity

Since the mathematicians have invaded the theory of relativity, I do not understand it myself any more.
—Albert Einstein

In this lesson you will learn more about congruency and how to use reflections, rotations, and translations to determine if two figures are congruent. You will also learn about similar figures and how to use dilations, along with the other transformations, to determine if two figures are similar.

STANDARD PREVIEW

Reproducing images through transformations is a common tool used in interior design, architecture, and advertising. One example is the repetition of objects that are slightly altered in some way. This is a technique that artists or builders have been using for centuries to create appealing images or stable structures. Andy Warhol's iconic silkscreens of repeated images of Marilyn Monroe or Campbell Soup cans are great examples of congruent images that have been replicated and translated to create popular art. The famous luxury brands Gucci and Coco Chanel have both marketed their products with logos that reflect and translate congruent letters. ***Standard 8.G.A.2 asks students to recognize that a figure is congruent to another when one shape can produce the other shape through a sequence of rotations, reflections, and translations.*** Rather than doing these transformations on a coordinate plane, we're going to focus on how these transformations change the appearance of objects while preserving their key attributes. ***Standard 8.G.A.4 adds dilations to the mix as students are asked to apply all four types of transformations in order to determine if two figures are similar.***

What Is Congruence?

When two shapes are **congruent**, it means that they have the same appearance, size, and relationship between their different parts. For example, if a tire on a small car goes flat, a tire that is *congruent* to the damaged tire is needed to replace it. An oversized tire from a tractor could not be used as a replacement, nor could a thin tire from a bicycle. The replacement tire would need to have the same thickness, height, and overall shape; it would need to be congruent to the original tire.

How Are Translations Used with Congruence?

When a first image can be rotated, reflected, and/or shifted (translated) to fit directly on top of a second image, it is said that the two images are congruent. If two figures are congruent, it is appropriate to say that "figure 1 maps onto figure 2" after a specific series of transformations (such as a horizontal reflection followed by a translation). Look for that language in the upcoming examples.

TIP: Two-dimensional figures *A* and *B* are **congruent** if Figure *B* can be obtained from Figure *A* through a series of rotations, reflections, and/or translations.

Read on to understand how transformations are used to determine congruency between 2-dimensional figures:

Example: Use transformations to determine if ΔELF is congruent to ΔANT:

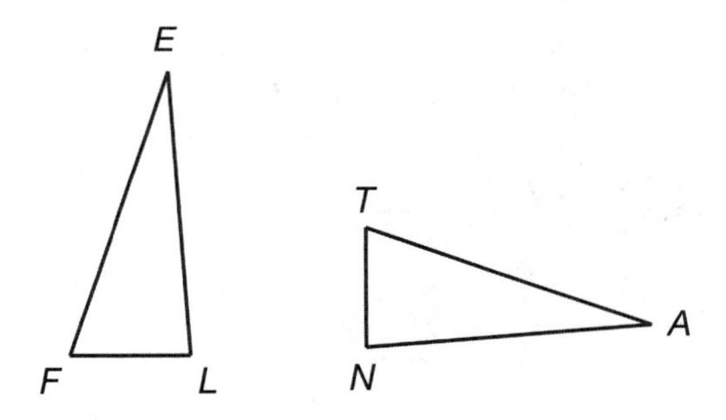

Step 1: Rotate $\Delta\, E\, L\, F$ clockwise to get $\Delta\, E'\, L'\, F'$

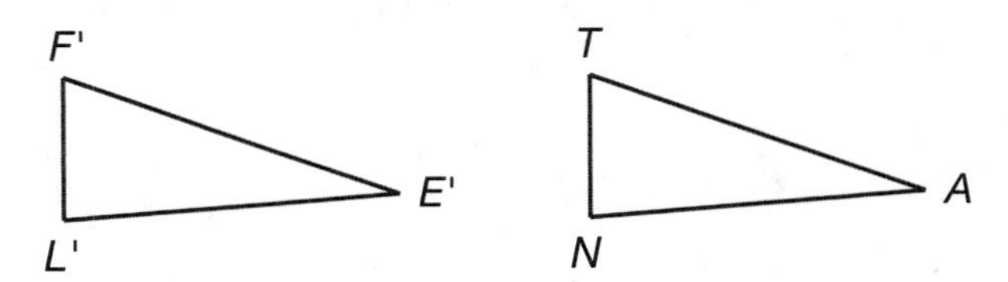

Step 2: Translate $\Delta\, E'\, L'\, F'$ so the corresponding sides and angles map onto $\Delta\, A\, N\, T$

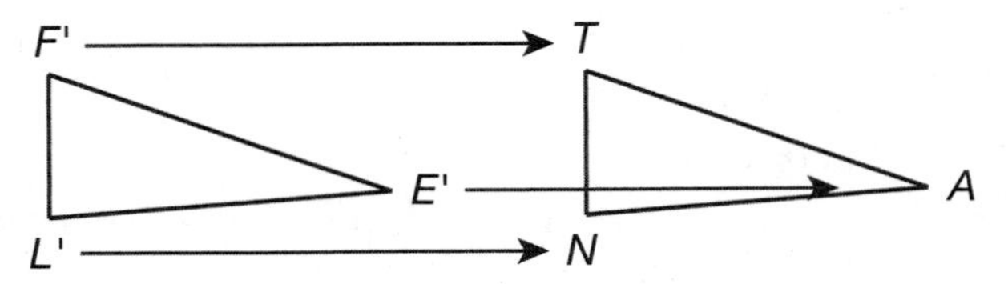

Therefore, it can be determined that ΔELF is congruent to ΔANT because a rotation followed by a translation maps ΔELF onto ΔANT.

Let's look at another example in which the figures are mirror images of each other:

> **Example:** Belicia Bradley uses the following logo for her curriculum consulting business. One *B* is a white letter with shaded accents, and the other *B* is shaded with white accents. Use transformations to determine if the white *B* is congruent to the shaded *B*.

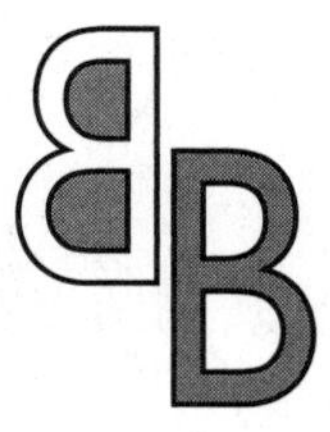

Step 1: Reflect the white *B* on the left over a vertical line to make white *B*'

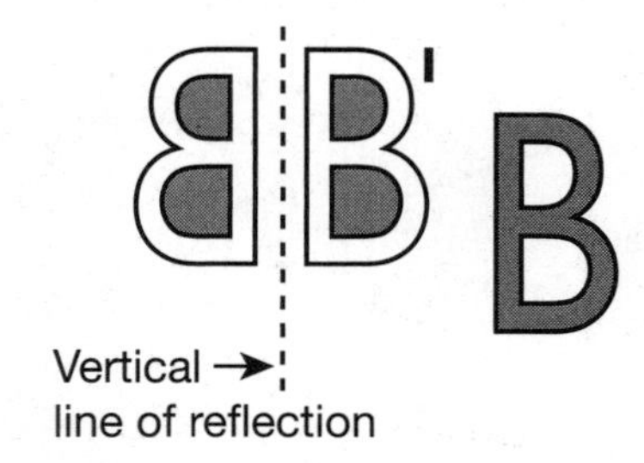

Step 2: Translate white *B*' so that it maps onto shaded *B*:

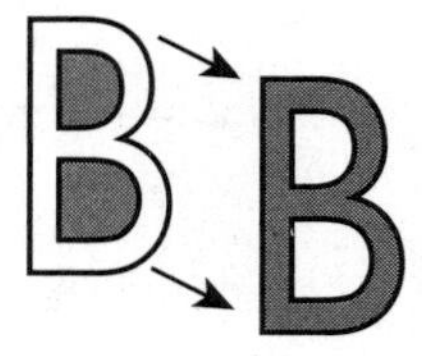

Since a reflection followed by a translation maps the white *B* onto the shaded *B*, it can be determined that they are congruent.

Sometimes when rotations are involved, it's more challenging to tell if two figures are congruent. The following is an example of a trickier investigation:

Example: A and B are images that a gaming company is thinking of using as graphics. Use transformations to determine if A is congruent to B.

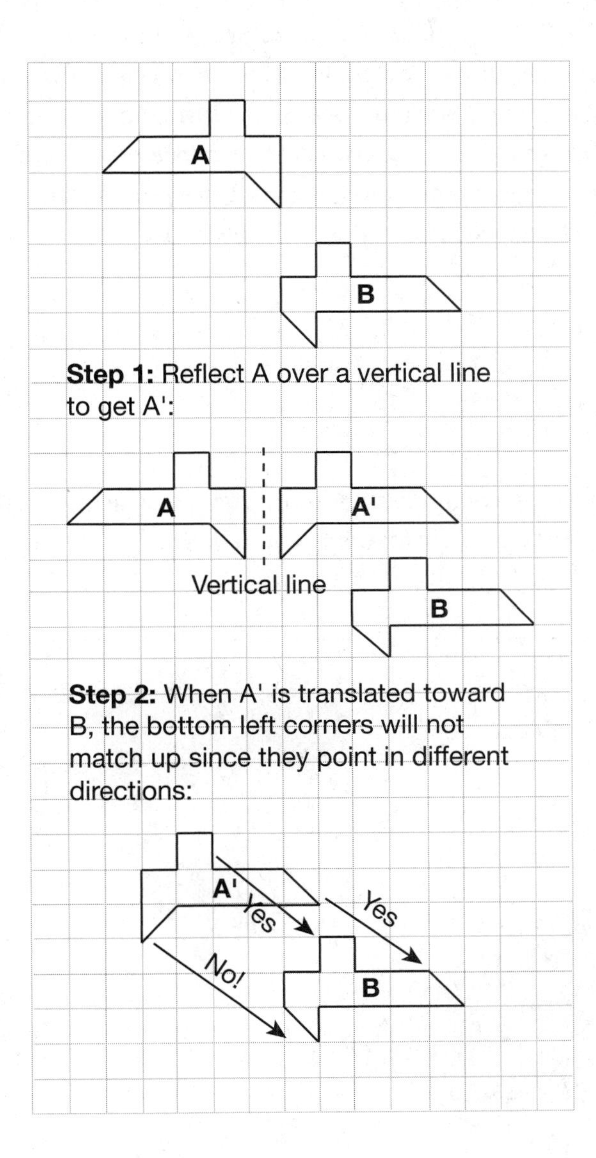

Although rotating and translating figure A matches up most of the corresponding parts, the figures do not map onto each other perfectly, since the bottom left corners don't match exactly. Therefore, these two figures are not congruent.

STANDARD ALERT!

If you understand that a 2-dimensional figure is congruent to another figure if it maps onto the second figure after a sequence of rotations, reflections, and translations, then you are well on your way to mastering Standard 8.G.A.2. This standard also requires students to describe the sequence of translations that determines the congruence between two given figures. You will demonstrate both of these skills in this first set of practice questions.

Practice 1

Determine if the two figures are congruent by using transformations. Name what specific transformations would be used, and justify your answer clearly.

1.

2.

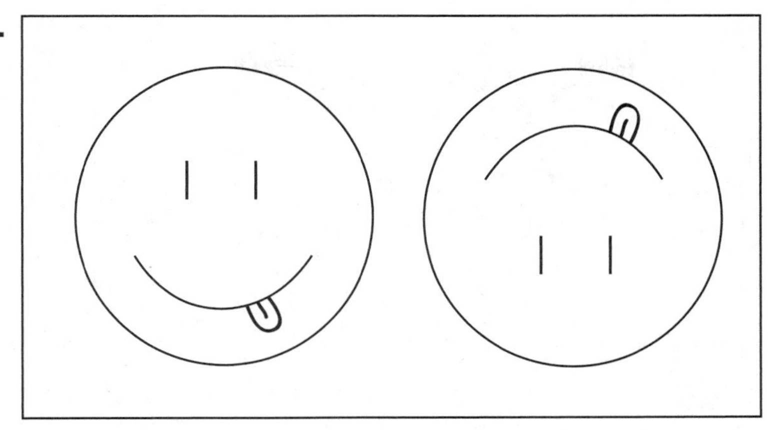

3.

4.

5.

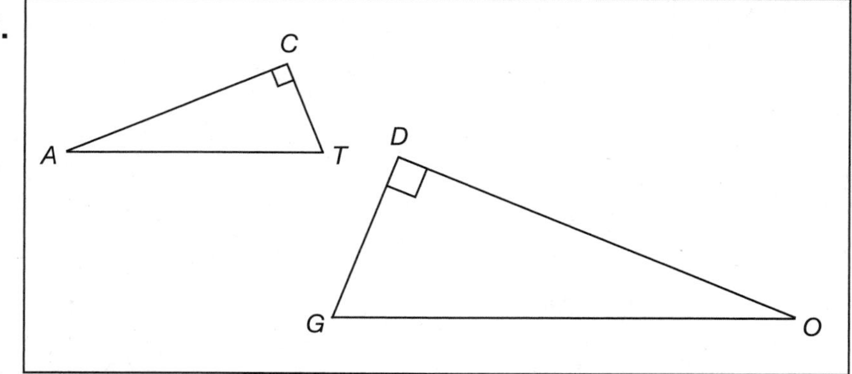

For questions 6 through 8, use the preimage and the given transformations to sketch the congruent image:

6. Reflect over a horizontal line and then reflect over a vertical line.

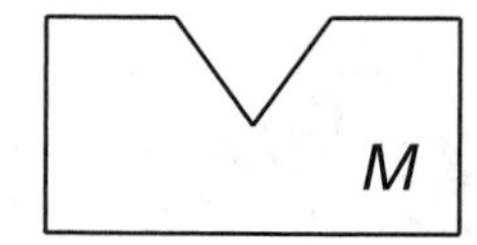

7. Rotate 90 degrees counterclockwise and then reflect over a horizontal line.

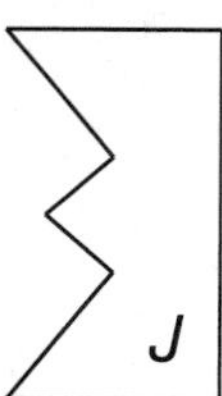

8. The following image is 10 units wide by 10 units tall. Reflect it to the right over a vertical line. Then translate it using $T_{-5,-5}$.

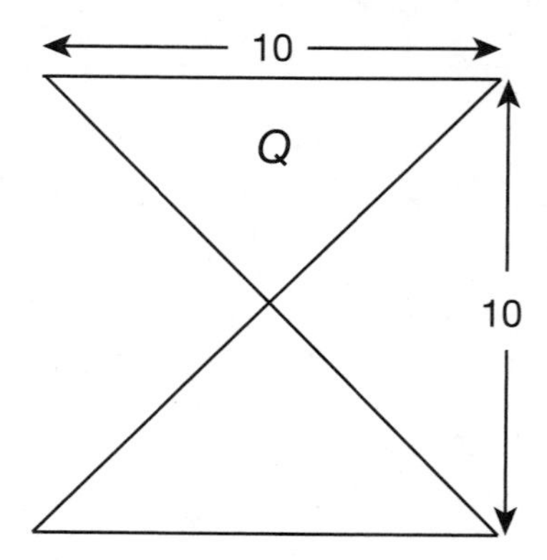

What Is Similarity?

Similarity is a bit like a relaxed version of congruence in that two figures look exactly alike but are not the same size. For example, consider two laptop computers made by the same company but with different screen sizes. A 15" laptop might have the same keyboard and overall appearance as a 13" laptop by the same manufacturer, but the 13" will be a smaller version of the 15" model. The same might go for two pairs of jeans made by the same designer but offered in different sizes. Even though the jeans are the same style and have the same relative appearance, one pair is just a smaller version of the other. This concept in math is referred to as "similarity." When two figures are the same overall shape but are different sizes, they are said to be **similar**.

How Are Translations Used with Similarity?

You already learned that two figures are congruent if one figure can map onto another figure after a series of rotations, reflections, and/or translations. Since similar figures are different-sized versions of each other, dilation is also needed to determine the similarity between two figures. When a first image can be rotated, reflected, translated, *and dilated* to fit directly on top of the second image, it is said that the two images are similar.

TIP: Two-dimensional figures *A* and *B* are **similar** if figure *B* can be obtained from figure *A* through a series of rotations, reflections, translations, and/or dilations.

In order to illustrate how transformations are used to determine similarity, let's return to two triangles that were presented in the first set of practice questions: ΔCAT and ΔDOG:

Example: Use transformations to determine if ΔCAT is similar to ΔDOG:

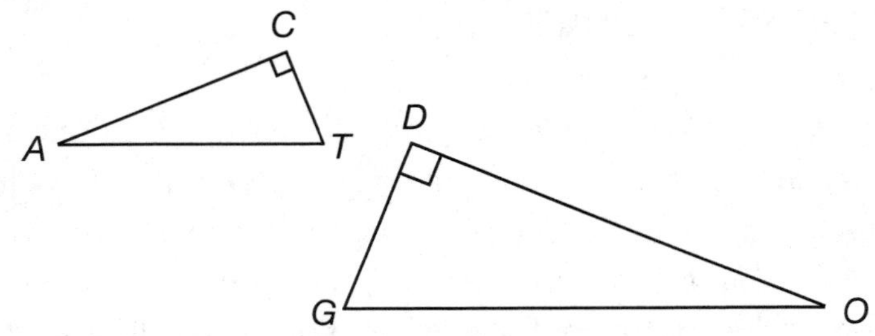

Step 1: Reflect Δ *CAT* over a vertical line to get *C'A'T'*:

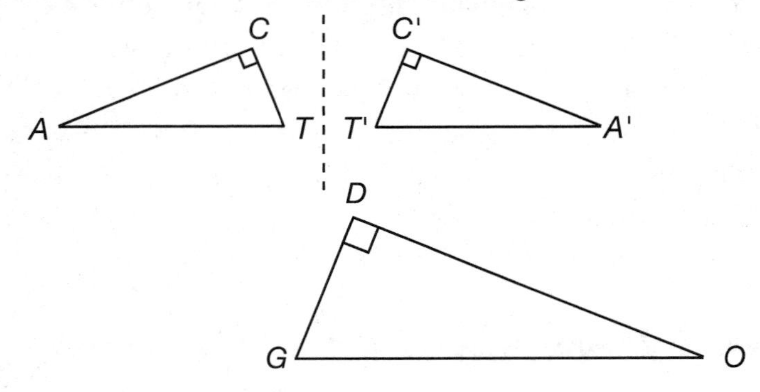

Step 2: Dilate Δ *C'A'T'* with a scale factor of 2 to get *C"A"T"*:

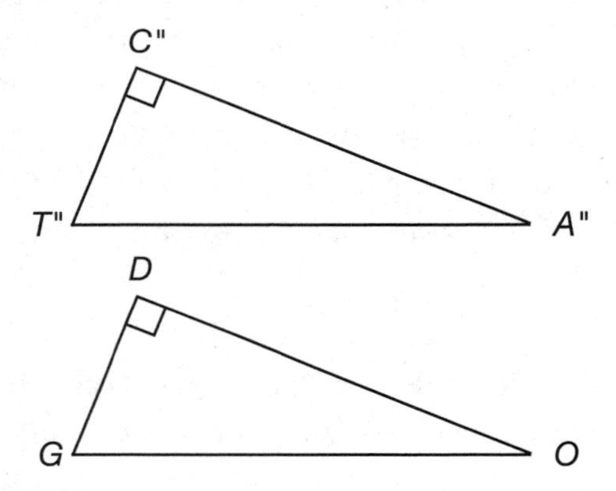

Step 3: Translate Δ *C"A"T"* so it maps onto Δ *DOG*:

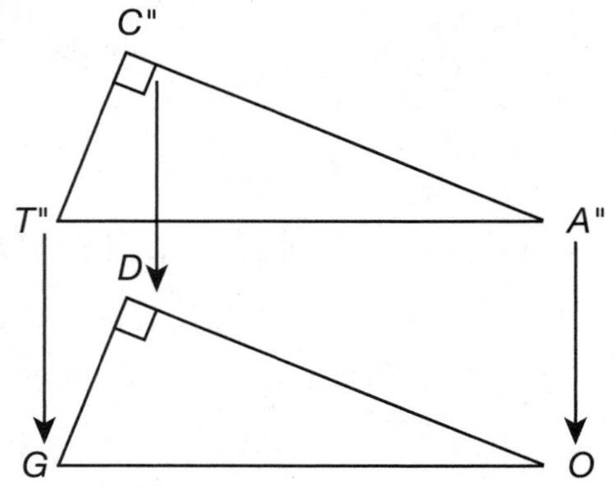

Since ΔCAT maps onto ΔDOG after a reflection, dilation, and translation, it follows that ΔCAT is similar to ΔDOG.

Since determining similarity has only added *dilation* to the list of transformations used, we think you're ready to try some questions on your own with this next practice set.

STANDARD ALERT!

If you now know that a pair of 2-dimensional figures are similar if one figure maps onto the second figure after a sequence of rotations, reflections, translations, and dilations, then you are doing well with Standard 8.G.A.4. This standard also requires student to describe the sequence of translations that determines the similarity between two given figures. This second set of practice questions gives you the opportunity to show your comfort with performing these skills.

Practice 2

Determine if the two figures are similar by using transformations. Name what transformations would be used, and justify your answer clearly.

1.

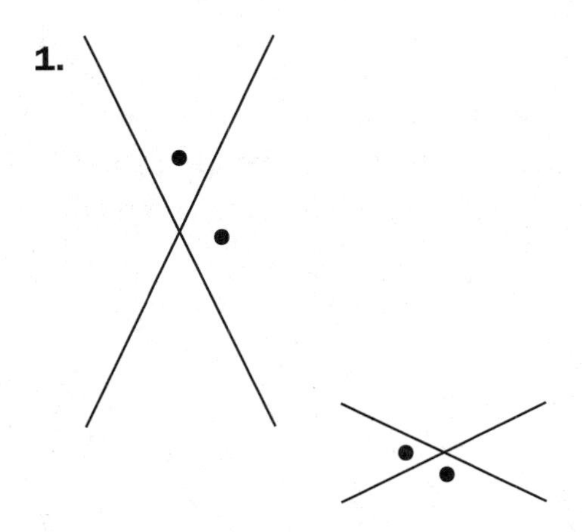

2.

3.

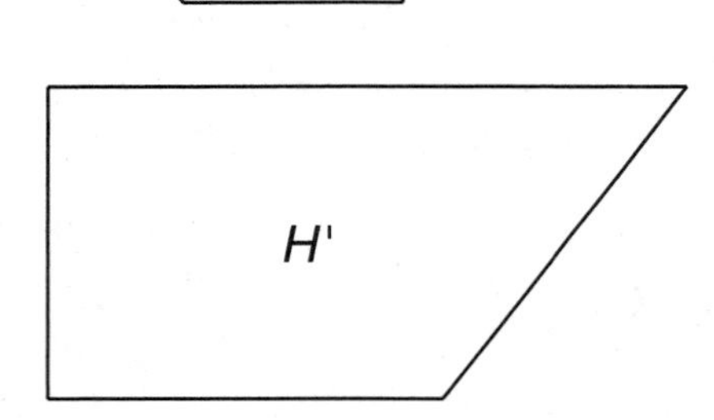

4. Emmy is a graphic designer who loves using transformations in her designs. She came up with this image for a charter-jet company. Determine if the *V* in "Vistair" is similar to the *A* in "Aviation" by using transformations.

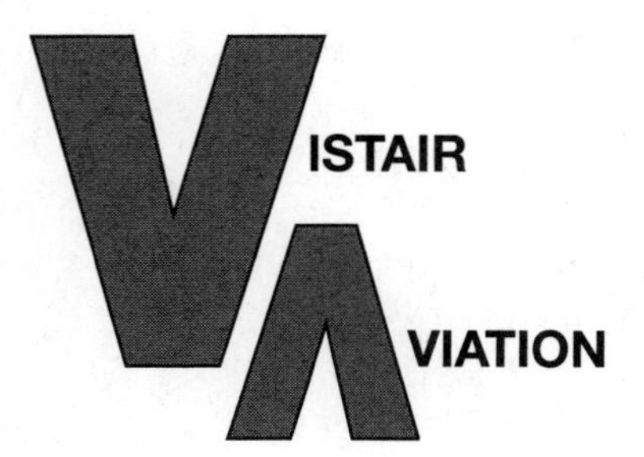

5. Asad and Amara are out flying kites! How could transformations show similarity between their two kites?

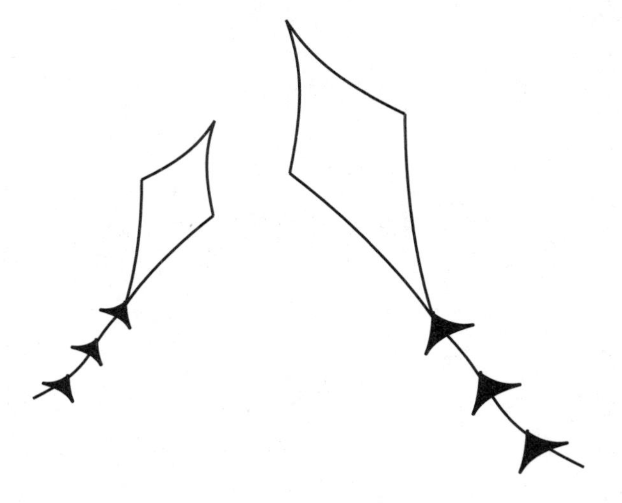

For questions 6 through 8, use the preimage and the given transformations to sketch the congruent image.

6. Rotate the preimage 90° clockwise and dilate using a scale factor of $\frac{1}{2}$. Would this be a good logo for Ned Zanders?

7. Reflect the preimage down over a horizontal line. Then reflect it to the right over a vertical line. Finally, dilate it using a scale of 1.5.

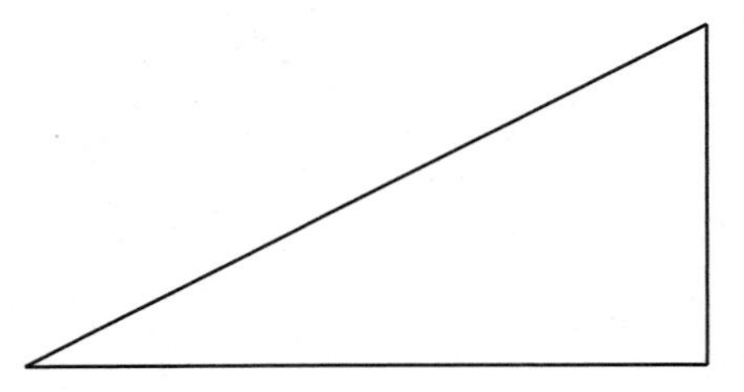

8. Reflect to the right over a vertical line and dilate using a scale factor of 2.

Answers

Practice 1

1. Yes, these figures are congruent by clockwise rotation followed by translation.
2. No, these figures are not congruent, since after the image on the right is rotated 180°, the tongue is on the opposite side as in the preimage on the left.
3. Yes, these figures are congruent by reflection over a vertical line followed by translation.
4. No, these figures are not congruent because the square inside the image on the right is located in the shorter side of the *L*, while the square inside the preimage on the left is located in the longer side of the *L*.
5. These two triangles are not congruent. Although they have the same shape, the image is larger than the preimage.
6.
7.

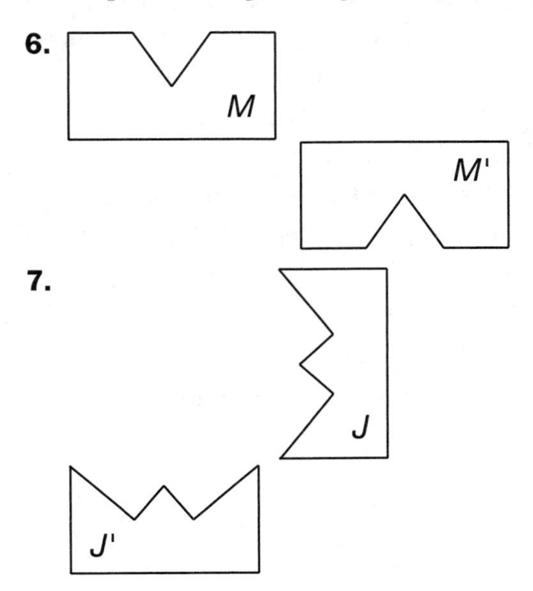

8.

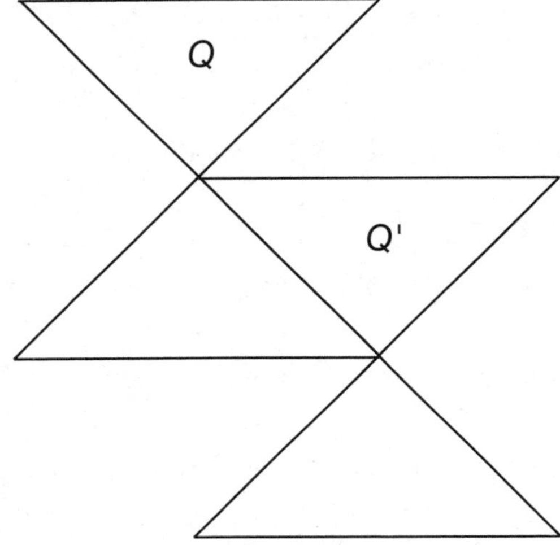

Practice 2

1. Yes, these figures are similar by a 90° clockwise rotation followed by a reflection over a vertical line, a dilation with a scale factor less than 1, and finally a translation. These figures used all four transformations!
2. No, these flowers are not similar, since after reflecting the preimage over a vertical line, they are both leaning the same direction, but the solo petal is on the opposite side of the stem.
3. Yes, these trapezoids are similar by a reflection over a vertical line, followed by a dilation with a scale factor greater than 1 and then by a translation.
4. Yes, the *V* and the *A* are similar, since the *A* is made out of a *V* that has been reflected, dilated, and translated.
5. These are similar by a reflection, dilation, and translation.

6.

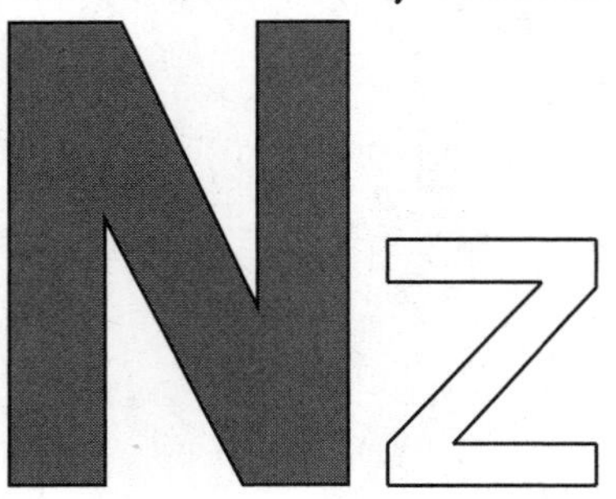

7.

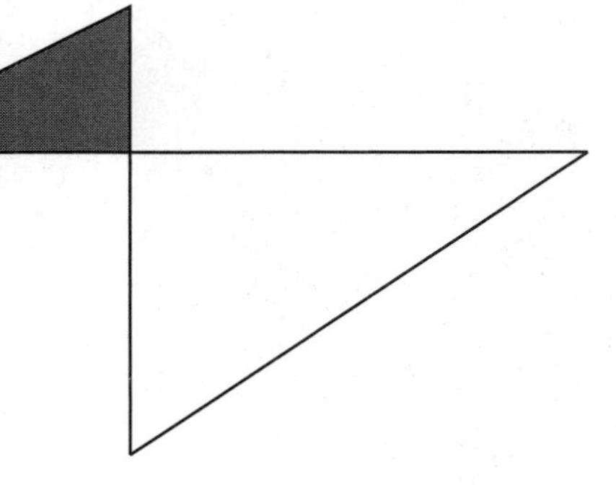

8.

22 Applying Proportions to Similar Polygons

Let no one ignorant of geometry enter here.
—Inscription above Plato's Academy

In this lesson you will learn what it means for polygons to be similar and about the special postulates that relate to similar triangles. You will also apply what you learned about ratios and proportions to solving problems involving similar polygons.

STANDARD PREVIEW

You have learned the building blocks of similarity, and in this lesson we will apply those to further investigate similarity of triangles and of polygons. You will first have an opportunity to get more practice with ***Standard 7.G.A.1 as you solve for missing side lengths in polygons***. Then you will cover ***Standard 8.G.A.5 as you come to understand that as long as triangles have two congruent angles, they are similar***. In this lesson we include some extensions of triangle similarity that are not required by the middle school Common Core State Standards, but they will give you a useful preview of material you will cover in high school.

How to Determine Similarity

In the last lesson you learned some of the basics of similarity, and in this lesson you will build upon that as we cover the last required standard for middle school geometry and give you a preview of some content you will tackle in high school. The first of the two conditions that must be met in order to determine similarity relates to angle measurements. *In order for two polygons to be similar, their corresponding angles must be congruent.* You can see in the figure below that $\angle C \cong \angle P$, $\angle U \cong \angle I$, $\angle T \cong \angle G$, and $\angle E \cong \angle S$, so this condition is met for polygons *CUTE* and *PIGS*.

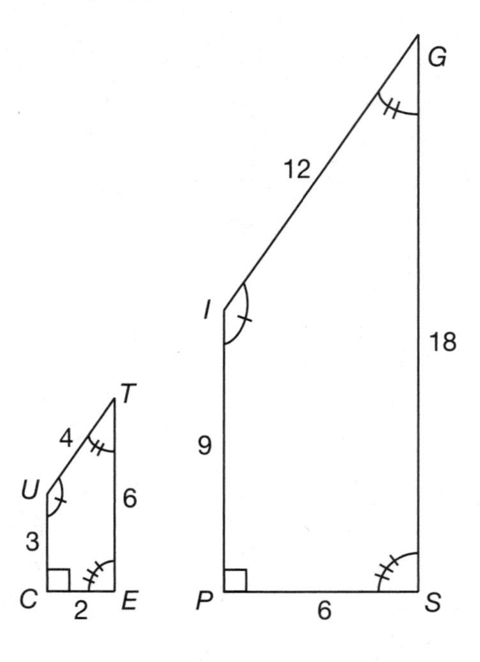

Next look at the polygons' corresponding sides. *The corresponding sides of similar polygons must be proportional.* This means that all the ratios of their corresponding sides must be equal. You can see here that the ratios of the corresponding sides all reduce to $\frac{1}{3}$.

$$\frac{\overline{CU}}{\overline{PI}} = \frac{3}{9} = \frac{1}{3}$$

$$\frac{\overline{UT}}{\overline{IG}} = \frac{4}{12} = \frac{1}{3}$$

$$\frac{\overline{TE}}{\overline{GS}} = \frac{6}{18} = \frac{1}{3}$$

$$\frac{\overline{EC}}{\overline{SP}} = \frac{2}{6} = \frac{1}{3}$$

Since both requirements for similarity have been satisfied, it can be concluded that *CUTE* is similar to *PIGS*. The symbol used to represent similarity is "~" so their similarity is represented as *CUTE ~ PIGS*.

TIP: If two polygons have congruent corresponding angles and all of their corresponding sides are proportional, then they are **similar polygons.**

Naming Similar Polygons

When dealing with similar polygons, the order that the vertices are written in is very important. You must write the corresponding vertices of each polygon in the same order. This way, the relationship between corresponding sides is clear. Consider the two similar polygons in the next figure.

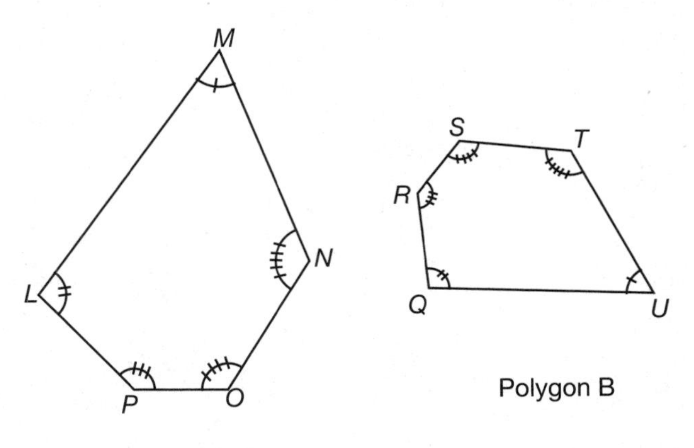

Polygon A

Polygon B

In this figure, we will name Polygon A by starting with $\angle M$ and moving in a counterclockwise direction: $MLPON$. In Polygon B, $\angle U$ is congruent to $\angle M$, so we will begin naming Polygon B with $\angle U$. If we were to again move in a counterclockwise direction, the name of Polygon B would be $UTSRQ$. This would imply that $\angle T$ is congruent to $\angle L$, since they both are the second vertex in the name. Looking at the hash marks of the angles, you can see that $\angle L$ is not congruent to $\angle T$, but instead $\angle L \cong \angle Q$. Therefore, to name Polygon B correctly, we must begin with vertex U and move in a clockwise fashion. When written that $UQRST \sim MLPON$, the congruent angles match up with one another. Now, the corresponding sides of these two polygons exist between the pairs of congruent angles. Therefore, we can conclude that $\overline{RS}$ is corresponding to $\overline{PO}$.

Practice 1

Use the preceding figure to answer questions 1 and 2.

1. $\angle O$ corresponds to what angle in Polygon B?
2. $\overline{LP}$ corresponds to what side in Polygon B?
3. True or false: The ratio of two pairs of sides of a polygon is sometimes equal to the ratio of the two corresponding pairs of a similar polygon.
4. True or false: When naming similar polygons, it is necessary to name both shapes by listing their vertices in the same direction—either clockwise or counterclockwise.

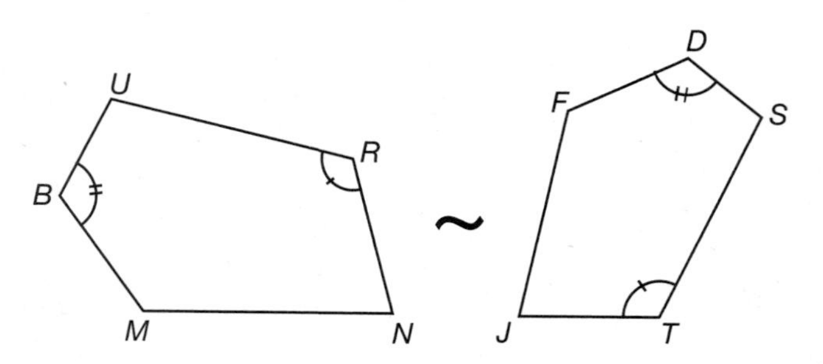

Use the similar polygons in the preceding figure to answer questions 5 through 10.

5. Beginning with $\angle M$, name the two polygons correctly to represent their similarity.

6. $\angle S$ corresponds to what angle?

7. $\angle N$ corresponds to what angle?

8. What is the corresponding side to $\overline{UR}$?

9. What is the corresponding side to $\overline{FJ}$?

10. Complete the following proportion: $\frac{\overline{BM}}{\overline{DF}} = \frac{?}{\overline{JT}}$

Using Proportions with Similar Polygons

Now that you know how to name similar polygons and identify their corresponding sides, you are ready to learn how to use proportions to solve for missing information in similar polygons. Consider the two triangles in the following figure.

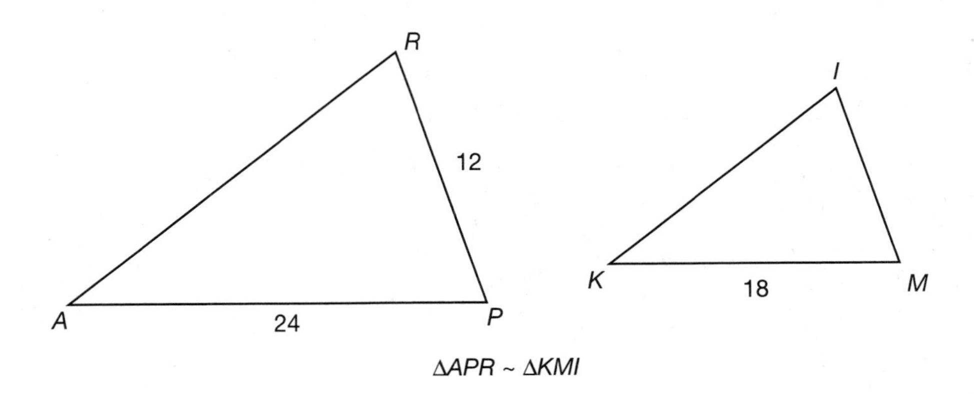

$\Delta APR \sim \Delta KMI$

The following proportion can be written about the corresponding sides of similar triangles *APR* and *KMI*:

$$\frac{\text{Medium side of Large } \Delta}{\text{Medium side of Small } \Delta} = \frac{\text{Small side of Large } \Delta}{\text{Small side of Small } \Delta}$$

Notice that this proportion compares the corresponding sides of the two triangles in each ratio. Fill in the names of the side lengths, and then substitute in the corresponding values from the illustration:

$$\frac{\overline{AP}}{\overline{KM}} = \frac{\overline{RP}}{\overline{IM}} \rightarrow \frac{24}{18} = \frac{12}{IM}$$

In Lesson 13 we learned to use cross multiplication to solve a proportion:

$$24(\overline{IM}) = 18(12)$$
$$24(\overline{IM}) = 216$$
$$(\overline{IM}) = \frac{216}{24}$$
$$\overline{IM} = 9$$

This is the method used to solve for missing information in similar polygons.

You probably remember that if the cross products from a proportion are not equal, then the proportion is false. This fact can be applied to similar polygons. If the proportion of corresponding sides of a polygon is false, then the polygons are not similar. For example, in the next figure, ΔCAT is not similar to ΔDOG since the cross products of the proportion are not equal.

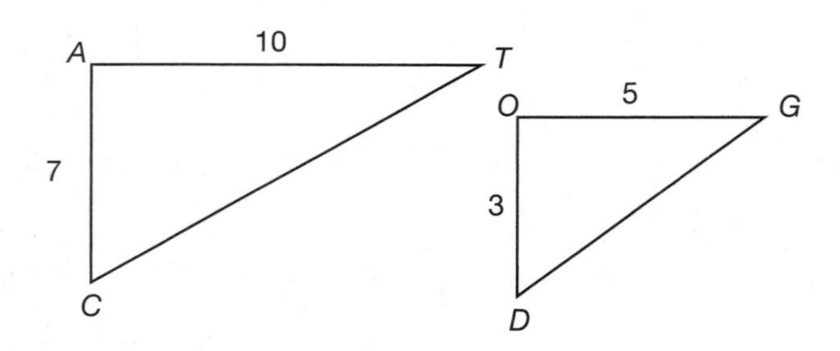

$$\frac{\overline{AT}}{\overline{OG}} \stackrel{?}{=} \frac{\overline{AC}}{\overline{OD}}$$
$$\frac{10}{5} \stackrel{?}{=} \frac{7}{3}$$
$$10(3) \stackrel{?}{=} 5(7)$$
$$30 \neq 35$$

Practice 2

Use this figure to answer questions 1 through 3.

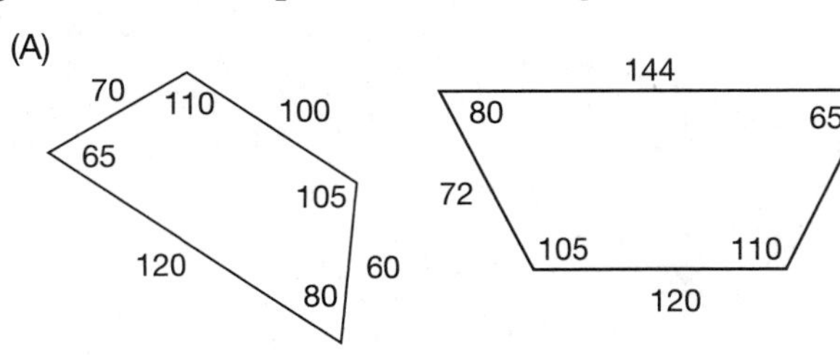

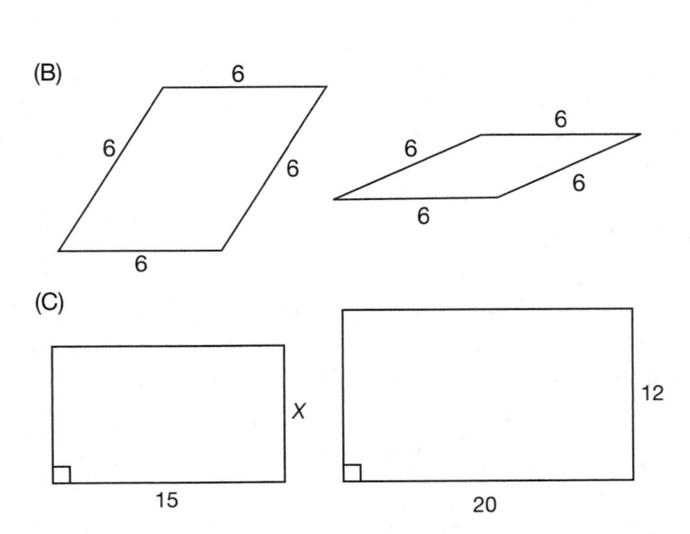

1. Explain why the quadrilaterals in part A are or are not similar.

2. Explain why the quadrilaterals in part B are or are not similar.

3. What would the length of side x be in order to make the rectangles similar in part C?

4. Polygon *WERT* is similar to polygon *VBNM*; $ER = 18$, $BN = 25.2$, and $WT = 26$. What is the length of side *VM*?

The two following trapezoids are similar. Use them to answer questions 5 through 8.

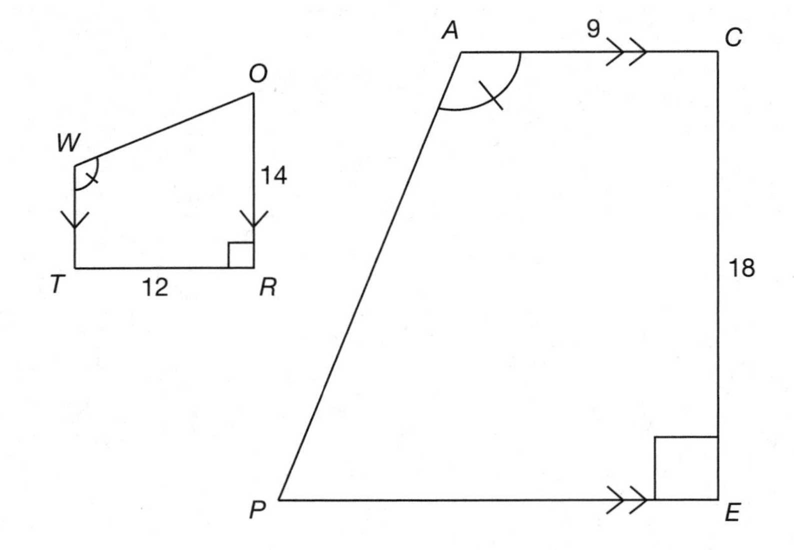

5. ______ ~ ______ Name the trapezoids to show their similarity.

6. Set up and solve a proportion to find the length of $\overline{PE}$.

7. Set up and solve a proportion to find the length of $\overline{WT}$.

8. What is the area of trapezoid *WORT*?

Similarity of Triangles

Triangles are unique in that both similarity requirements do not need to be simultaneously met in order to determine that two triangles are similar. There are actually three separate tests (also called *postulates*) that can be used to prove that two triangles are similar.

STANDARD ALERT!

Pay close attention to the first test here, since this is part of Standard 8.G.A.5. The rest of the material after Test 1 is an extension of similarity and proportion and is not required by the middle school standards.

Test 1. Angle-Angle

If two triangles have two congruent interior angles, then they are similar. (Since the interior angles of a triangle sum to 180 degrees, as long as two angles are congruent, it can be identified that the third angles in each triangle are also congruent.)

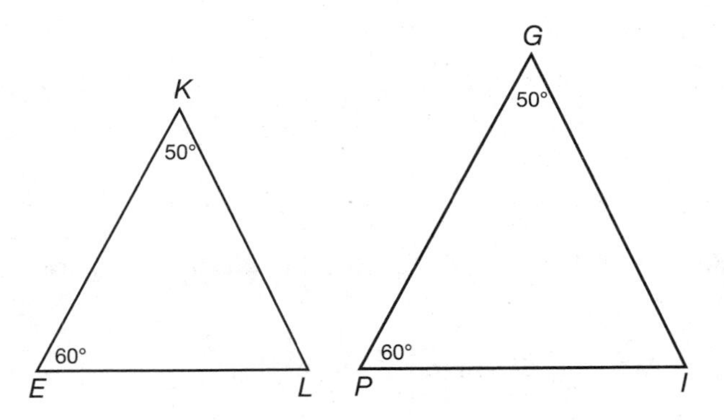

Test 2. Side-Angle-Side

If two pairs of corresponding sides of triangles are proportional, and their *included* angles are congruent, then the triangles are similar. The *included* angle is the angle formed by the two given sides. When checking to see if sides are proportional in triangles, be certain to compare the smallest sides, largest sides, and middle-length sides of each triangle to each other.

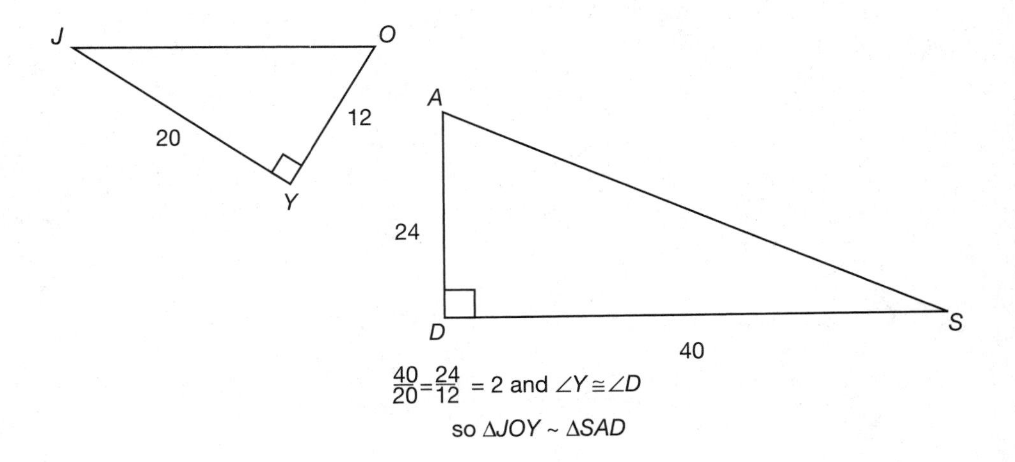

$\frac{40}{20} = \frac{24}{12} = 2$ and $\angle Y \cong \angle D$

so $\Delta JOY \sim \Delta SAD$

Test 3. Side-Side-Side

If the lengths of all pairs of corresponding sides of two triangles are all equally proportional, then the triangles are similar.

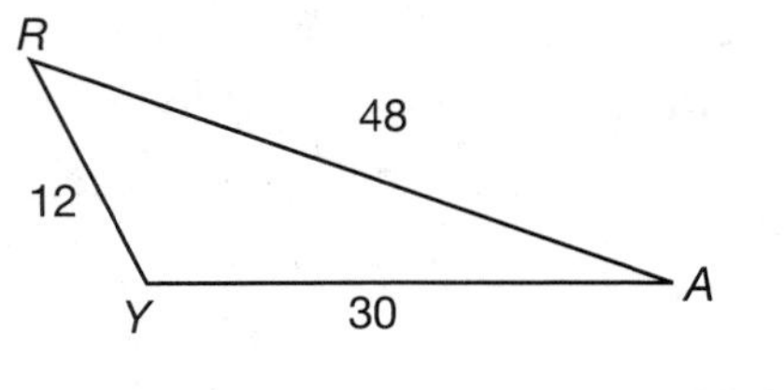

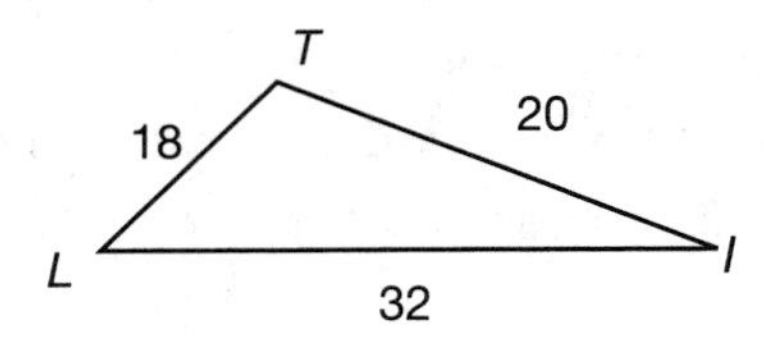

$$\frac{48}{32} = \frac{30}{20} = \frac{18}{12} = \frac{1.5}{1}$$

$$\Delta RAY \sim \Delta LIT$$

TIP: The following postulates can be used to determine that two triangles are similar:

1. **Angle-Angle:** Two pairs of congruent angles.
2. **Side-Angle-Side:** Two pairs of corresponding sides are proportional and the included angles are congruent.
3. **Side-Side-Side:** Three pairs of corresponding sides are proportional.

Practice 3

State whether the triangles are similar and which postulate supports your answer.

1.

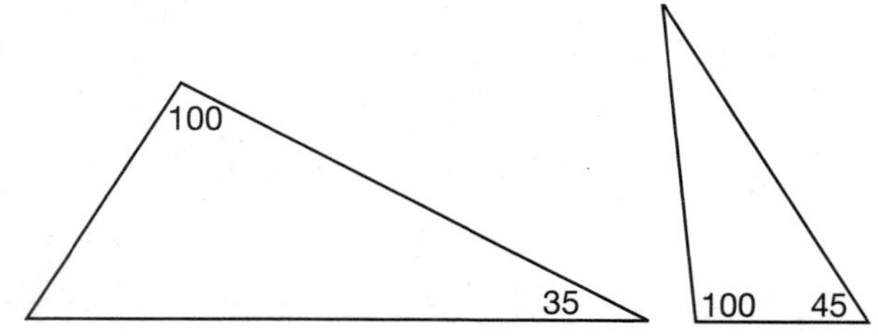

2.

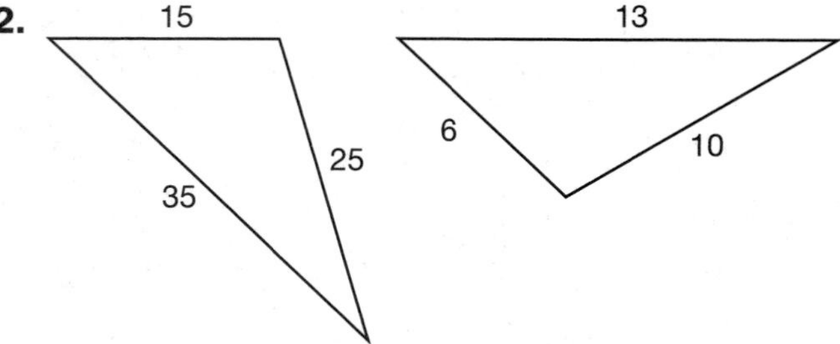

Which postulate, if any, could you use to prove that the triangles are similar?

3.

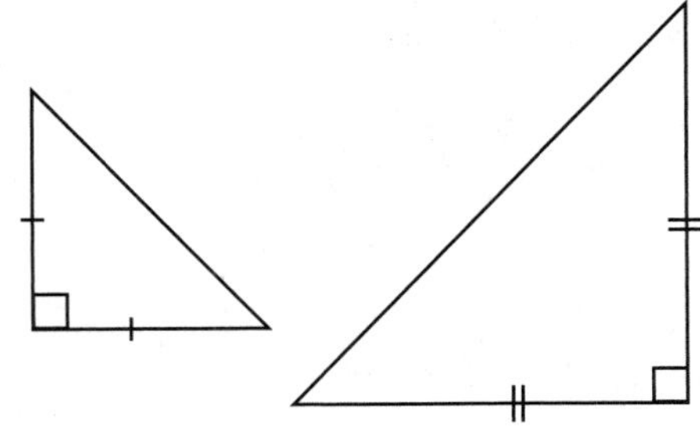

4.

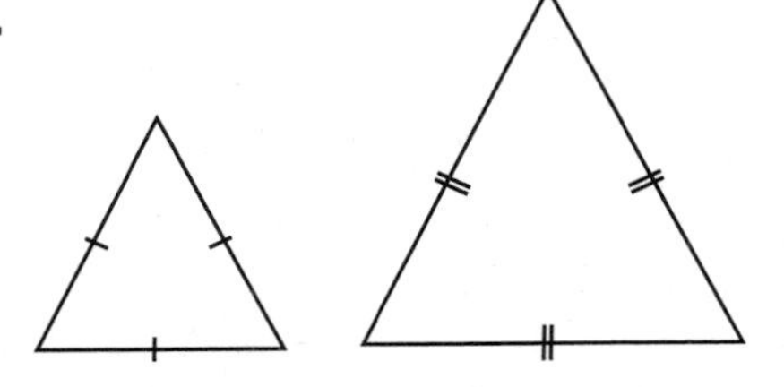

5.

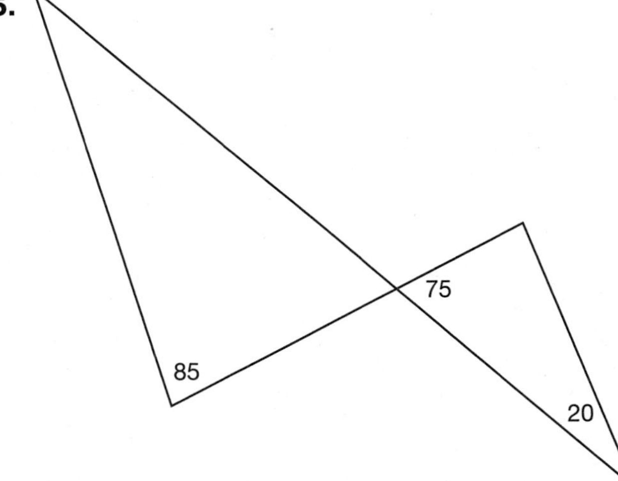

6.

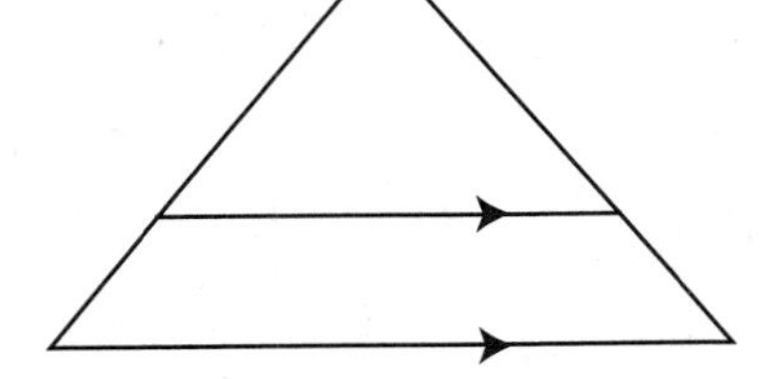

7.

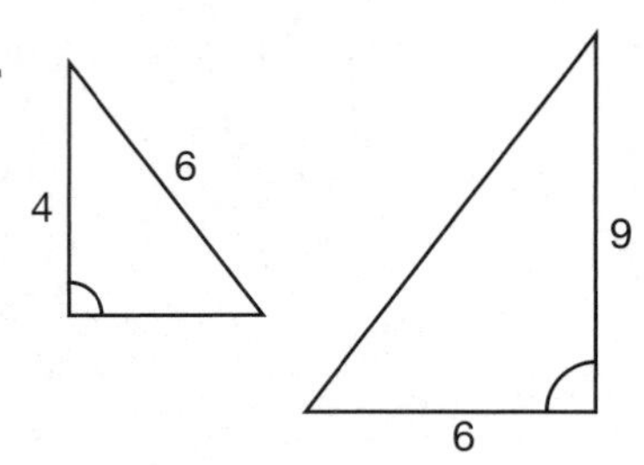

8.

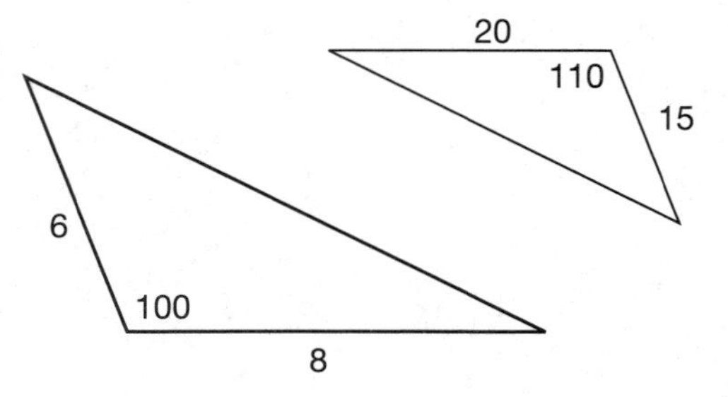

9.

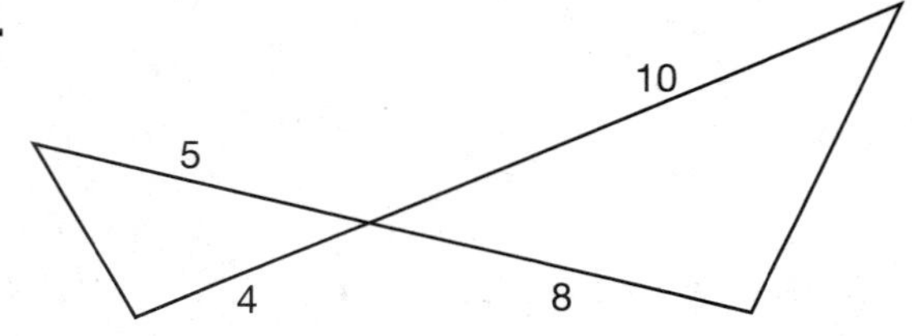

10.

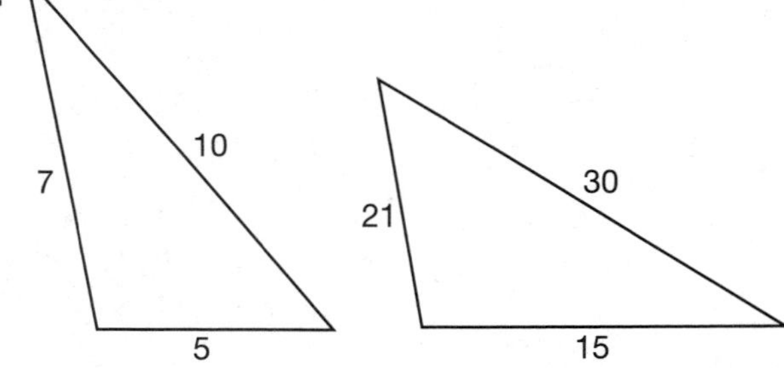

Answers

Practice 1

1. $\angle S$
2. $\overline{QR}$
3. False. The ratio of two pairs of sides of a polygon is *always* equal to the ratio of the two corresponding pairs of a similar polygon.
4. False. The direction is not important—what matters is that the congruent angles from each polygon are listed in the same order.
5. $MBURN \sim FDSTJ$ or $MNRUB \sim FJTSD$
6. $\angle U$
7. $\angle J$
8. $\overline{ST}$
9. $\overline{MN}$
10. $\frac{\overline{BM}}{\overline{DF}} = \frac{\overline{NR}}{\overline{JT}}$ (It would not be correct to use $\overline{RN}$.)

Practice 2

1. $\frac{72}{60} = \frac{84}{70} = \frac{120}{100} = \frac{144}{120} = 1.2$. These quadrilaterals are similar because the ratios of corresponding sides of the larger quadrilateral to the smaller quadrilateral are all equal to 1.2.
2. Although these polygons appear to be similar, because no information is given about their angles, we can't determine if they are indeed similar. A rhombus and a square can both have side lengths of 6, but different angle measures would keep them from being similar.
3. $\frac{15}{20} = \frac{x}{12}$

   $20(x) = 15(12)$

   $20x = 180$

   $x = \frac{180}{20}$

   $x = 9$
4. $\frac{\overline{ER}}{\overline{BN}} = \frac{\overline{WT}}{\overline{VM}}$

   $\frac{18}{25.2} = \frac{26}{\overline{VM}}$

   $25.2(26) = 18(\overline{VM})$

   $655.2 = 18(\overline{VM})$

   $\frac{655.2}{18} = (\overline{VM})$

   $\overline{VM} = 36.4$

5. $WORT \sim APEC$ or any other name that corresponds angles W and A; O and P; R and E; and T and C.
6. $\overline{PE} = 21$
7. $\overline{WT} = 6$
8. Area ($WORT$) = 120 units^2

Practice 3

1. Yes. Angle-Angle. (Solve for the missing angle in each triangle, and you will see that all three pairs of angles are congruent.)
2. No. $\frac{35}{13} \stackrel{?}{=} \frac{15}{6}$
   $210 \neq 195$
3. Yes. Side-Angle-Side
4. Yes. Angle-Angle; Side-Angle-Side; or Side-Side-Side
5. Yes. Angle-Angle (using vertical angles)
6. Yes. Angle-Angle
7. No. The proportional sides are not corresponding.
8. No
9. Yes. Side-Angle-Side
10. Yes. Side-Side-Side

23 Slope of a Line

The things of this world cannot be made known without a knowledge of mathematics.
—Roger Bacon

In this lesson you will learn what the slope of a line is and how slope is measured. You will learn about the slopes of horizontal, vertical, parallel, and perpendicular lines.

STANDARD PREVIEW

Linear equations and slope of lines are both most certainly part of the middle school math curriculum, but these two topics do not fall under the geometry standards. However, the geometry skills you have learned in previous lessons have given you the tools to successfully work with linear equations and their slopes. So although slope of lines is not found in the geometry standards, understanding slope requires much of the previous material in this book and we have decided to include it here for your enjoyment. You are encouraged to thoroughly read through this lesson because you will need to demonstrate an understanding of slope in your middle school math classes.

What Is Slope?

Any two points on a coordinate grid can be connected to form a line that will have a slope. Similar to how different ski runs at a winter resort have different trails of varying steepness, so do the lines on a coordinate grid. The **slope** of a line is its measure of steepness. Slope is read like words—from left to right. If a line is angled upward as you look at it from left to right, then it has a positive *slope*. Conversely, if a line is angled downward when read from left to right, it has a negative slope. In the next figure, $\overleftrightarrow{MJ}$ has a *steep positive slope* and $\overleftrightarrow{SF}$ has a *moderate negative slope*.

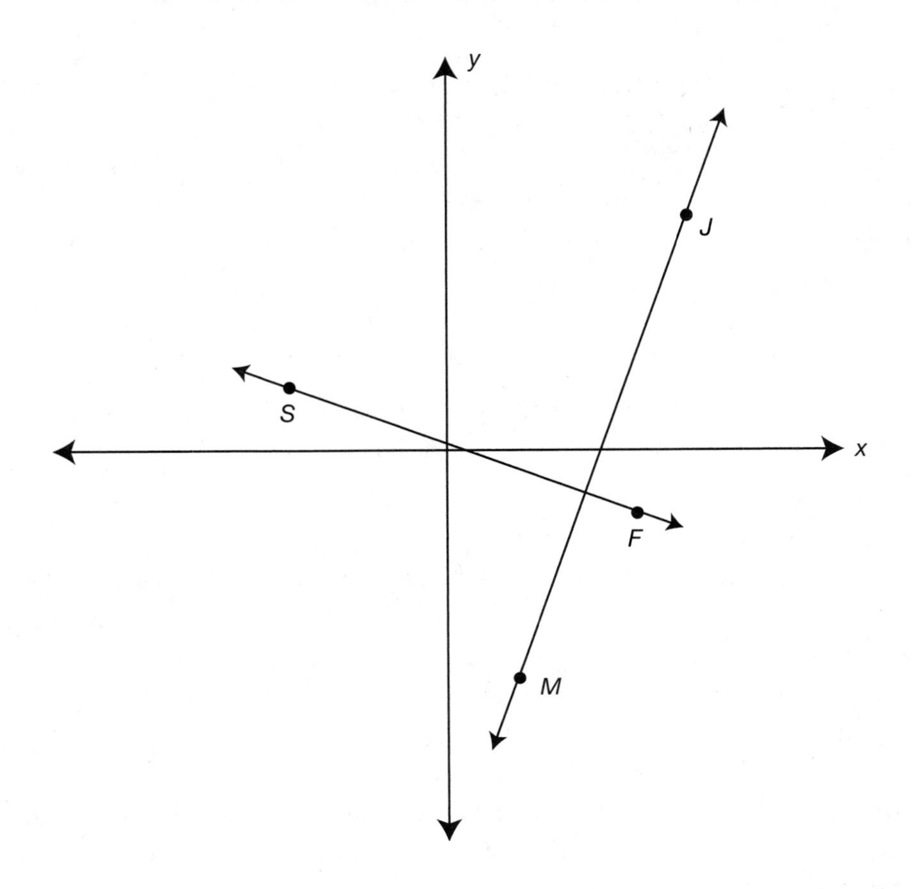

How Is Slope Measured?

Sometimes, people have different opinions of what is steep. One runner might think that the hill on Palms Avenue is terribly steep, but another runner might think it is only a moderate hill. When we are talking about the steepness of lines, we must use a specific and uniform method to measure slope. By exact definition, slope is the ratio of the vertical change between two points divided by the horizontal change between those same two points. This is commonly said, "change in ***y***, over change in ***x***." An accepted shorthand way of writing slope is with the Greek symbol delta (Δ), which means change: $\frac{\Delta y}{\Delta x}$. The letter that is always used to represent slope in coordinate geometry is ***m***.

TIP: The slope, m, between any two points (x_1,y_1) and (x_2,y_2) is $m = \frac{y_2 - y_1}{x_2 - x_1}$.

Another way that people refer to slope is $\frac{\text{rise}}{\text{run}}$, which is said "rise over run." This is a helpful method to remember that the vertical change, or y-shift, is in the numerator of the slope, and that the horizontal change, or x-shift, is in the denominator. The most common mistake that students make is putting the x-coordinates in the numerator of the fraction and the y-coordinates in the demoninator.

TIP: Avoid the common mistake of putting the *x*-coordinates in the numerator by always thinking of $\frac{\text{rise}}{\text{run}}$ when working with slope.

An important difference between the distance formula and the slope formula is that order is very important with the slope formula. Recall that with the distance formula, the differences $(x_2 - x_1)$ and $(y_2 - y_1)$ will be squared, so any negative differences will cancel out. This is not the case with the slope formula, so it is very important to make sure that the x- and y-coordinates you begin with in the numerator and denominator are from the same (x,y) point. So although it is not really important which y-coordinate you begin with in the numerator, you must be certain that you begin with the corresponding x-coordinate in the denominator.

Example: Find the slope between the points (–3,7) and (9,4).
Solution: Let (–3,7) be (x_1,y_1); and let (9,4) be (x_2,y_2).

$$m = \frac{y_2 - y_1}{x_2 - x_1}$$

$$m = \frac{4 - 7}{9 - -3}$$

$$m = \frac{-3}{9 + 3} = \frac{-3}{12} = \frac{-1}{4}$$

Notice in the example above that the slope of $\frac{-3}{12}$ was reduced to $\frac{-1}{4}$. It is best to reduce slopes to lowest terms.

A slope of $m = \frac{-1}{4}$ means that for every one unit down, the line moves 4 units to the right. Since slope represents a ratio of change between the y-coordinates and the x-coordinates, it is best to keep it in fractional form. Improper fractions are fine, but mixed fractions should never be used for slope, nor should decimals. When you're writing a negative slope, the negative sign is typically written in front of the fraction or with the numerator, but these three slopes all have the same meaning: $\frac{-1}{2}, \frac{1}{-2}, -\frac{1}{2}$

TIP: Slope should be written in simplest fractional form, and should be kept as improper fractions. Mixed fractions and decimals should never be used to represent slope.

Practice 1

1. True or false: A line can have a negative slope or positive slope, depending on which way you look at it.

2. True or false: Slope is the change in x-coordinates divided by the change in y-coordinates.

3. What are two abbreviated ways to represent slope, other than the algebraic formula?

4. Which of the following slopes is different from the rest?
   a. $\frac{-4}{1}$
   b. $\frac{4}{-1}$
   c. -4
   d. $-\frac{4}{1}$

For questions 5 through 8, find the slope between the two pairs of points.

5. (0,5) and (–1,7)

6. (9,6) and (–3,–4)

7. (–2,–9) and (7,–6)

8. (–7,13) and (–4,4)

The Slope of Vertical Lines

Both vertical lines and horizontal lines have special slopes. When two different points have the same x-coordinate, the line connecting them is vertical. Look at the two points in the next figure and what the resulting slope is.

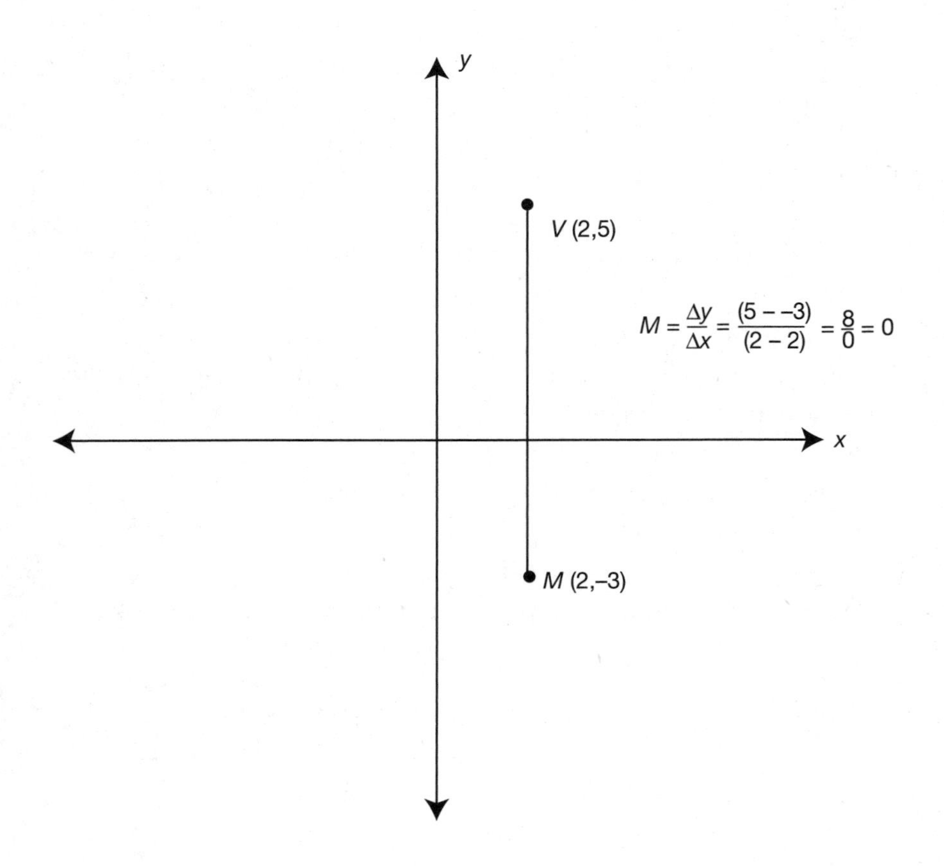

As you can see, the change in the y-coordinates is 8, but the change in the x-coordinates is 0. Any fraction that has zero in the denominator is undefined, since it is impossible to divide any number by zero. (Can 8 brownies be divided properly by zero people? No! There is no one to eat the brownies, so this situation is undefined!) In order to remember that a vertical line has an undefined slope, I imagine walking down a ski slope that goes straight down. This would be impossible to do, so I know that a vertical line's slope is undefined.

TIP: Vertical lines have an undefined slope.

The Slope of Horizontal Lines

Next, let us investigate the slope of a horizontal line. In this case, any two points on the line will have the same y-coordinate. Look at the two points in the next figure.

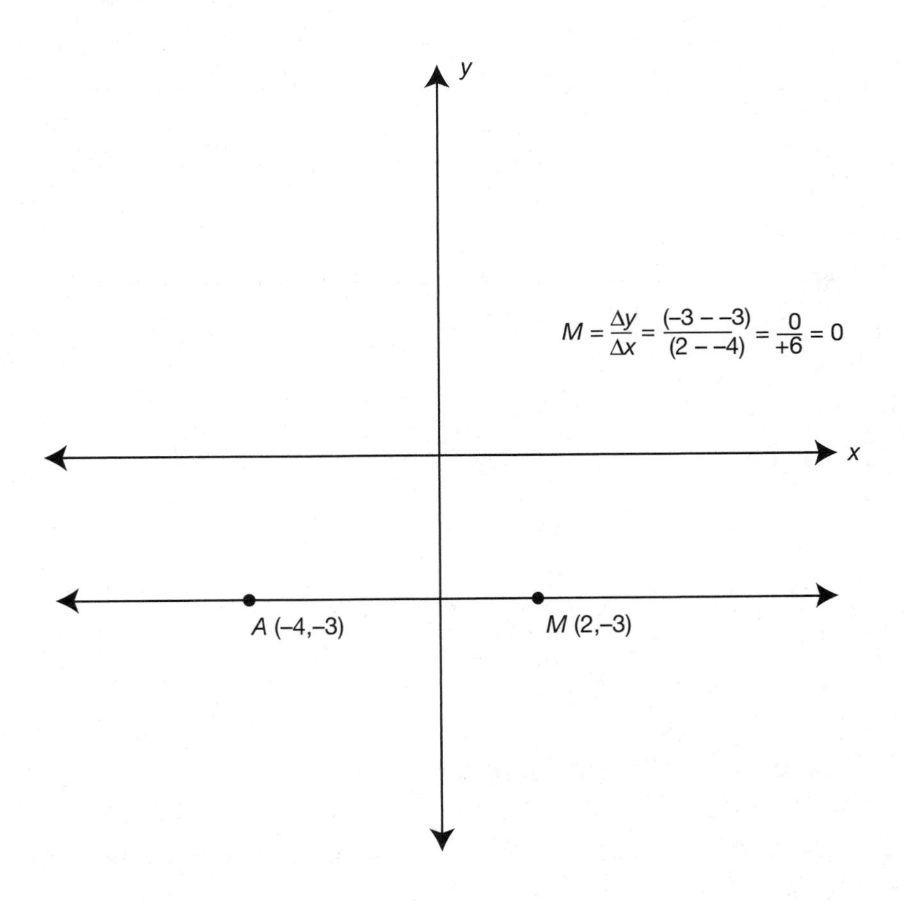

As you can see, the change in the x-coordinates is 6, but the change in the y-coordinates is 0. Any fraction that has zero in the numerator is equal to zero; if zero brownies are shared by 6 people, every person gets zero brownies. I still think of the ski slope likeness to help me remember that horizontal lines have a slope of zero: If I were to walk on a completely flat trail, I would be able to tell a friend that it had no incline or decline.

TIP: Horizontal lines have a slope of zero.

Practice 2

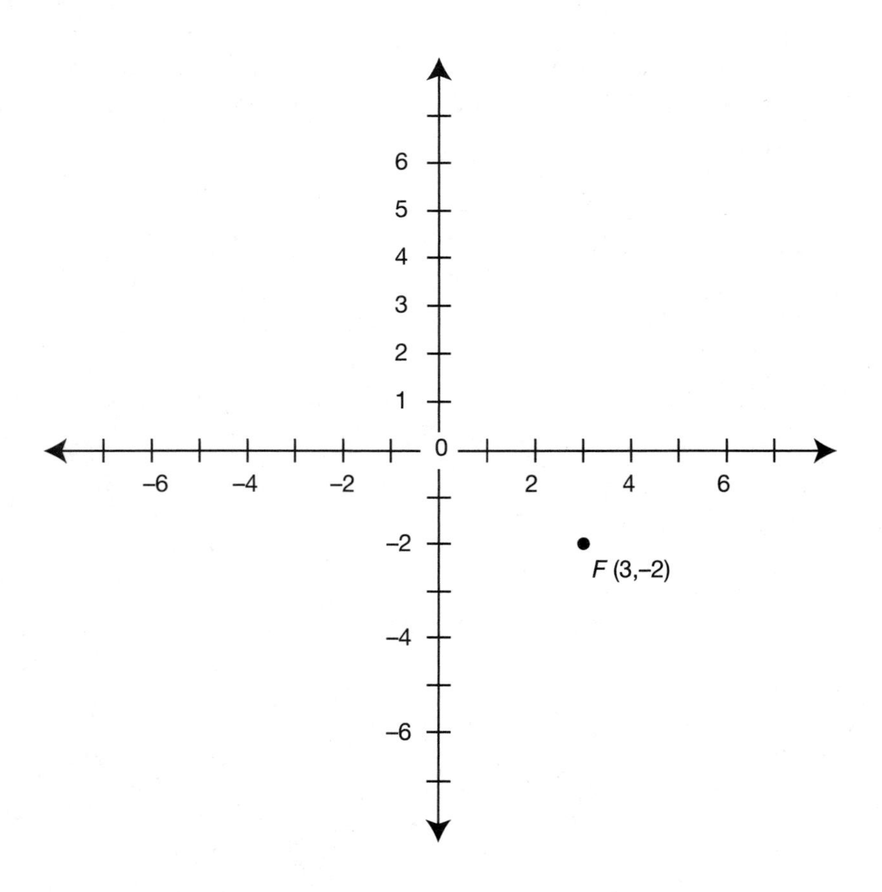

Starting with point F in the coordinate plane provided in this figure, plot and label points A through H so that the requirements for each question are met. (Do not use the origin as one of your points until you are instructed to do so in question 6.)

1. Plot and label point A in Quadrant III, so that the line connecting A and F has a negative slope.

2. Plot and label point B in Quadrant I, so that the line connecting B and F has a negative slope.

3. Plot and label point C in Quadrant I, so that the line connecting C and F has a positive slope.

4. Plot and label point D, so that the line connecting D and F has an undefined slope.

5. Plot and label point E, so that the line connecting E and F has a slope of zero.

6. Plot and label point G at the origin. Connect the line going from G to F and find the slope of this line.

7. Plot and label point H at (–4,–6). Connect the line going from G to H, and find the slope of this line.

8. What do you notice about the intersection of the two lines from questions 6 and 7? What kind of angle do they seem to form?

9. What do you notice about the slopes from questions 6 and 7?

10. $\overline{GH}$ and $\overline{GF}$ are perpendicular line segments. If you had to make up a rule for the slopes of perpendicular lines, what would it be?

Slopes of Perpendicular Lines

Two lines are perpendicular when they form right angles. In the preceding three practice questions, you began to investigate the slopes of perpendicular lines. The **slopes of perpendicular lines** will always be **opposite reciprocals**. "Opposite" means having opposite signs and a "reciprocal" is when the numerator and denominator of a fraction switch places. For example, the opposite reciprocal of $\frac{c}{d}$ will be $\frac{-d}{c}$. When you're looking for opposite reciprocals, remember that "–3" is equivalent to $\frac{-3}{1}$, $\frac{3}{-1}$, or $-\frac{3}{1}$. Always reduce slopes into simplest terms before identifying their relationship to one another.

TIP: Perpendicular lines have slopes that are opposite reciprocals.

Slopes of Parallel Lines

When two lines are parallel, it means they will never intersect. The reason they will not intersect is because they have the same rate of change. Therefore, the *slopes of parallel lines* will always be *equal.*

TIP: Parallel lines have slopes that are the same.

In the next figure, look at the examples of perpendicular and parallel lines and how their slopes are related.

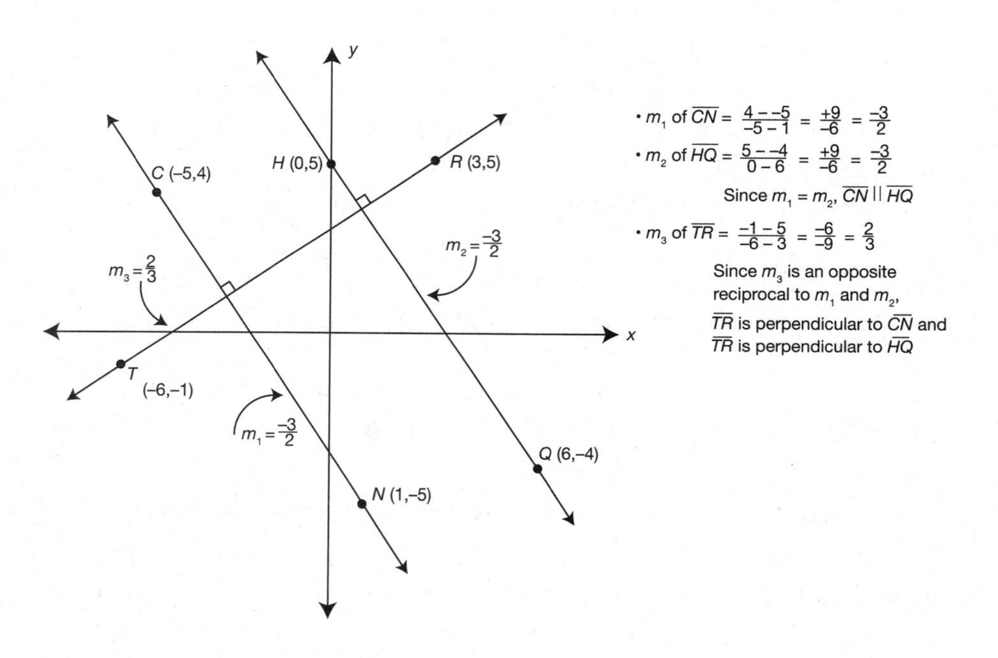

Practice 3

m_1 and m_2 represent the slopes of two different lines. Identify whether these lines are parallel, perpendicular, or neither, and explain your answer.

1. $m_1 = \frac{-4}{7}$; $m_2 = \frac{7}{-4}$

2. $m_1 = \frac{0}{-3}$; $m_2 = \frac{0}{2}$

3. $m_1 = 6; m_2 = \frac{1}{6}$

4. $m_1 = \frac{-12}{4}; m_2 = \frac{3}{9}$

5. $m_1 = -9; m_2 = \frac{9}{1}$

6. $m_1 = \frac{1}{2}; m_2 = -2$

7. $m_1 = \frac{0}{5}; m_2 = \frac{8}{0}$

Answers

Practice 1

1. False: Lines are read from left to right and cannot have both a positive and a negative slope.
2. False: Slope is the change in *y-coordinates* divided by the change in *x-coordinates.*
3. $\frac{\Delta y}{\Delta x}$ and $\frac{\text{rise}}{\text{run}}$ are two abbreviated ways to represent slope.
4. None of the listed slopes is different from the rest. Choices **a** through **d** all represent a slope of –4.
5. $m = \frac{y_2 - y_1}{x_2 - x_1} = \frac{5-7}{0--1} = \frac{-2}{1} = -\frac{2}{1} = \frac{-2}{1}$
6. $m = \frac{y_2 - y_1}{x_2 - x_1} = \frac{6--4}{9--3} = \frac{10}{12} = \frac{5}{6}$
7. $m = \frac{y_2 - y_1}{x_2 - x_1} = \frac{-9--6}{-2-7} = \frac{-3}{-9} = \frac{1}{3}$
8. $m = \frac{y_2 - y_1}{x_2 - x_1} - \frac{13-4}{-7--4} = \frac{9}{-3} = -\frac{-3}{1}$

Practice 2

1. Any point A with a negative x-coordinate and a y-coordinate less than 0 and greater than –2 will be correct.
2. Any point B with an x-coordinate greater than 0 and less than 3 and a positive y-coordinate will be correct.
3. Any point C with an x-coordinate greater 3 and a positive y-coordinate will be correct.
4. Any point D with an x-coordinate of 3 will be correct.
5. Any point E with a y-coordinate of –2 will be correct.
6. Point G will be at (0,0) and the slope of $\overline{GF}$ will be $\frac{-2}{3}$.
7. The slope of $\overline{GH}$ will be $\frac{-6}{-4} = \frac{3}{2}$.
8. $\overline{GH}$ and $\overline{GF}$ are perpendicular and form a right angle.
9. The slopes of $\overline{GF}$ and $\overline{GH}$ are opposite signs and are reciprocals of each other.
10. If two slopes are opposite reciprocals, then the lines are perpendicular.

Practice 3

1. Neither: the slopes are reciprocals, but both have the same sign
2. Parallel: both of these slopes equal 0
3. Neither: the slopes are reciprocals, but both have the same sign
4. Perpendicular: $m_1 = \frac{-12}{4} = \frac{-3}{1}$; and $m_2 = \frac{3}{9} = \frac{1}{3}$, so they are opposite reciprocals
5. Parallel: both of these slopes equal –9
6. Perpendicular: $m_1 = \frac{1}{2}$; and $m_2 = \frac{-2}{1}$, so they are opposite reciprocals
7. Perpendicular: $m_1 = 0$, so it is a horizontal line; m_2 is undefined, so it is a vertical line

24

Introduction to Linear Equations

I'm sorry to say that the subject I most disliked was mathematics.
I have thought about it. I think the reason was
that mathematics leaves no room for argument.
If you made a mistake, that was all there was to it.
—Malcolm X

In this lesson you will learn how to identify a linear equation. You will also learn how to see if a point lies on a line and how to graph a line when you're given a linear equation.

STANDARD PREVIEW

As stated in the previous lesson, linear equations are not part of the Common Core State Standards for middle school geometry, but they are a critical component of the middle school CCSS for other areas in math. Linear equations are extremely important when modeling real-life relationships in many professional fields, including finance, sports, and retail. The study of linear equations is paramount in middle school math and it's a natural extension of the work you did on slope in Lesson 23, so we feel this is the best skill to present as the culminating lesson of this book. We hope you pay close attention to this lesson, since you will be expected to demonstrate a comprehensive understanding of linear equations in your middle school math classes.

Why Are Linear Equations Important?

Linear equations are used to model many real-life situations. Consider placing an online order of T-shirts. There may be a constant shipping cost that applies to all orders. On top of that shipping charge, with every T-shirt ordered your bill total will increase at a constant rate. A linear equation will show exactly how the total cost of your order increases with the number of T-shirts you purchase. Linear equations are helpful for making predictions and, when graphed, help us understand the relationship between two things.

What Does a Linear Equation Look Like?

Linear equations have a standard form $Ax + By = C$, where A, B, and C are all numbers and x and y remain as variables. When you're trying to determine if an equation is linear, the signs of the terms do not matter, nor does the order that the terms occur in the equation. An equation can still be linear if the y-term comes before the x-term, or if the y-term is on the opposite side of the equation from the x-term. Sometimes, A or B will be zero and will cancel out the x or y variable. This is okay and the resulting equation will still be linear. Examples of other forms a linear equation can take are:

$Ax + C = By$
$-By - Ax = C$
$C = Ax$ (in this case, $B = 0$)
$Ax = -By$ (in this case, $C = 0$)
$y = -5$ (in this case, $A = 0$ and $C = -5$)

An important factor in linear equations is that x and y do not have exponents. Any equation with x^2 or y^2 will *not* be a linear equation. Also, x and y cannot be multiplied by each other in linear equations. Take a look at the next figure for examples of linear and nonlinear equations.

Linear Equations	**Nonlinear Equations**
$2x + 3y = 7$	$2x^2 + 3y = 7$
$x = -3$	$\frac{-3}{y} = x$
$y = \frac{2}{3}x + 4$	$y^2 = \frac{2}{3}x + 4$
$y = 7$	$xy = 7$

TIP: The standard form of a linear equation is $Ax + By = C$, where A, B, and C are all constant numbers and x and y are variables.

Practice 1

State whether each equation represents a linear or a nonlinear equation.

1. $4x + 8 = -3y$

2. $x = 7y^2$

3. $\frac{3}{y} = -2x$

4. $x^2 + 4 = 8y$

5. $-9y - \frac{2}{3}x = 0$

6. $\frac{7}{x} = 9$

7. $y = 0.5$

8. $x \div y = 4$

Graphing Linear Equations

In order to graph a linear equation, you first need to find three points that each make the equation work. Although only two points are *needed* to define a line, finding a third point helps confirm that there were no arithmetic mistakes made while you were finding the points. (If the three points do not sit on the same line, and instead make a triangle, then an error has been made.) Therefore, three points should always be used. After finding three points, plot them on a coordinate grid. As you learned in the last section, linear equations can be presented in many different ways. What is the easiest way to find points that will work as a linear equation? First, isolate the y variable in the equation. Look at the following example, where the equation $-6x + 3y = 9$ is manipulated using algebraic principles, to get y on its own.

Example: Isolate y in the following equation: $-6x + 3y = 9$
Solution:

$-6x + 3y = 9$

$\underline{+6x \qquad\quad + 6x}$

$3y = 9 + 6x$

$\underline{\div 3 \qquad \div 3}$ divide by 3 to *both* terms on right side of equation

$y = 3 + 2x$

The line $y = 3 + 2x$ is the same as the original line, $6x + 3y = 9$, but it is just in a different format. Once y is by itself, it is easy to find three points that will work in the equation. Simply plug in three different values for x and then evaluate the expression $3 + 2x$ for y. (It is usually best to plug in x-values between –3 and 3. Then your points will be easier to graph, since they will not be too far from the origin.)

Column 1	Column 2	Column 3
let x = 0	let x = 1	let x = −2
y = 3 + 2x	y = 3 + 2x	y = 3 + 2x
y = 3 + 2(0)	y = 3 + 2(1)	y = 3 + 2(−2)
y = 3	y = 5	y = −1

You now know that the following three points will lie on the line $-6x + 3y = 9$: (0,3); (1,5); and (−2,−1). To graph the line, plot these points on a graph and connect them. In the next figure the points have been plotted for you. Now connect them to make the line for $-6x + 3y = 9$. (Remember that this can also be called line $y = 3 + 2x$.)

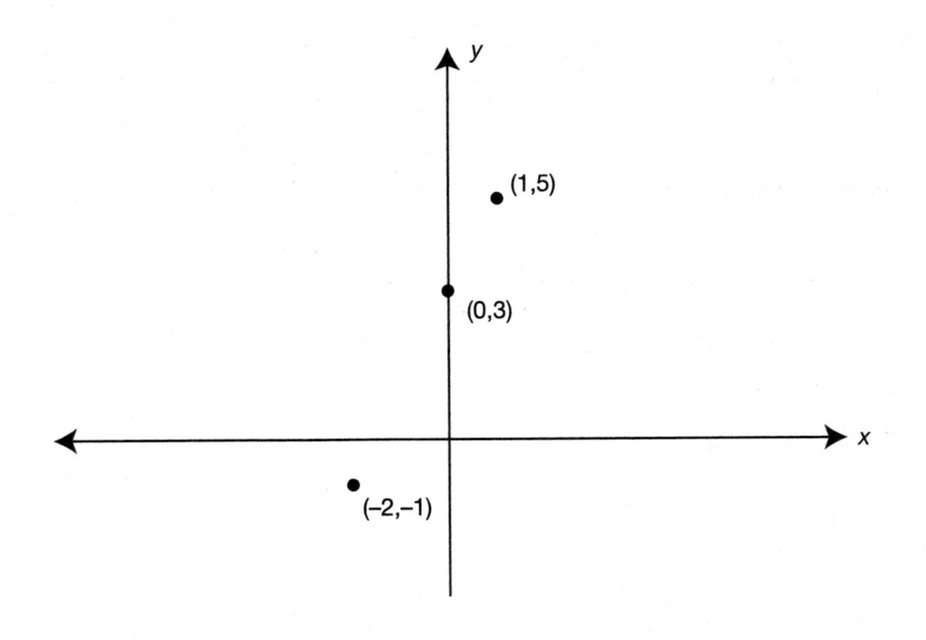

TIP: To graph a linear equation:

1. **Get the *y* variable by itself.**
2. **Substitute three different values in for *x* and solve for *y* in order to find three coordinate pairs that satisfy the equation.**
3. **Plot the three coordinate pairs on a graph and connect them.**

One tricky thing to look out for is that the coefficient of x will often be in fractional form, such as in $y = \frac{3}{4}x - 2$. When this is the case, use values for x that are divisible by the denominator of the fraction being multiplied by x. It is also helpful to use zero for x.

Example: What x-values would be easiest to solve for in the linear equation $2x - 3y = -6$?
Solution:

$$\begin{array}{rl} 2x - 3y &= -6 \\ \underline{-2x \qquad} & \underline{\quad -2x} \\ -3y &= -6 - 2x \\ \underline{\div -3} & \underline{\quad \div -3} \\ y &= 2 + \frac{2}{3}x \end{array}$$

In this case, you would want to pick values for x that will be divisible by 3, such as $x = 0, 3, 6, -3, -6, 9$, and so on.

Practice 2

For questions 1 through 4, use algebra to get the y-variable by itself.

1. $12 = -4x + 3y$

2. $9y - \frac{2}{3}x = 6$

3. $7x = 5y - 10$

4. $\frac{1}{2}y + (-6) = x$

For questions 5 through 8, find three points that would work in the given linear equation.

5. $3y - 6x = -12$

6. $2x = 3y - 3$

7. $7y + 7x = -14$

8. $\frac{2}{3}y = x + \frac{10}{3}$

Determining if Points Are on Linear Equations

Now that you know how to find points on a line and how to graph linear equations, this next task will be easy for you. Sometimes, you will be asked if a particular coordinate pair sits on a given line, or "satisfies the equation." In order to do this, simply plug the x and y values into the linear equation, and see if the result is a true statement. You will be happy to learn that for this method, you do not have to change the format of the line.

Example: Do the points (–4,–10) and (5,4) satisfy the line $-3x + 2y = -8$?

Solution:

Plug in (–4,–10)
$-3(-4) + 2(-10) = -8$
$12 + -20 = -8$
$-8 = -8$
This is a true statement, so the point (–4,–10) satisfies the equation and must lie on the line $-3x + 2y = -8$.

Plug in (5,4)
$-3(5) + 2(4) = -8$
$-15 + 8 = -8$
$-7 \neq -8$
Since –7 does not equal –8, the point (5,4) does *not* satisfy the equation and does not lie on the line $-3x + 2y = -8$.

Equations for Horizontal and Vertical Lines

Horizontal and vertical lines have unique equations that at first might not look like they are in the standard form $Ax + By = C$. With a horizontal line, the y-coordinate will be the same for every point, and only the x-coordinate will be changing. That is why horizontal lines have the form $y = C$. The x-coordinate can be an infinite number of values without affecting the y-value, so x does not need to be in the equation.

TIP: Horizontal lines are written in the form $y = C$, where C is any constant number.

On the other hand, vertical lines have a constant x-value. The value of y does not affect the value of x in vertical lines, so y does not need to be in the equation for vertical lines. Vertical lines have a standard form of $x = C$.

TIP: Vertical lines are written $x = C$, where C is any constant number.

Practice 3

Identify whether the coordinate pairs in questions 1 through 3 satisfy the linear equation $2x - 5y = 10$.

1. (3,–1)

2. $(\frac{5}{2},-1)$

3. (–5,4)

Identify whether the coordinate pairs in questions 4 through 6 satisfy the linear equation $\frac{1}{2}y - \frac{3}{2}x = 4$.

4. (0,8)

5. (1,11)

6. (2,–2)

Identify whether the linear equations in questions 7 through 10 represent horizontal, vertical, or sloped lines.

7. $y = -7$

8. $2 = 3x$

9. The line between points (–5,0) and (0,–5)

10. The line between points (–2,8) and (–2,–7)

Answers

Practice 1

1. linear
2. nonlinear (cannot have y^2)
3. nonlinear (cannot have yx, which is what will happen when both sides are multiplied by y)
4. nonlinear (cannot have x^2)
5. linear
6. linear (will be $7 = 9x$ when both sides are multiplied by x)
7. linear
8. linear (In order to change $x \div y = 4$ into a standard linear format, multiply both sides by y to get $x = 4y$, and then divide by 4 to end with $y = \frac{1}{4}x$.)

Practice 2

1. $y = 4 + \frac{4}{3}x$
2. $y = \frac{2}{3} + \frac{2}{27}x$
3. $y = \frac{7}{5}x + 2$
4. $y = 2x + 12$
5. Any points that work in: $y = 2x - 4$. Examples include: (0,–4); (1,–2); (2,0); (–1,–6)
6. Any points that work in: $y = \frac{2}{3}x + 1$. Examples include: (0,1); (3,3); (6,5); (–3,–1)
7. Any points that work in: $y = -2 - x$. Examples include: (0,–2); (1,–3); (2,–4); (–1,–1)
8. Any points that work in: $y = \frac{3}{2}x + 5$. Examples include: (0,5); (2,8); (4,11); (–2,2)

Practice 3

1. No. $11 \neq 10$
2. Yes. $10 = 10$
3. No. $-30 \neq 10$

For questions 4 through 6, make sure you substituted the x- and y-values into the equation correctly—the order of x and y in the given equation $\frac{1}{2}y - \frac{3}{2}x = 4$ was flipped.

4. Yes. $4 - 0 = 4$
5. Yes. $5.5 - 1.5 = 4$
6. No. $-1 - 3 \neq 4$
7. $y = -7$ is horizontal
8. $2 = 3x$ is the same as $x = \frac{2}{3}$, so it is vertical.
9. The line between points (–5,0) and (0,–5) is sloped.
10. The line between points (–2,8) and (–2,–7) is vertical with the equation $x = -2$.

Posttest

The 24 lessons you have just read have given you the skills necessary to ace the posttest. The posttest has 30 questions. The questions are similar to those in the pretest, so you can see your progress and improvement since taking the pretest.

The posttest can help you identify which areas you have mastered and which areas you need to review a bit more. Return to the lessons that cover the topics that are tough for you until those topics become your strengths.

Posttest

1. What is the least number of points needed to define a plane?

a. 0
b. 1
c. 2
d. 3

2. C H I L E

$\overline{LE} = 6$ cm

$\overline{IL} = \frac{2}{3}\overline{LE}$

$\overline{HI}$ is twice as long as $\overline{IL}$

If $\overline{CI}$ is equal to 22 cm, how long is $\overline{HI}$?

a. 10 cm
b. 12 cm
c. 14 cm
d. 16 cm

3. $\angle P$, $\angle R$, and $\angle Q$ are supplementary angles. If $m\angle Q = 120°$ and $\angle P$ is half the measure of $\angle R$, what is the $m\angle P$?

a. 60°
b. 30°
c. 40°
d. 20°

4. In the next figure, $m\angle AOT = 52°$ and $m\angle POG = 137°$. What is the $m\angle POA$?

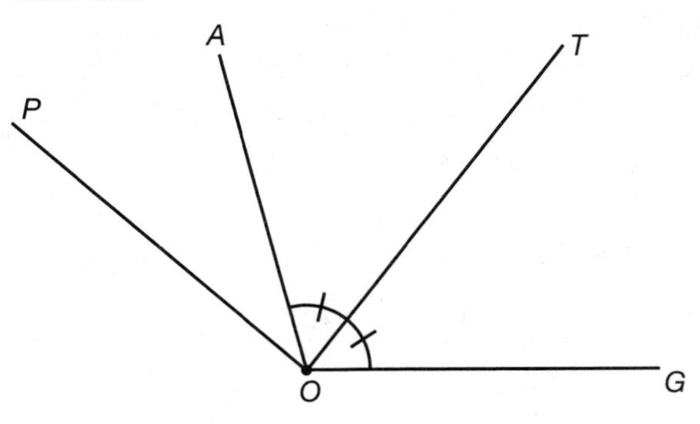

a. 33°
b. 104°
c. 52°
d. 34°

5. In the next figure, $q \parallel r$. Solve for x.

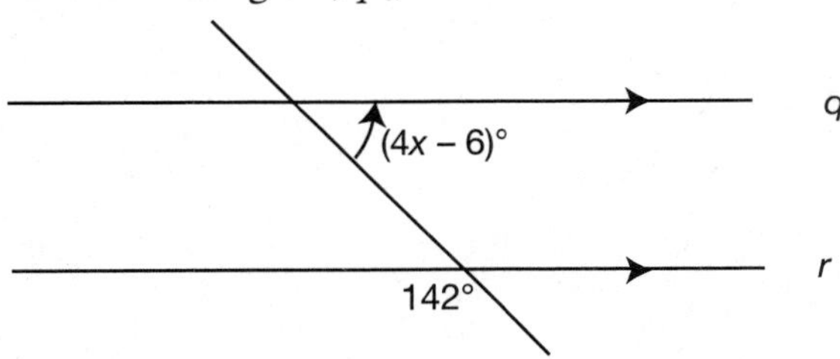

a. 38
b. 48
c. 11
d. 14

6. In ΔABC, find the $m\angle C$ if $m\angle A = (2x + 7)°$, $m\angle B = 28°$, and $m\angle C = (3x)°$.

a. 29°
b. 87°
c. 65°
d. 180°

7. In ΔPIG, $\overline{PI} = 6$, $\overline{PG} = 8.5$, and $\overline{IG} = 6$. What kind of triangle is ΔPIG?

a. Scalene
b. Right
c. Isosceles
d. Equilateral

8. The area of $\Delta THE = 620\text{ cm}^2$. If the base of the triangle is 40 cm, what is the measure of the altitude of ΔTHE?

a. 15.5 cm
b. 80 cm
c. 31 cm
d. 62 cm

9. Find the missing side length in the given right triangle.

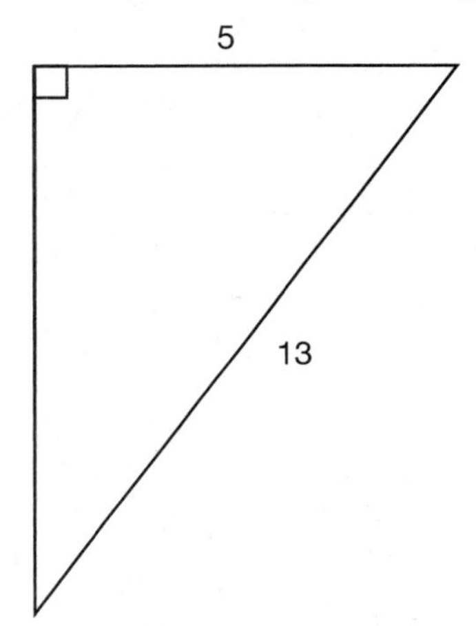

a. 8
b. 9
c. 11
d. 12

10. ΔCAB is a 45°-45°-90° triangle. If one of its legs measures 9 inches, what are the measures of the other two sides?

a. 9 inches and $9\sqrt{2}$ inches
b. 9 inches and $9\sqrt{3}$ inches
c. 9 inches and 9 inches
d. Cannot be determined with the information given

11. Given parallelogram *ABLE* that follows, find the $\angle EAL$.

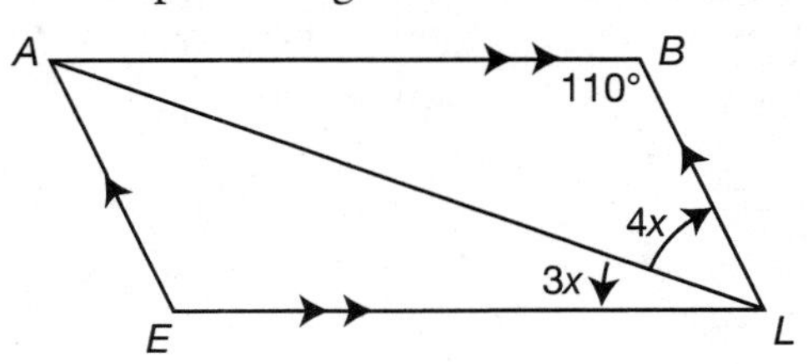

a. 30°
b. 40°
c. 70°
d. 80°

12. Find the area of the following lawn:

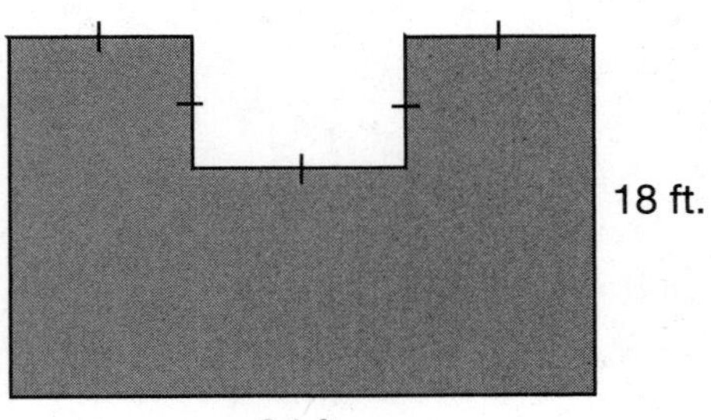

a. 432 feet^2
b. 408 feet^2
c. 100 feet^2
d. 368 feet^2

13. Pete receives a box of fruits and vegetables from a local farm that contains the following items: 4 apples, 5 peppers, 2 mangoes, 3 cucumbers, 6 onions, and 1 cantaloupe. What is the ratio of fruits to vegetables that Pete received?

a. $\frac{1}{3}$
b. $\frac{1}{2}$
c. $\frac{2}{3}$
d. 2

14. A circle is circumscribed within a square so that its edge touches all sides of the square, such as in the next figure. If the area of the square is 36 m², what is the circumference of the circle, in terms of π?

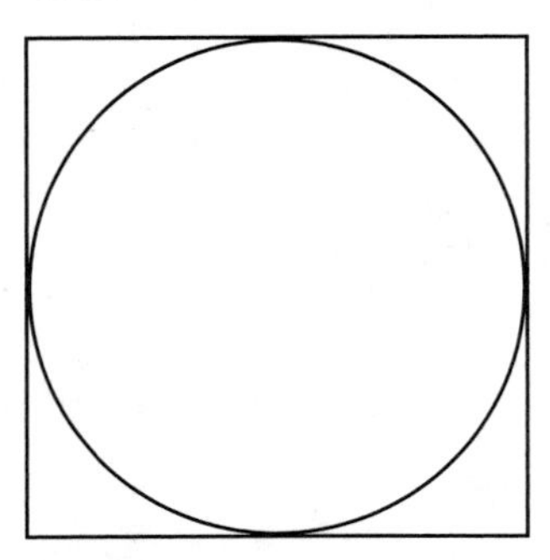

a. 6π m
b. 9π m
c. 12π m
d. 18π m

15. Using the same figure from question 14, if the perimeter of the square is 32 cm, what is the area of the circle, in terms of π?
a. 8π cm²
b. 64π cm²
c. 16π cm²
d. 32π cm²

16. The shaded area in the figure of a fountain represents a grassy area surrounding a 5-yard by 10-yard fountain. Kaoru is purchasing grass seed to plant. How many square feet of grass will she be growing in the shaded area around the fountain?

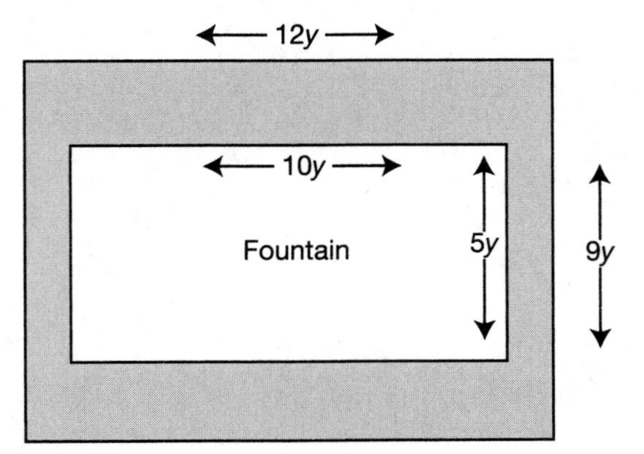

a. 108 yd^2
b. 50 yd^2
c. 62 yd^2
d. 58 yd^2

17. What is the surface area of a cylinder that has a diameter of 8 inches and a height of 14 inches? Round your answer to the nearest tenth.
a. 418.6 $in.^2$
b. 427.0 $in.^2$
c. 452.2 $in.^2$
d. 542.2 $in.^2$

18. What is the volume of the triangular prism in the next figure?

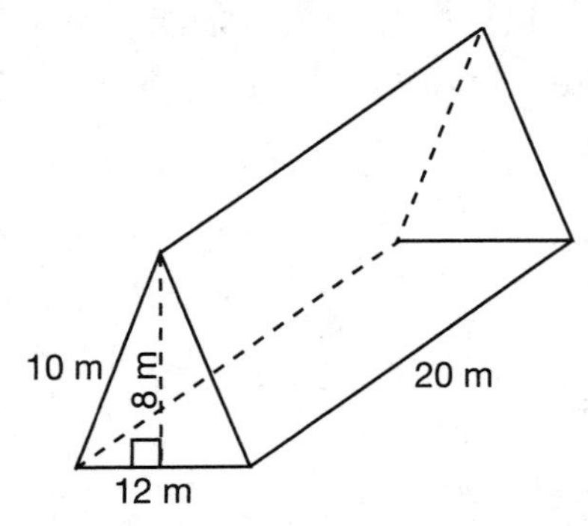

a. 2,400 m^2
b. 1,640 m^2
c. 1,200 m^2
d. 960 m^2

19. What is the distance d, between $J(-6,1)$ and $P(4,6)$?
a. 15
b. $5\sqrt{5}$
c. $\sqrt{27}$
d. $\sqrt{51}$

20. What transformation could be used to create image point $N'(-5,-9)$ from preimage point $N(9,-5)$?
a. A reflection over the x-axis.
b. A translation using $T_{-4,4}$.
c. A clockwise rotation of 90°.
d. A counterclockwise rotation of 90°.

21. Which transformations preserve the general shape, angle measurements, and size of a figure?
a. translation, rotation, dilation
b. dilation, rotation, reflection
c. reflection, translation, rotation
d. none of the above

22. The two triangles below can best be described as being

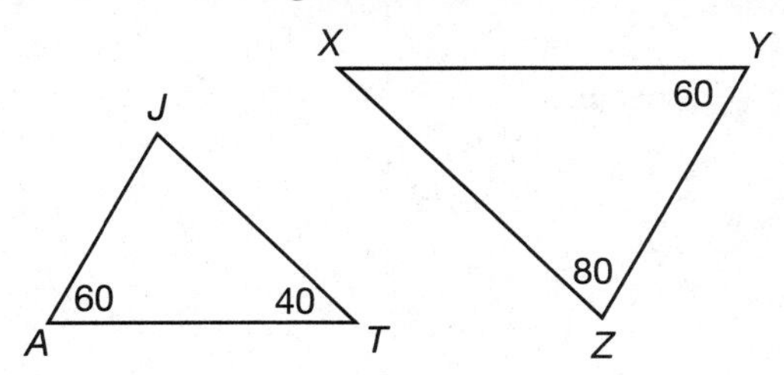

a. Similar
b. Congruent
c. Regular
d. Corresponding

23. If a line has a slope of $\frac{-1}{5}$, a line *perpendicular* to it will have a slope of
a. 5
b. –5
c. $\frac{1}{5}$
d. Cannot be determined with the given information

24. Which coordinate pair is not on the line $\frac{2}{3}x - \frac{1}{2}y = -4$?
a. (–3,4)
b. (–9,–4)
c. (3,12)
d. (0,–8)

Answers

For any questions that you may have missed, go back to review the associated lesson.

1. **d.** Lesson 1
2. **c.** Lesson 2
3. **d.** Lesson 3
4. **a.** Lesson 4
5. **c.** Lesson 5
6. **b.** Lesson 6
7. **c.** Lesson 7
8. **c.** Lesson 8
9. **d.** Lesson 9
10. **a.** Lesson 10
11. **b.** Lesson 11
12. **d.** Lesson 12
13. **b.** Lesson 13
14. **a.** Lesson 14
15. **c.** Lesson 15
16. **d.** Lesson 16
17. **c.** Lesson 17
18. **d.** Lesson 18
19. **b.** Lesson 19
20. **d.** Lesson 20
21. **c.** Lesson 21
22. **a.** Lesson 22
23. **a.** Lesson 23
24. **d.** Lesson 24

Reference Section A: Proving the Pythagorean Theorem

Standard 8.G.B.6 asks students to explain a proof of the Pythagorean theorem.

Pythagoras proved this over 2,000 years ago, so why do we still need to go over this? The proof of this important theorem requires the applications of some of the most fundamental geometric principles with some fundamental algebraic principles.

Hopefully, you remember that the Pythagorean theorem is used for finding unknown side lengths in right triangles. The theorem is $a^2 + b^2 = c^2$, where a and b represent the lengths of the legs and c represents the length of the hypotenuse. Although proving that this is true for all right triangles might sound like an impossible task, hopefully all that you've learned in the previous lessons has prepared you to take this on. Two key required formulas for this proof are the area formula for triangles and the area formula for a square. It is also necessary to know how the angles in triangles and straight angles relate to one another. These principles can be combined with a little algebra to present a proof that isn't too hard to follow:

> Begin with a right triangle with legs a and b and hypotenuse c. Notice that the $m\angle C = 90°$ and recall that this means that the smaller acute angle, $\angle A$, is complementary to the larger acute angle, $\angle B$.

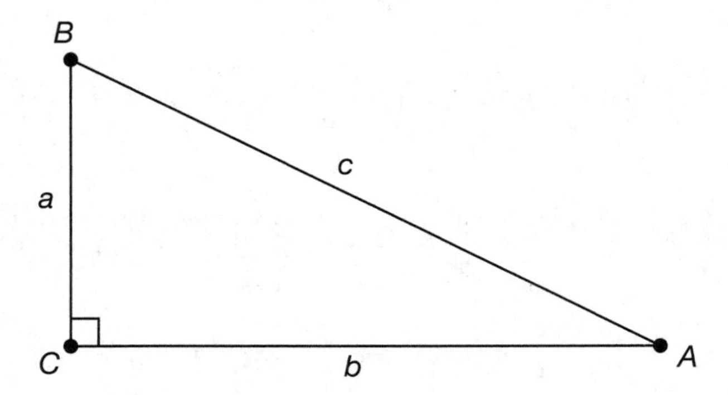

Use four copies of this triangle to construct the following quadrilateral. We know the larger quadrilateral is a square because its four corners are all right angles C, and its side lengths are congruent since they all equal $a + b$.

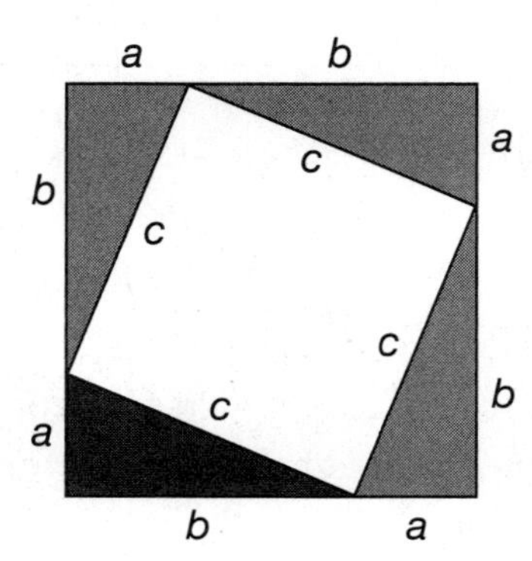

It can also be determined that the non-shaded quadrilateral in the center must be a square since its side lengths are all C and its angles are all 90°.

(We know its angles are right angles since the straight angle minus complementary angles A and B leaves 90°.)

Therefore, it follows that the area of the whole square must be equal to the area of the four right triangles plus the area of the inner square. Write this relationship as:

$$Area_{\text{(Whole Square)}} = Area_{\text{(4 Triangles)}} + Area_{\text{(Inner Square)}}$$

Use the following formulas to calculate the areas of the squares and triangles:

$$Area_{\text{(Square)}} = s^2 \text{ and } Area_{\text{(Triangle)}} = \tfrac{1}{2}(base)(height)$$

$$Area_{\text{(Whole Square)}} = s^2 = (a + b)(a + b) = a^2 + 2ab + b^2$$

$$Area_{\text{(4 Triangles)}} = 4 \times \tfrac{1}{2}(base)(height) = 4 \times \tfrac{1}{2}(b)(a) = 2ab$$

$$Area_{\text{(Inner Square)}} = s^2 = c^2$$

Now plug each of these area values into the statement above:

$$Area_{\text{(Whole Square)}} = Area_{\text{(4 Triangles)}} + Area_{\text{(Small Square)}}$$

$$a^2 + 2ab + b^2 = 2ab + c^2$$

Subtracting $2ab$ from both sides leaves:

$$a^2 + b^2 = c^2$$

And there we have it—a proof for the Pythagorean theorem! Come back to this proof a few more times until you feel comfortable enough with it to explain this to a friend. Then you will have satisfied **Standard 8.G.B.6.**

**Note:* The side length of the whole square is $a + b$, so that must be multiplied by itself.

Reference Section B: Slicing Three-Dimensional Figures

Standard 7.G.A.3 asks students to describe the two-dimensional figures that result from slicing three-dimensional figures such as rectangular prisms, cubes, and right rectangular pyramids.

One of the purposes of this lesson is that you'll see that a 3-dimensional figure is actually composed of an infinite number of 2-dimensional figures. Think of how a loaf of bread is a rectangular prism made of up square pieces of bread. (Of course in this example the pieces of bread are also prisms, but hopefully you can understand that those pieces of bread could be further sliced to create flat squares, that when pressed together, make the rectangular loaf of bread.)

When a plane intersects a 3-dimensional figure, a 2-dimensional shape called a *cross section* results from this intersection. Let's start our investigation of what 2-dimensional cross sections are made when a plane cuts a rectangular prism. Let's picture a stick of butter as our 3-dimensional rectangular prism and a knife as our plane. The angle at which you use the knife to cut the butter will determine the shape made. Typically butter is cut in a way that will reveal a square cross section:

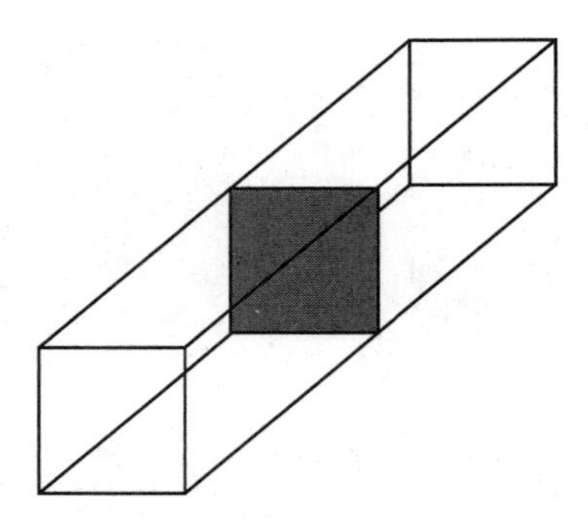

What type of cross section would be made if the stick of butter were cut lengthwise? Hopefully, you can see that it would make a rectangle. It definitely takes a keen spatial awareness and a strong visual mind to determine the cross sections that are created by planes that intersect the stick of butter at an angle, rather than in a manner that is parallel to the front or side faces. We've envisioned two different cuts. Let's see how many more ways we can use our plane (knife) to slice the prism (butter). You might want to head to the kitchen and grab some butter and a butter knife to work through the following simulations of a plane slicing a right prism.

Let's cut off one of the corners by slicing directly downward at a 45-degree angle to each side of the butter. Your knife should look like this:

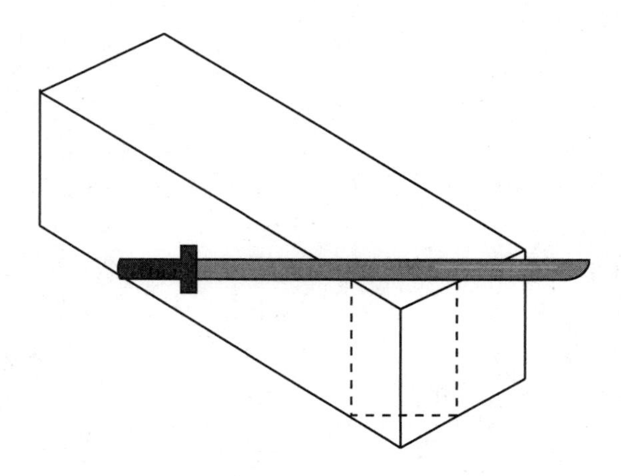

Once you make that cut you will notice that the cross section you just created was a rectangle:

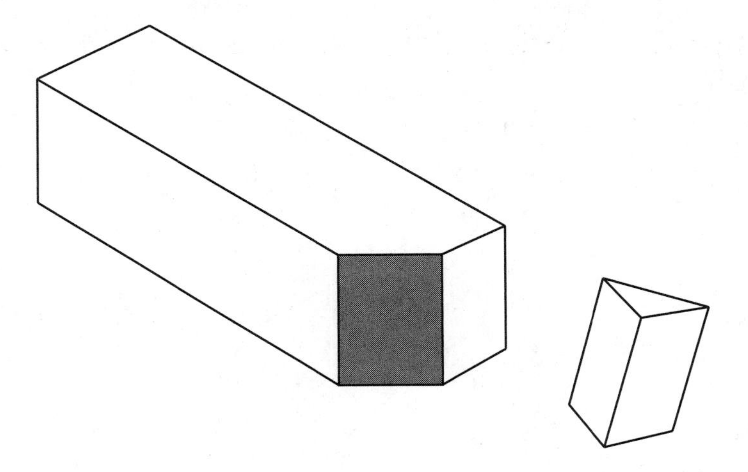

Turn your stick of butter around so you have a fresh corner, and cut the corner off so that you make a little pyramid of butter. Position your knife like this:

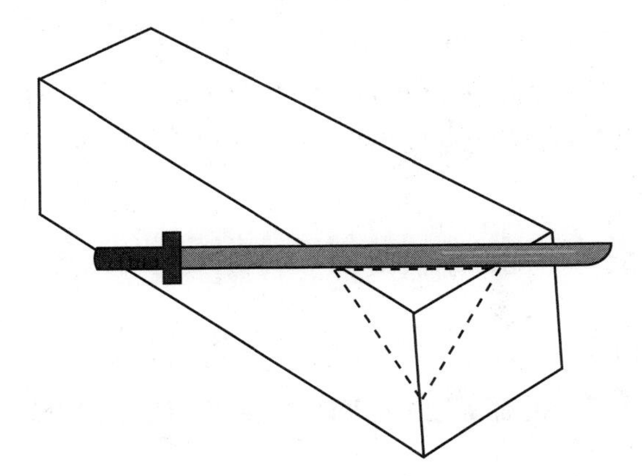

Let's cut off one of the corners by slicing directly downward at a 45-degree angle to each side of the butter.

What type of cross section will this slice reveal? Look at the equilateral triangle below:

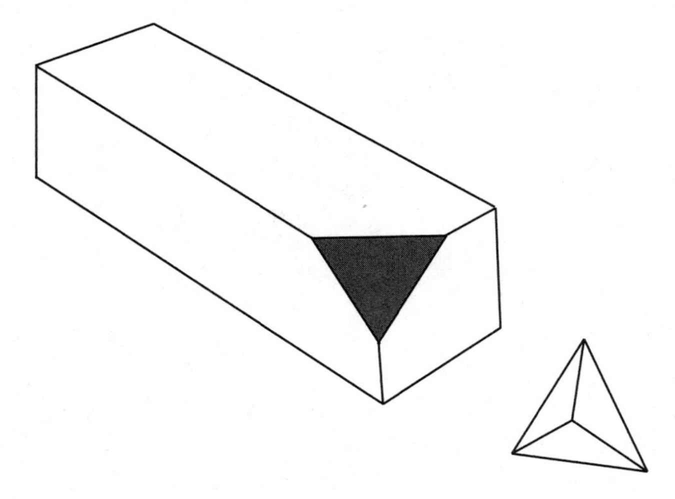

How could you slice one of the remaining corners to create a triangle that isn't equilateral? All you need to do is shift the angle of your butter knife so that it's making a triangular slice at the top of the butter that is scalene:

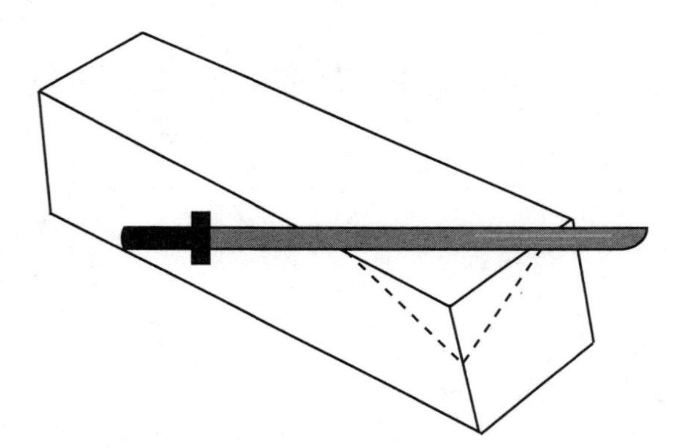

This type of slice will create the following triangular cross section:

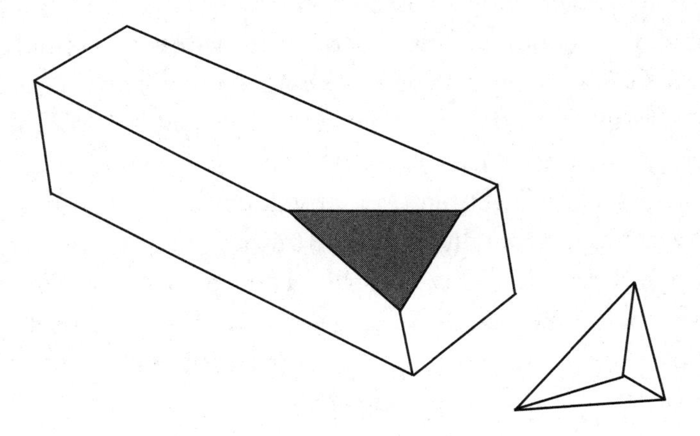

The last cross section we will illustrate is how to create a parallelogram. For this maneuver, you'll need to line your butter knife diagonally across the top of the butter and cut diagonally toward the front of the butter. Here is the pre- and post-image of this slice:

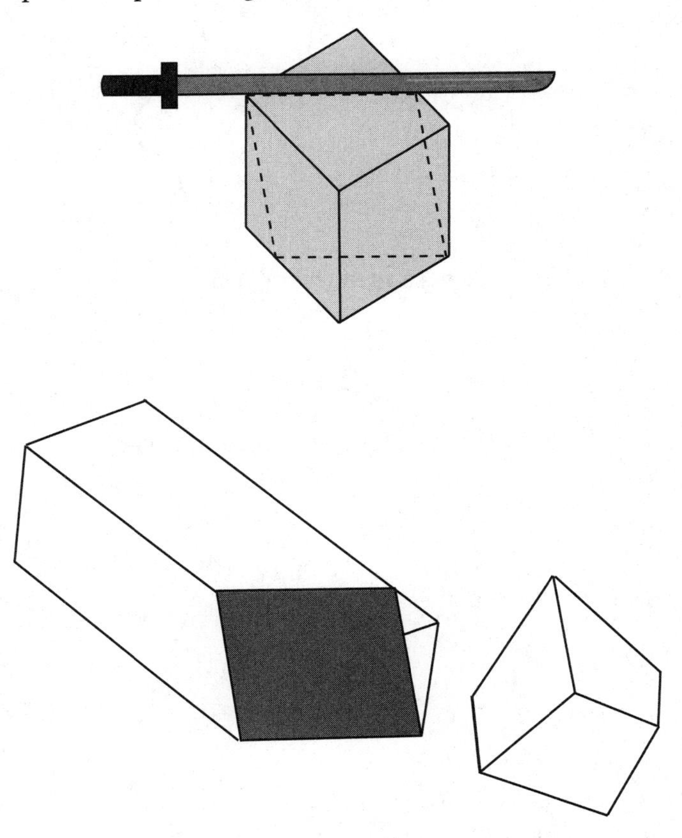

If you have enough butter you should be able to experiment to see how you could make a pentagon and a hexagon!

Now that we're done with our stick of butter, we need to remind you that **Standard 7.G.A.3** also wants you to be able to describe the 2-dimensional figures that result from slicing a right rectangular pyramid. We encourage you to use these two illustrations to make some hypotheses about the types of cross sectional 2D figures that will be made when the following pyramids are cut horizontally, vertically through their vertices, and vertically through their sides. How will the 2D shapes change if the pyramid has a square base versus a rectangular base? We'll give you the following illustrations to help you get started; then, look at the next page to see the cross sections that have been illustrated.

Square Pyramid

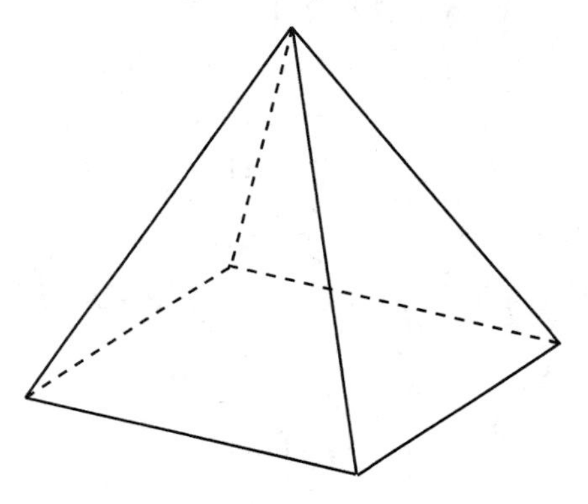

Rectangular Pyramid

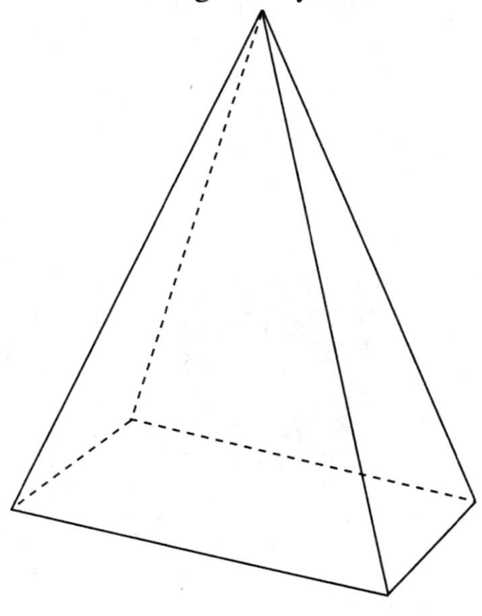

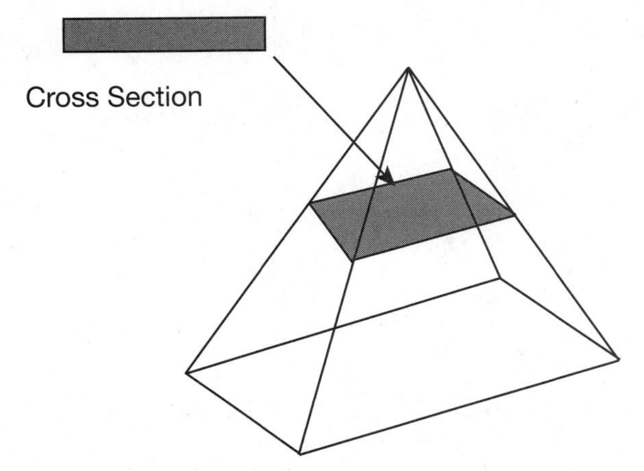

A horizontal slicing of a rectangular pyramid reveals a rectangle.

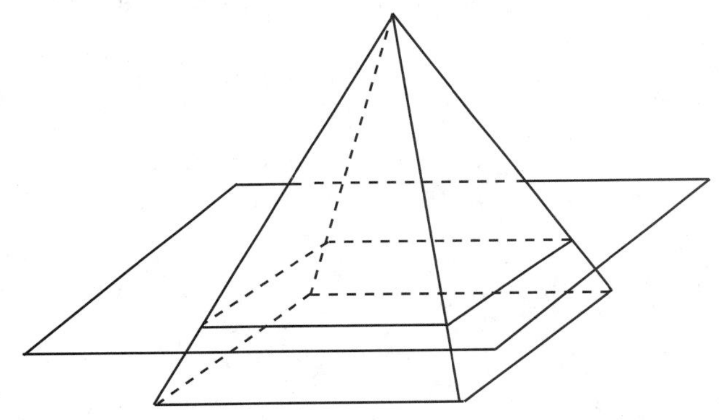

A horizontal slicing of a square pyramid reveals a square.

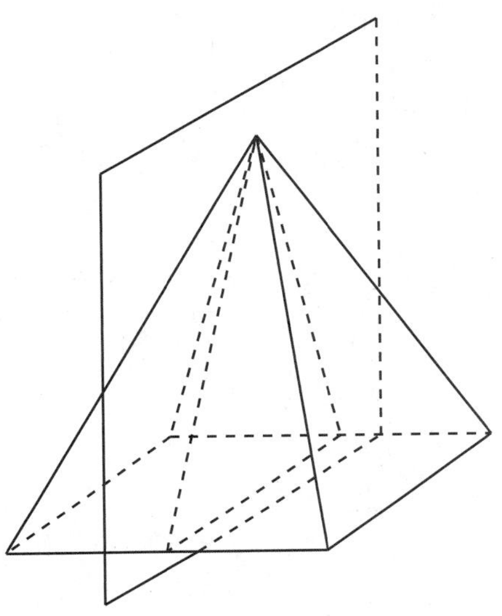

A vertical slicing of a pyramid through the vertex reveals a triangle. How will it change if the pyramid has a rectangular base?

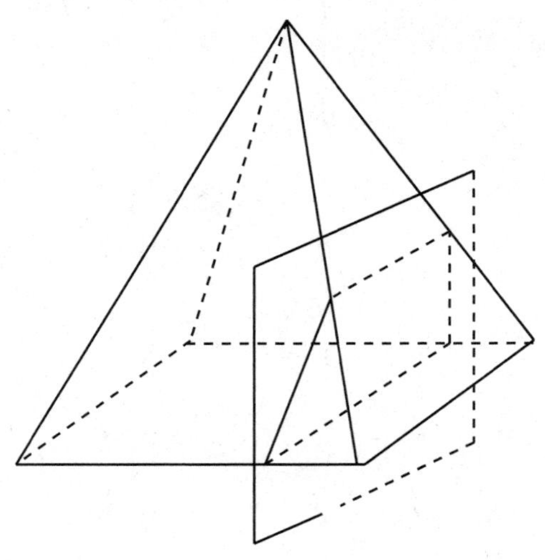

A vertical slicing of a pyramid through the side reveals a trapezoid.

Glossary

adjacent angles angles that share a common vertex, have one common side, and have no interior points in common

alternate exterior angles When a transversal crosses a pair of lines, the alternate exterior angles are the angles outside the pair of lines, on opposite sides of the transversal. If the two lines crossed by the transversal are parallel, then the alternate exterior angles will be congruent.

alternate interior angles When a transversal crosses a pair of lines, the alternate interior angles are the angles inside the pair of lines, on opposite sides of the transversal. If the pair of lines crossed by the transversal is parallel, then the alternate interior angles will be congruent.

altitude of a triangle the line that passes through the vertex of a triangle, forming a right angle with the opposite side, which is defined as the base of that triangle. All triangles have three altitudes.

angle bisector a line, ray, or segment that goes through the vertex of an angle, dividing it into two angles of equal measure

arc a curved section of the outside of the circle that is defined by two points on the circle

base of a prism the two congruent ends of a prism

bisector a line, ray, or segment that goes through a line segment's midpoint, dividing the segment into two equal parts

central angle an angle formed by two radii in a circle. Its vertex is the center of the circle and its endpoints sit on the circle.

chord any line segment in a circle that extends from one point on the circle to another point on the circle

circumference the distance around a circle, defined by $2\pi r$ or πd

complementary angles two angles whose measures sum to 90°

coordinate plane a plane that contains a horizontal number line (the x-axis) and a vertical number line (the y-axis). Every point in the coordinate plane can be named by a pair of numbers written in the form (x,y).

corresponding angles When a transversal crosses a pair of lines, the corresponding angles are the angles at the same relative position at each intersection. If the two lines crossed by the transversal are parallel, the corresponding angles will be congruent.

decagon a polygon with 10 sides

degree the unit by which angles are measured. A circle contains 360 degrees.

diameter of a circle a chord in a circle that passes through the center of the circle. Any diameter will be twice the length of a circle's radius.

dilation a transformation, written D_k, that will either expand or shrink a shape, depending on the scale factor being used

equiangular polygon a polygon with all angles equal in measure

equiangular triangle a triangle with all angles of equal measure. Its sides will also be equal in length.

equilateral polygon a polygon with all sides equal in length

equilateral triangle a triangle with all sides of equal length. Its angles will also be equal in measure.

heptagon a polygon with seven sides

hexagon a polygon with six sides

inscribed angle an angle in a circle whose sides are two chords within the circle, and whose vertex sits on the circle. The measure of an inscribed angle is half the arc it forms.

intercepted arc the minor arc formed by a central angle in a circle. Its measure is equal to the measure of the central angle that forms it.

isosceles trapezoid a trapezoid with one pair of congruent sides

legs of an isosceles triangle the two congruent sides in an isosceles triangle

legs of a trapezoid the two non-parallel sides in a trapezoid

linear equation an equation whose graph is a line. Linear equations have a standard form $Ax + By = C$, where A, B, and C are all numbers and x and y are variables.

major arc the longer arc connecting any two points on a circle

median of a triangle the line extending from the vertex of a triangle to the midpoint of the opposite side. All triangles have three medians.

median of a trapezoid the line segment that joins the midpoints of the legs of a trapezoid

midpoint a point that divides a line segment into two segments of equal length

mid-segment of a triangle a segment that connects the midpoints of any two sides of a triangle

minor arc the shorter arc connecting any two points on a circle

net a 2-dimensional figure that can be folded to make a 3-dimensional figure

nonagon a polygon with nine sides

octagon a polygon with eight sides

origin the point in a coordinate plane where the x-axis and y-axis intersect. The coordinates of the origin are (0,0).

pentagon a polygon with five sides

plane a collection of points that make a flat surface, which extends continuously in all directions, like an endless wall

prism a three-dimensional figure that has two congruent polygon ends and parallelograms as the remaining sides

pyramid a three-dimensional shape with a regular polygon base, and congruent triangular sides that meet at a common point

Pythagorean theorem For any right triangle, the square of the length of the hypotenuse is equal to the sum of the squares of the lengths of the legs. Written as: $a^2 + b^2 = c^2$.

reflection a transformation that flips a point or object over a given axis or line

rotation a type of translation that swivels a point or polygon around a fixed point, such as the origin

same-side interior angles When a transversal crosses a pair of lines, the same-side interior angles are the angles inside the pair of lines, on the same side of the transversal. If the two lines crossed by the transversal are parallel, then the same-side interior angles will be supplementary.

sector of a circle a part of a circle bound by two radii and their intercepted arc

scale the ratio between the measurements of a model and the measurements of the real-life equivalent it is representing

similar Polygons are similar when all their corresponding sides are proportional to one another.

supplementary angles two angles whose measures sum to 180°

surface area the combined area of all of a prism's outward-facing surfaces

translation a transformation that slides a point or object horizontally, vertically, or in both directions, which creates a diagonal shift. The notation $T_{h,k}$ is used to give translation instructions.

transformation a general movement or manipulation of a point or shape on a coordinate graph. Three forms of transformations are translations, reflections, and rotations. The original point or shape is called the *preimage* and the transformed object is called the *image*.

vertex the intersection of two sides of a polygon; also the intersection of two line segments or rays that form an angle

vertical angles a pair of non-adjacent angles formed when two rays or lines intersect

volume the amount of space enclosed by a 3-dimensional object. Volume is expressed in cubic units.

***x*-axis** the horizontal number line that is the scale in a coordinate plane. Values to the right of the origin are positive and those to the left are negative.

***y*-axis** the vertical number line that is the scale in a coordinate plane. Values above the origin are positive and those below it are negative.

Additional Online Practice

Using the codes below, you'll be able to log in and access additional online practice materials!

Your free online practice access codes are:
FVE034720WX42N8B2V2O
FVE2U5DQMD464B1I74J1
FVEA53PBSLHDHD5F6O42

Follow these simple steps to redeem your codes:

- Go to **www.learningexpresshub.com/affiliate** and have your access codes handy.

If you're a new user:

- Click the **New user? Register here** button and complete the registration form to create your account and access your products.
- Be sure to enter your unique access codes only once. If you have multiple access codes, you can enter them all—just use a comma to separate each code.
- The next time you visit, simply click the **Returning user? Sign in** button and enter your username and password.
- Do not re-enter previously redeemed access codes. Any products you previously accessed are saved in the **My Account** section on the site. Entering a previously redeemed access code will result in an error message.

If you're a returning user:

- Click the **Returning user? Sign in** button, enter your username and password, and click **Sign In**.
- You will automatically be brought to the **My Account** page to access your products.
- Do not re-enter previously redeemed access codes. Any products you previously accessed are saved in the **My Account** section on the site. Entering a previously redeemed access code will result in an error message.

If you're a returning user with new access codes:

- Click the **Returning user? Sign in** button, enter your username, password, and new access codes, and click **Sign In**.
- If you have multiple access codes, you can enter them all—just use a comma to separate each code.
- Do not re-enter previously redeemed access codes. Any products you previously accessed are saved in the **My Account** section on the site. Entering a previously redeemed access code will result in an error message.

If you have any questions, please contact LearningExpress Customer Support at LXHub@LearningExpressHub.com. All inquiries will be responded to within a 24-hour period during our normal business hours: 9:00 A.M.–5:00 P.M. Eastern Time. Thank you!

NOTES

NOTES

NOTES

NOTES

NOTES

NOTES